BARRON'S

SAT® SUBJECT TEST MATH LEVEL 1

3RD EDITION

Ira K. Wolf, Ph.D.
President, PowerPrep, Inc.
Former High School Math Teacher,
College Professor of Mathematics, and
University Director of Teacher Preparation

About the Author

Dr. Ira Wolf has had a long career in math education. In addition to teaching math at the high school level for several years, he was a professor of mathematics at Brooklyn College and the Director of the Mathematics Teacher Preparation program at SUNY Stony Brook.

Dr. Wolf has been helping students prepare for the PSAT, SAT, and SAT Subject Tests in Math for 35 years. He is the founder and president of PowerPrep, a test preparation company on Long Island that currently works with more than 1,000 high school students each year.

All inquiries should be addressed to:
Barron's Educational Series, Inc.
250 Wireless Boulevard
Hauppauge, New York 11788
www.barronseduc.com

ISBN-13: 978-0-7641-4355-7 (book only)
ISBN-10: 0-7641-4355-7 (book only)
ISBN-13: 978-0-7641-9683-6 (book/CD-ROM package)
ISBN-10: 0-7641-9683-9 (book/CD-ROM package)

ISSN 1938–5722

PRINTED IN THE UNITED STATES OF AMERICA

9 8 7 6 5 4 3 2 1

Contents

What You Need to Know About SAT Subject Tests

The importance of the College Board's new Score Choice policy

- What Are SAT Subject Tests?
- How Many SAT Subject Tests Should You Take?
- How Are SAT Subject Tests Scored?
- How Do You Register for an SAT Subject Test?

Since you are reading this book, it is likely that you have already decided to take the SAT Subject Test in Math Level 1; at the very least, you are seriously considering taking it. Therefore, you probably know something about the College Board and the tests it administers to high school students: PSAT, SAT, and SAT Subject Tests. In this short introductory chapter, you will learn the basic facts you need to know about the Subject Tests. In the next chapter, you will learn everything you need to know about the Math Level 1 test in particular.

In 2009, the College Board instituted a Score Choice policy for all SAT Subject Tests, as well as for the SAT. What this means is that at any point in your high school career you can take (or even retake) any Subject Tests you want, receive your scores, and then choose whether or not the colleges to which you eventually apply will ever see those scores. In fact, you don't have to make that choice until your senior year when you are actually sending in your college applications. Suppose, for example, that you take the Biology test one year and the Chemistry test the following year. If you earn very good scores on both exams, then, of course, you can send the colleges both scores; if, however, your Chemistry score is much better than your Biology score, you can send the colleges only your Chemistry score and the colleges won't even know that you took the Biology test. Similarly, if you take the Math Level 1 test in June and retake it in November, you can send the colleges just your higher score and they will never know that you took it twice.

WHAT ARE SAT SUBJECT TESTS?

Each SAT Subject Test is an hour-long exam designed to test your knowledge of one specific course that you studied in high school. The following chart lists all the SAT Subject Tests that the College Board offers.

Subject	Tests
English	Literature
Social Studies	World History
	United States History
Mathematics	Math Level 1
	Math Level 2
Science	Biology
	Chemistry
	Physics

Subject	Tests
Foreign Language	French
	German
	Hebrew
	Italian
	Latin
	Spanish
	Chinese
	Japanese
	Korean

Each of these tests consists entirely of multiple-choice questions. The number of questions ranges from 50 on the Math 1 and Math 2 tests to 95 on World History.

Why Should You Take SAT Subject Tests?

Not every college and university requires you to submit SAT Subject Test scores as part of the admissions process. So if you knew with certainty that you were applying only to schools that do not require their applicants to take Subject Tests, you would not have to take any.

However, when you are in ninth-, tenth-, or even eleventh-grade, it is impossible for you to know exactly which schools you will be applying to in the fall of your senior year. Also colleges and universities that don't currently insist that applicants submit scores from SAT Subject Tests may change their policy. In the past few years, many colleges that previously had not required applicants to take Subject Tests have begun requiring them. Therefore, most students—and certainly all good students—should plan on taking some Subject Tests.

Another reason for taking SAT Subject Tests is that even colleges that do not require them for admissions may use them for placement purposes. Often, if you have a good score on a Subject Test, you may be exempted from taking an introductory course in that area and be able to take a more interesting elective.

Finally, remember that because of Score Choice, you are at *no risk*. If you take a Subject Test and don't get a score you are happy with, you never have to submit it.

How Many SAT Subject Tests Should You Take?

No college requires applicants to submit scores from more than three SAT Subject Tests, but many schools—including almost all of the most competitive ones—do ask for two or three. Consequently, most students should plan to take at least two Subject Tests and very strong students should take at least three. You should know, however, that many students take more than three, some as many as six or seven.

They do this because the Score Choice policy allows them to send whichever scores they like. So they can pick their best two or three scores from among all the tests they have taken. Or, if they have really good scores on more than three tests, they can try to really impress the admissions officers by submitting scores from four, or five, or even more tests. A good guideline is that you should take an SAT Subject Test in any subject in which you feel you can earn a high score. On any test date, you may take one, two, or three Subject Tests, but you may not take any in the month you take the SAT.

When Should You Take SAT Subject Tests?

Most SAT Subject Tests are given six times per year: in January, May, June, October, November, and December. By far the most common month in which to take a Subject Test is June, at the end of the year in which you study the subject on that test. For example, you certainly should not take a history or science Subject Test in December or January of the year you are taking the course—at that point you will have covered less than half the year's work. Also, taking one of those tests in the fall after your course is over makes no sense when you have not looked at the subject for several months. The exceptions to this general rule are the Math Level 1 test, which you can take any time after you have completed three years of high school math, and the foreign language exams, which you should put off until you have completed as many years of study as possible.

HOW ARE SAT SUBJECT TESTS SCORED?

Two types of scores are associated with SAT Subject Tests: raw scores and scaled scores. Your raw score, which you do not receive, is calculated by giving you 1 point for a correct answer and deducting $\frac{1}{4}$ point for an incorrect answer. Suppose, for example, that when you take the Math Level 1 test you answer 42 of the 50 questions and omit the other 8. If, of the 42 questions you answer, 38 are correct and 4 are incorrect, your raw score will be 37 (38 points for the 38 correct answers minus $4 \times \frac{1}{4} = 1$ point for the 4 wrong answers). Your raw score is then converted to a scaled score between 200 and 800. Only the scaled score is reported to you (and to the colleges to which you apply). Each test has its own conversion chart. See page 8 for a conversion chart for the Math Level 1 test.

The method of scoring described in the preceding paragraph is the basis for understanding when you should guess while taking a Subject Test. Be sure you read the explanation in the next chapter of when to guess on the Math Level 1 test. In fact, read it twice. It is critically important that you know why wild guessing does not hurt you and why educated guessing can improve your score dramatically.

What Is a Good Score on an SAT Subject Test?

Obviously, different students will have different answers to this question. Many students would be thrilled with any score above 600. Others might not want to take a test if they felt they could not earn at least a 650 or 700. For most Subject Tests, the average score is between 580 and 600. On those tests, therefore, any score in the 600s is well above average and scores in the 700s are excellent. The average scores for Physics and Math 2 are somewhat higher—in the mid 600s. Remember, if your score on a particular test isn't as high as you would have liked, because of Score Choice you don't have to submit it, as long as you have at least two or three that are higher.

How Can You Tell How Well You Will Do?

Of course, you cannot know for sure. However, here is a way to get a good sense of your potential. About six weeks before the test, get a copy of the Barron's review book for that subject and study it for several weeks before the test date. You should also buy a copy of the College Board's book *The Official Study Guide for all SAT Subject Tests* and take the practice test in each subject for which you are planning to take a Subject Test. Give yourself exactly one hour for each exam. Reviewing the subject matter and taking practice tests should enable you to raise your score by 50–100 points or even more and help you to reach your goal.

HOW DO YOU REGISTER FOR AN SAT SUBJECT TEST?

You can get a registration form in your school's guidance office and mail it in. However, most students register online. To do that, just go to the College Board's web site—*www.collegeboard.com*—and follow the simple directions.

If you register by mail, an admissions ticket will be sent to you between ten days and two weeks before your test. If you register online, you can print out your admissions ticket as soon as you have completed your registration.

What Should You Bring to the Test Center?

The night before you are scheduled to take any SAT Subject Test, assemble the following materials:

- Admission ticket
- Photo ID
- Several sharpened No. 2 pencils with erasers (do not assume there will be a pencil sharpener in the test room)
- Your calculator—if you are taking either Math Level 1 or Math Level 2
- Spare batteries or a backup calculator
- An easy-to-read watch or small clock to keep on your desk during the test. (You may *not* use the clock or stop watch on your cell phone.)

What You Need to Know About the SAT Subject Test in Math Level 1

- What Topics Are Covered on the Math Level 1 Test?
- How Many Questions Should You Answer?
- When Should You Guess?
- Should You Use a Calculator on the Math Level 1 Test?

There are two SAT Subject Tests in math: Level 1 and Level 2. Although there is some overlap in the material covered on the two tests, basically, the Level 1 material is less advanced than the Level 2 material. The Level 1 test is based on the math that most students learn in their first three years of high school, whereas many of the questions on the Level 2 test are on material normally taught in a fourth year of math (usually precalculus and trigonometry).

WHAT TOPICS ARE COVERED?

The Math 1 test consists of 50 multiple-choice questions. If you have completed three years of high school math, you have likely learned all the topics covered on the test. In fact, you almost surely have learned more than you need. At most, two or three questions, and possibly none, should seem completely unfamiliar to you. The following chart lists the topics included on the Math 1 test and indicates how many questions you should expect on each topic.

Topic	Percent of Test	Number of Questions
Algebra	30%	15
Plane geometry	20%	10
Solid geometry	6%	3
Coordinate geometry	12%	6
Trigonometry	8%	4
Functions	12%	6
Statistics and sets	6%	3
Miscellaneous	6%	3

The information in this chart can help you guide your study. If you start your preparation early enough, you should plan on reviewing most, if not all, of the material in this book. If, however, your time is limited because you waited until right before your test date to start studying, you should concentrate on the topics that are heavily tested—algebra and plane geometry—and spend little or no time on solid geometry, statistics, and trigonometry.

The numbers in the chart above are approximations, because the percentages can vary slightly from test to test and also because some questions belong to more than one category. For example, you may need to solve an algebraic equation to answer a geometry question or you may need to use trigonometry to answer a question in coordinate geometry.

In the math review part of this book, you will specifically learn which facts you need to know for each topic. For example, you will learn that you do *not* need to know most of the trigonometry you were taught in school: only the most basic trigonometry is tested. The more advanced topics in trigonometry appear on the Math Level 2 test. You will also learn within each topic which facts are more heavily emphasized on the test. For example, to answer the 10 questions on plane geometry, you need to know several facts about triangles, quadrilaterals, and circles. However, some facts about circles are much more important than others, and more questions are asked about triangles than about quadrilaterals.

What Formulas Do You Have to Memorize?

You need to know well over a hundred facts and formulas to do well on the Math 1 test. However, many of them you have known for years, such as the formulas for the areas of rectangles, triangles, and circles. Others you learned more recently, such as the laws of exponents and the quadratic formula. In the math review chapters, each essential fact is referred to as a KEY FACT, and you should study and memorize each one that you do not already know.

If you have already taken the PSAT or SAT, you may recall that 12 facts about geometry are provided for you in a reference box on the first page of each math section. For the Math 1 test, you need to know these formulas (and many more), but they are not given to you.

There are five formulas, however, that you do *not* have to memorize. They are provided for you in a reference box on the first page of the test. All five concern solid geometry and are explained in Chapter 11. It is unlikely that more than one of the 50 questions on any Math 1 test would require you to use one of these formulas, and it is possible that none of them will. So don't worry if you are not familiar with them. The five formulas appear in the box to the right.

REFERENCE INFORMATION

Here are formulas for the volumes of three solids and the areas of two of them. Although they will be provided on the test itself, memorizing them can save you time.

For a sphere with radius r:

- $V = \frac{4}{3}\pi r^3$
- $A = 4\pi r^2$

For a right circular cone with radius r, circumference c, height h, and slant height l:

- $V = \frac{1}{3}\pi r^2 h$
- $A = \frac{1}{2}cl$

For a pyramid with base area B and height h:

- $V = \frac{1}{3}Bh$

HOW MANY QUESTIONS SHOULD YOU ANSWER?

This seems like a strange question. Most students, especially good students, try to answer all the questions on a test. Occasionally, they might have to leave out a question because they get stuck, but they never start a test planning to pace themselves in such a way as to omit 10, 15, or 20 percent of the questions intentionally. Surprisingly, this is precisely what many students should do on the Math 1 test. The biggest mistake most students make when taking this test is trying to answer too many questions. It is far better to go slowly, answering fewer questions and getting most of them right than to rush through the test answering all the questions but getting many of them wrong.

Because nothing lowers one's score more than making careless mistakes on easy questions and because a major cause of careless errors is rushing to finish, take the test slowly enough to be accurate, even if you don't get to finish.

The best way to increase your score on the Math 1 test is to attempt fewer questions.

So exactly how many questions should you answer? Obviously, the answer to this question depends on your goal. If you are an outstanding math student and your goal is to get an 800, then not only do you have to answer all 50 questions, you have to get all of them right. If, on the other hand, your goal is to earn a 650, then, as you can see from the conversion chart on page 8, you could answer fewer than 40 questions and even miss a few.

To see why this is so, consider the following situation. Suppose Bob took the Math 1 test, answered all 50 questions, and got 34 right and 16 wrong. Then his raw score would be 30 (34 points for the 34 right answers minus $\frac{16}{4} = 4$ points for the 16 wrong answers), and his scaled score would be 600. Probably among the 16 questions he missed were a few that he just didn't know how to solve. It is also likely that several of his mistakes were careless. Especially during the last 10 or 15 minutes, he probably went too fast trying to finish and missed questions he could have gotten right had he worked more slowly and more carefully. A likely scenario is that in the first 30 questions, when he was not rushing, he got about 26 right and 4 wrong. On the last 20 questions, in contrast, when he was going too fast, he got about 8 right and 12 wrong.

What if he had worked as slowly and as accurately at the end of the test as he had at the beginning of the test? He would have run out of time. *However, his score would have been higher.* Suppose in the last 20 questions he omitted 8, answering only 12, but getting 10 right and 2 wrong. Then in total he would have had 36 right answers and 6 wrong ones. His raw score would have been 35 and his scaled score a 650. By slowing down and answering fewer questions, his score would have increased by 50 points!

SAMPLE MATH 1 CONVERSION CHART

Raw Score	Scaled Score	Raw Score	Scaled Score	Raw Score	Scaled Score
50	800	30	600	10	400
49	790	29	590	9	390
48	780	28	580	8	380
47	770	27	570	7	370
46	760	26	560	6	360
45	750	25	550	5	350
44	740	24	540	4	340
43	730	23	530	3	330
42	720	22	520	2	320
41	710	21	510	1	310
40	700	20	500	0	300
39	690	19	490	–1	290
38	680	18	480	–2	280
37	670	17	470	–3	270
36	660	16	460	–4	260
35	650	15	450	–5	250
34	640	14	440	–6	240
33	630	13	430	–7	230
32	620	12	420	–8	220
31	610	11	410	–9 or less	200

Which Questions Should You Answer?

Every question has the same raw score value, 1 point. You get the same 1 point for a correct answer to the easiest question on the test, which you could answer in less than 30 seconds, as you do for a correct answer to the hardest question, which might take you more than three minutes to answer. Therefore, if you are not going to answer all the questions, then you should answer the easy and moderately difficult ones and leave out the hardest ones.

Of course, to follow this advice, you need to know which questions are easy and which ones are hard. Fortunately, that is not a problem. The first ones are the easiest, the last ones are the hardest. In general, the questions on the Math 1 test go in order from easy to difficult.

On a recent actual Math 1 test, on questions 1–10, the average percentage of students answering a question correctly was 82 percent, and on questions 41–50 the average percentage of students answering a question correctly was 28 percent. Of questions 1–27, every question was answered correctly by more than 60 percent of the students taking the test; of questions 28–50, not one question was answered correctly by at least 60 percent of the students.

You may not find question 30 to be harder than question 26—especially if you are better in algebra than geometry and question 30 is on algebra and question 26 is on geometry. However, you will definitely find questions 10–19 to be easier than questions 20–29, which in turn will be significantly easier than questions 30–39.

SHOULD YOU GUESS ON THE MATH 1 TEST?

The simple answer is "YES." In general, it pays to guess. To be fair, however, that answer was a little too simple. There are really two types of guessing—wild guessing and educated guessing—and they should be handled separately.

How Does Wild Guessing Affect Your Score?

Suppose that when you take the Math 1 test you work slowly and carefully and answer only 40 of the 50 questions but get them all right. First of all, is that good or bad? Well, probably on a math test in school that would not be very good—you probably wouldn't be happy with a grade of 80. On the Math 1 test, however, those 40 right answers give you 40 raw score points, which convert to a very respectable 700!

Now comes the big question. Should you take your last 10 seconds and quickly bubble in an answer to the last 10 questions without even looking at them? In other words, should you make 10 wild guesses? The answer is that it probably won't matter. Since there are 5 answer choices to each question, the most likely outcome is that you will get $\frac{1}{5}$ of them right. So if you guess on those last 10 questions, you will probably get 2 right and 8 wrong. For the 2 right answers you will earn 2 points and for the 8 wrong answers you will lose $\frac{8}{4} = 2$ points.

If that happens, your score remains the same—your raw score is still 40 and your scaled score is still 700. Of course, you might be unlucky and get only 1 right answer or really unlucky and get none correct, in which case your score would drop to 690 or 680. On the other hand, you might be lucky and get 3 or 4 right, in which case your score would increase to 710 or even 730. On average, however, *wild guessing does not affect your score*, so whether you make wild guesses or not is completely up to you.

How Does Educated Guessing Affect Your Score?

Educated guessing is very different from wild guessing. Sometimes, even though you don't know how to solve a problem, you are sure that some of the answer choices are wrong. When that occurs, you eliminate everything you know is wrong and guess among the remaining choices. This use of the process of elimination is called educated guessing and, unlike wild guessing, can increase your score significantly.

TIP

Educated guessing can increase your score dramatically.

To see why educated guessing is so important, consider a scenario slightly different from the one in our discussion of wild guessing. Suppose now that you have time to answer all 50 questions, but you are sure of only 40 of them. On the other 10 you are able to eliminate 3 choices, say A, B, and C, but have no idea whether D or E is the correct answer. Should you guess at these 10 questions and risk getting some wrong, or should you leave them out? If you omit these questions, your raw score will remain at 40 and your scaled score will still be 700. Now, however, if you guess, since you have a 50-50 chance of guessing correctly, you will probably get about 5 right and 5 wrong. How will that affect your score? For the 5 you get right, you will earn 5 points; for the 5 you get wrong, you will lose $\frac{5}{4}=1.25$ points. This is a net gain of 3.75 points. Your raw score would go from 40 to 43.75, which would get rounded up to 44, and your scaled score would go from 700 to 740, which is a tremendous improvement. You cannot afford to give up those 40 points because you are afraid to guess.

When Should You Guess?

You should be able to make an educated guess on most of the questions you attempt. As you will see in the next chapter on tactics for taking the Math 1 test, there are strategies for dealing with almost all of the questions on the Math 1 test that you do not know how to do or get stuck on. Incredibly, when properly used, some of these tactics are guaranteed to get you the right answer. Others will enable you to eliminate choices. Whenever you can eliminate one or more choices, you must guess.

Basically, if you attempt a question, you should almost always answer it: either you will know how to do it or you should be able to make an educated guess. Certainly, you should omit very few, if any, of the first 25 questions, which make up the easier half of the test.

When Should You Omit Questions?

There are two reasons for omitting a question on the Math 1 test:

- You absolutely do not understand what the question is asking. You do not know how to answer it and have no basis for making a guess.
- You do not get to that question. Most students who pace themselves properly, going slowly enough to avoid careless errors, do not have enough time to answer every question. If you run out of time, you may omit the remaining questions—or, if you like, you can make a few wild guesses.

SHOULD YOU USE A CALCULATOR ON THE MATH 1 TEST?

You must bring your own calculator to the test. None will be available at the test center, and you are absolutely forbidden from sharing a calculator with a friend.

On the PSAT and SAT, using a calculator is optional. Although almost all students bring one to the test and use it on at least a few questions, there isn't a single question that requires the use of a calculator. On the Math 1 test, the situation is very different. At least 20 percent of the questions on the Math 1 test *require* the use of a calculator (to evaluate $\sin 40°$, $\sqrt[3]{200}$, $\log 17$, or $(1.08)^{20}$, for example). On another 20–30 percent of the questions, a calculator might be helpful. So it is absolutely mandatory that you bring a calculator with you when you take the test.

What Calculator Should You Use?

Basically, you have two options—a scientific calculator or a graphing calculator. The decision is really quite simple: you should bring a calculator with which you are very comfortable. This is probably the calculator you are currently using in your math class.

Do not go out and buy a new calculator right before you take the Math 1 test. If, for any reason, you want a new calculator, get it now, become familiar with it, and use it as you go through this book and especially as you do all the model tests.

The College Board recommends that if you are comfortable with both a scientific calculator and a graphing calculator, you bring a graphing calculator. This is perfectly good advice because there is no disadvantage to having a graphing calculator, but the advantages are small.

One advantage is that in the larger window of a graphing calculator, you can see the answers to your last few calculations, so you may not have to write down the results of intermediate steps in a problem whose solution requires a few steps.

Suppose, for example, that you are asked to find the area of ΔABC in the figure below.

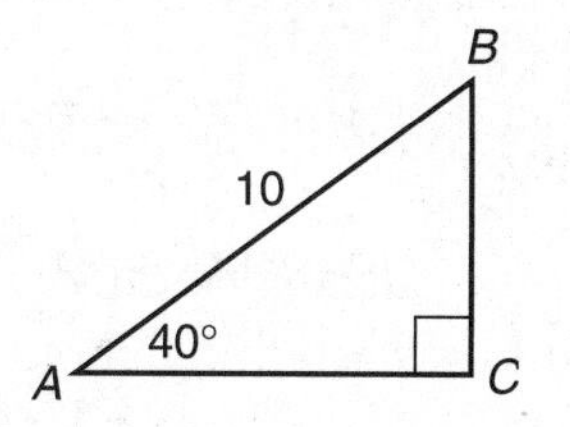

The straightforward way to answer this question is to use the area formula $A = \frac{1}{2}bh$. The area of $\Delta ABC = \frac{1}{2}bh = \frac{1}{2}(AC)(BC)$. Now make three calculations.

Step 1:

$$\cos 40° = \frac{AC}{10} \Rightarrow AC = 10 \cos 40° = 7.66$$

On most scientific calculators, the value will disappear as soon as you start your next calculation, so you would have to write 7.66 in your exam booklet. On a graphing calculator, "$10 \cos 40° = 7.66$" remains visible in the screen when you do step 2:

$$\sin 40° = \frac{BC}{10} \Rightarrow BC = 10 \sin 40° = 6.43$$

On a graphing calculator, both values are still there when you need to do step 3:

$$A = \frac{1}{2}(6.43)(7.66) = 24.62$$

$$or\ A = 0.5(\text{ANS})7.66 = 24.62$$

A second advantage of a graphing calculator is the obvious one—it can graph. However, this is not as big an advantage as you might think. As you will see in the next chapter, occasionally if you get stuck on a question and cannot come up with the correct mathematical solution, looking at a graph may help you to get the right answer or at least make an educated guess. However, this is not a common situation, and no question on the Math 1 test requires the use of a graphing calculator.

To summarize, there is absolutely no reason not to use a graphing calculator if you own one and are comfortable with it, but the advantages of using it are small and do not warrant buying one just for this test.

By the way, you may bring two calculators and use whichever you prefer on any question. In fact, the College Board recommends that you bring batteries and/or a backup calculator to the test center. Remember, if your calculator fails during the test, you may not borrow or share anyone else's and the test center won't have any to lend you.

What Else Do I Need to Know About Calculators?

In the next chapter, "Tactics," you will receive very important advice about when to use and when not to use your calculator. Be sure to read that chapter—it is critical for learning good test-taking skills.

This discussion of calculators concludes with a few miscellaneous bits of advice.

- As you will see in Chapter 14, all angles on the Math 1 test are measured in degrees. You do not have to know anything about radians, so keep your calculator in degree mode.
- If you are using a graphing calculator, you do not have to clear its memory. Therefore, you can store any formulas you like and even program your calculator, if you know how. You should know, however, that this is usually not advisable. If you have a program to solve quadratic equations, for example, you may very well spend more time searching for it and running it than it would take just to solve the equation in your test booklet.
- If your calculator fails during the test and you do not have a backup and if you immediately tell the proctor, you may cancel your math test without canceling any other SAT Subject Tests you are taking that day. (Normally, if you want to cancel a test, you must cancel all the tests you take that day.)

TEST-TAKING STRATEGIES

Important Tactics

CHAPTER 1

- Use of the Calculator
- Backsolving
- Extra Variables
- Proper Use of Diagrams
- Roman Numeral Problems
- Eliminating Choices

As a general rule, students should take SAT Subject Tests in those subjects in which they excel and avoid taking them in subjects that are difficult for them. Consequently, almost all students who take the Math 1 test have good averages in math (typically at least a B+).

CALCULATOR TIPS

You know, of course, that you are allowed to use a calculator whenever you like on the Math 1 test. (See pages 10–12 for a discussion of calculator use.) The College Board classifies about 10 of the 50 questions on the Math 1 test as "calculator active." For these questions, a calculator is absolutely required. No one can evaluate $\tan 23°$ or $\sqrt[5]{100}$ or 21^{25} without a calculator. Another 10 to 15 questions are classified as "calculator neutral." On those questions, the use of a calculator is optional. You can surely evaluate 2^7 or $\frac{1}{5}+\frac{2}{3}$ or $\frac{2}{7}$ of 168 or $987 - 789$ without a calculator, but why should you? You have a calculator, so use it.

If there is any chance that you will make a mistake adding or subtracting negative numbers, use your calculator. A general rule of thumb is this: do not do arithmetic in your exam booklet. If you cannot do it in your head, use your calculator. In particular, never do long multiplication or long division; do not find common denominators; do not simplify radicals; do not rationalize denominators. All these things you should do on your calculator.

THROUGHOUT THIS BOOK, WHENEVER THE USE OF A CALCULATOR IS REQUIRED OR RECOMMENDED, A CALCULATOR ICON APPEARS IN THE MARGIN.

The College Board considers the other 25 to 30 questions on the Math 1 test to be "calculator inactive." These are questions about which they say, "There is no advantage, perhaps even a disadvantage, to using a calculator." The discussion following TACTIC 1, on page 16, shows you that even on some of these calculator inactive questions, if you get stuck, there is a way to use your calculator to get the right answer.

THE INSIDE SCOOP FOR SOLVING PROBLEMS

Why do some students do so much better on the Math 1 test than others? Of course, A+ students tend to do better on the test than B+ or A– students. Among students with exactly the same grades in school, though, why do some earn significantly higher scores than others—perhaps 100 to 200 points higher? Those students are better test takers. Either instinctively or by having been taught, they know and use most of the tactics discussed in this chapter. If you master these strategies, you will be a much better test taker and will earn significantly higher scores, not only on the Math 1 but also on the PSAT, SAT, and other standardized math tests.

THE PROPER USE OF THE TACTICS EXPLAINED IN THIS CHAPTER CAN MAKE THE DIFFERENCE BETWEEN GOOD SCORES AND GREAT SCORES.

Tactic 1

Use your calculator even when no calculations are necessary

Often, if you get stuck on a calculator inactive question, you can use your calculator to get the right answer.

EXAMPLE 1: If $\left(x+\frac{1}{x}\right)^2 = a$, then $x^2 + \frac{1}{x^2} =$

(A) $a - 2$
(B) a
(C) a^2
(D) $\sqrt{a}$
(E) $a\sqrt{a}$

Solution: If you think that this is an algebra question for which a calculator would not be helpful, you would be only partly right. There is an algebraic solution that does not require the use of a calculator. However, if you don't see how to do it, you can plug in a number for x and then use your calculator. Let $x = 2$. Then $\left(x+\frac{1}{x}\right)^2 = \left(2+\frac{1}{2}\right)^2 = (2.5)^2 = 6.25$. So $a = 6.25$.

Now $x^2 + \frac{1}{x^2} = 2^2 + \frac{1}{2^2} = 4 + \frac{1}{4} = 4.25$.

When $a = 6.25$, only choice A, $a - 2$, is equal to 4.25.

In Example 1, you actually didn't need your calculator very much. You used it only to square 2.5. Example 2 looks easier because there is only one variable, but it actually requires a greater use of the calculator.

EXAMPLE 2: If $x > 0$ and $\left(x+\frac{1}{x}\right)^2 = 64$, what is the value of $x^2+\frac{1}{x^2}$?

(A) 4
(B) 16
(C) 60
(D) 62
(E) 64

Solution (using TACTIC 1): $\left(x + \frac{1}{x}\right)^2 = 64 \Rightarrow x+\frac{1}{x} = 8$. Now use your calculator to approximate x by guessing and checking. Since $8+\frac{1}{8} = 8.125$, x must be a little less than 8.

- $7.5+\frac{1}{7.5} = 7.63$ too small
- $7.8+\frac{1}{7.8} = 7.92$ just a little too small
- $7.9+\frac{1}{7.9} = 8.02$ just a little too big

So $7.8 < x < 7.9$.

Now evaluate $x^2+\frac{1}{x^2}$:

- $7.8^2+\frac{1}{7.8^2} = 60.86$
- $7.9^2+\frac{1}{7.9^2} = 62.42$

Only choice D, 62, lies between 60.86 and 62.42.

With a graphing calculator, you can find x by graphing $y = x+\frac{1}{x}$ and tracing along the graph, zooming in if necessary, until the y-coordinate is very close to 8. You could also look at $y = x+\frac{1}{x}$ in a table, using increments of 0.1 or even 0.01.

Mathematical Solution to Examples 1 and 2:

$$\left(x+\frac{1}{x}\right)^2 = \left(x+\frac{1}{x}\right)\left(x+\frac{1}{x}\right) = x^2 + x\left(\frac{1}{x}\right) + x\left(\frac{1}{x}\right) + \left(\frac{1}{x}\right)^2 = x^2+2+\frac{1}{x^2}$$

So $\left(x + \frac{1}{x}\right)^2$ is 2 more than $x^2+\frac{1}{x^2}$.

In Example 1, $\left(x + \frac{1}{x}\right)^2 = a$, so $x^2+\frac{1}{x^2} = a - 2$.

In Example 2, $\left(x + \frac{1}{x}\right)^2 = 64$, so $x^2+\frac{1}{x^2} = 62$.

The algebraic solution to the following example doesn't require the use of a calculator, but many students wouldn't find that solution when taking the test and would leave the question out. Of course, you would not do that. You would use TACTIC 1.

EXAMPLE 3: If $7^x = 2$ then $7^{3x} =$

(A) $\sqrt[3]{2}$
(B) 4
(C) 6
(D) 8
(E) $4\sqrt[3]{2}$

Solution (using TACTIC 1): First you need to find (or approximate) x. There are several ways to do that, all of which require a calculator. Here are four methods.

1. Guess and check

$7^1 = 7$	way too big
$7^{0.5} = 2.6$	still too big
$7^{0.4} = 2.17$	getting close
$7^{0.35} = 1.98$	a little too small, but close enough

2. Graph $y = 7^x$ and trace until y is very close to 2
3. Look at the table for $y = 7^x$ and scroll until you find a y value very near 2
4. Use logarithms: $\log 2 = \log 7^x = x \log 7 \Rightarrow x = \dfrac{\log 2}{\log 7} = 0.356$.

Using $x = 0.35$, the first value we got, we have $3x = 1.05$ and $7^{1.05} = 7.7$. Clearly, the correct answer is 8 (especially since we know that $x = 3.5$ is slightly too small).

Mathematical Solution: Of course, the solution using logarithms is a correct mathematical solution. If you carefully enter $7^{3\left(\frac{\log 2}{\log 7}\right)}$ into your calculator, you will get 8. The shortest and nicest solution does not require a calculator at all: $7^{3x} = (7^x)^3 = 2^3 = 8$.

EXAMPLE 4: What is the range of the function $f(x) = (x - 2)^2 + 2$?

(A) All real numbers
(B) All real numbers not equal to 2
(C) All real numbers not equal to −2
(D) All real numbers greater than or equal to 2
(E) All real numbers less than or equal to 2

The easiest, correct mathematical solution is to observe that since the square of a number can never be negative, $(x - 2)^2$ must be greater than or equal to 0. Therefore, $f(x) = (x - 2)^2 + 2$ must be greater than or equal to 2.

If you do not see that, however, and if you have a graphing calculator, you can graph $y = (x - 2)^2 + 2$ and see immediately that the graph is a parabola whose minimum value is 2—the turning point is at (2, 2).

Tactic 2

Backsolve

Backsolving is the process of working backward from the answers. When you backsolve, you test the five answer choices to determine which one satisfies the conditions in the given problem. This strategy is particularly useful when you have to solve for a variable and you are not sure how to do it. Of course, it can also be used when you do know how to solve for the variable but feel that it would take too long or that you might make a mistake with the mathematics.

Always test choice C first. On the Math 1 test, when the five answer choices for a question are numerical, they are almost always listed in either increasing or decreasing order. (The occasional exceptions occur when the choices involve radicals or π.) When you test a choice, if it is not the correct answer, it is usually clear whether the correct answer is greater or smaller than the choice tested. Therefore, if choice C does not work because it is too small, you can immediately eliminate three choices—C and the two choices that are even smaller (usually A and B). Similarly, if choice C does not work because it is too big, you can immediately eliminate three choices—C and the two choices that are even bigger (usually D and E).

Examples 5 and 6 illustrate the proper use of TACTIC 2.

EXAMPLE 5: For what value of n is $21^n = 3^5 \times 7^5$?

(A) 5
(B) 10
(C) 25
(D) 50
(E) 125

Solution (using TACTICS 1 and 2): Use your calculator to evaluate $3^5 \times 7^5 = 4{,}084{,}101$. Now test the choices, starting with C.

Is $21^{25} = 4{,}084{,}101$? No. $21^{25} = 1.13 \times 10^{33}$, which is way too big. Eliminate choice C and choices D and E, which are even bigger, and try something smaller. Whether you now test 5 (choice A) or 10 (choice B) does not matter. However, since your first attempt was ridiculously large, try the smaller value, 5. Is $21^5 = 4{,}084{,}101$? Yes, so the answer is A. NOTE: If after eliminating C, D, and E you tried B, you would have found that $21^{10} = 1.668 \times 10^{13}$, which is still much too big and you would have known that the answer is A.

Did you have to backsolve to answer this question? Of course not. You *never* have to backsolve. You can always get the correct answer to a question directly if you know the mathematics and if you do not make a mistake. You also did not need to use your calculator.

TIP 

Always start with choice C. Doing so can save you time.

Mathematical Solution: One of the laws of exponents states that for any numbers a, b, and n: $a^n b^n = (ab)^n$. So $3^5 \times 7^5 = (3 \times 7)^5 = 21^5$.

Note that in this case the mathematical solution is much faster. If, however, while taking a test, you come to a question such as Example 5 and you don't remember the laws of exponents or are unsure about how to apply them, you don't have to omit the question—you can use TACTIC 2 and be certain that you will get the correct answer.

EXAMPLE 6: Alice, Beth, and Carol divided a cash prize as follows. Alice received $\frac{2}{5}$ of it, Beth received $\frac{1}{3}$ of it, and Carol received the remaining $120. What was the value of the prize?

(A) $360
(B) $450
(C) $540
(D) $600
(E) $750

Solution (using TACTIC 2): Backsolve starting with C. If the prize was worth $540, then Alice received $\left(\frac{2}{5}\right)(\$540) = \$216$ and Beth received $\left(\frac{1}{3}\right)(\$540) = \$180$.

So, they received a total of $396, leaving $\$540 - \$396 = \$144$ for Carol. Since that is too much (Carol only received $120), eliminate choices C, D, and E and try B. If the prize was worth $450, then Alice received $\left(\frac{2}{5}\right)(\$450) = \$180$ and Beth received $\left(\frac{1}{3}\right)(\$450) = \$150$.

So, they received a total of $330, leaving $\$450 - \$330 = \$120$ for Carol, which is correct.

Mathematical Solution: Let x represent the value of the prize. Here are two ways to proceed.

1. Solve the equation: $\frac{2}{5}x + \frac{1}{3}x + 120 = x$ using the 6-step method from Chapter 6.

   Get rid of the fractions by multiplying each term by 15:

$$\overset{3}{\cancel{15}}\left(\frac{2}{\cancel{5}}\right)x + \overset{5}{\cancel{15}}\left(\frac{1}{\cancel{3}}\right)x + 15(120) = 15(x)$$

$$6x + 5x + 1{,}800 = 15x$$

Combine like terms: $11x + 1{,}800 = 15x$
Subtract $11x$ from each side: $1{,}800 = 4x$
Divide both sides by 4: $x = 450$

2. Add $\frac{2}{5}$ and $\frac{1}{3}$ to determine that together Alice and Beth received $\frac{11}{15}$ of the prize, leaving $\frac{4}{15}$ of the prize for Carol. So

$$\frac{4}{15}x = 120 \Rightarrow x = 120 \div \frac{4}{15} = 450$$

If you are comfortable with either algebraic solution and are confident you can solve the equations correctly, just do it, and save backsolving for a harder problem. If you start to do the algebra and you get stuck, you can always revert to backsolving. Note that unlike the situation in Example 5, in Example 6 the correct mathematical solutions are not much faster than backsolving.

Tactic 3

Plug in numbers whenever you have EXTRA variables

To use this tactic, you have to understand what we mean by *extra* variables. Whenever you have a question involving variables:

- Count the number of variables.
- Count the number of equations.
- Subtract these two numbers. This gives you the number of *extra* variables.
- For each extra variable, plug in any number you like.

If $x+y+z=10$, you have three variables and one equation. Hence you have two extra variables and can plug in *any* numbers for two of the variables. You could let x and y each equal 2 (in which case $z=6$); you could let $x=1$ and $z=11$ (in which case $y=-2$); and so on. You could not, however, let $x=1$, $y=2$, and $z=3$—you do not have three extra variables, and, of course, $1+2+3$ is *not* equal to 10.

If $x+y=10$, you have two variables and one equation. Hence you have one extra variable and can plug in *any one* number you want for x or y but not for both. You cannot let $x=2$ and $y=2$ since $2+2 \neq 10$. If you let $x=2$, then $y=8$; if you let $x=10$, then $y=0$; if you let $y=12$, then $x=-2$.

If $2x+4=10$, you have one variable and one equation. So you have no extra variables, and you cannot plug in a number for x. You have to solve for x.

If a question requires you to simplify $\frac{8m+6n}{3n+4m}$, you should recognize that you have two variables and *no* equations. Note that $\frac{8m+6n}{3n+4m}$ is *not* an equation; it is an expression. An equation is a statement that one expression is equal to another expression. Since you have two extra variables, you can let $m=1$ and $n=2$, in which case $\frac{8m+6n}{3n+4m}=\frac{8(1)+6(2)}{3(2)+4(1)}=\frac{8+12}{6+4}=\frac{20}{10}=2.$

Of course, since $\frac{8m+6n}{3n+4m}=\frac{2\cancel{(4m+3n)}}{\cancel{4m+3n}}=2$, this is the result you would get if you plugged in *any* numbers for m and n.

Look at Example 1 on page 16. Without saying so, TACTIC 3 was used. The given information was $\left(x+\frac{1}{x}\right)^2=a$. Two variables were given but only one equation. So, we had one *extra variable* and could have plugged in *any* number for either x or a. Clearly, it is easier to plug in for x and evaluate a than it would be to plug in a number for a and then have to solve for x. But we didn't have to replace x by 2; we could have used *any* number. For example, if we let $x=3$:

$$a=\left(x+\frac{1}{x}\right)^2=\left(3+\frac{1}{3}\right)^2=\left(\frac{10}{3}\right)^2=\frac{100}{9}=11\frac{1}{9}$$

$$\text{and } x^2+\frac{1}{x^2}=9+\frac{1}{9}=9\frac{1}{9}=a-2$$

Although all good test takers use TACTIC 3 when they want to avoid potentially messy algebraic manipulations, TACTIC 3 can also be used on geometry or trigonometry questions that contain variables. The basic idea is to

- replace each extra variable with an easy-to-use number;
- answer the question using those numbers;
- test each of the answer choices with the numbers you picked to determine which choices are equal to the answer you obtained.

If only one choice works, you are done. If two or three choices work, change at least one of your numbers, and test only the choices that have not yet been eliminated.

Now look at a few examples that illustrate the correct use of TACTIC 3.

EXAMPLE 7: If $a + a + a + a = b$, which of the following is equal to $4b - a$?

(A) 0
(B) $3a$
(C) $15a$
(D) $16a$
(E) $10a + 10$

Solution (using TACTIC 3): Since you have two variables and one equation, you have one extra variable, so let $a = 2$. Then

$$b = 2 + 2 + 2 + 2 = 8 \Rightarrow 4b = 32 \Rightarrow 4b - a = 32 - 2 = 30$$

Now replace a by 2 in each of the answer choices and eliminate any choice that does not equal 30.

- $0 \neq 30$ — Cross out A.
- $3(2) = 6 \neq 30$ — Cross out B.
- $15(2) = 30$ — C *could* be the correct answer, but you still have to test choices D and E.
- $16(2) = 32 \neq 30$ — Cross out D.
- $10(2) + 10 = 30$ — E *could* be the correct answer.

At this point, you know that the correct answer is either C or E. To break the tie, you have to choose another number for a, say 3. When $a = 3$, $b = 3 + 3 + 3 + 3 = 12$, and $4b - a = 4(12) - 3 = 45$. Now test choices C and E.

- $15(3) = 45$ — Choice C still works.
- $10(3) + 10 = 40 \neq 45$ — Now choice E does not work. Cross out E.

The answer is C.

Mathematical Solution:

$$a + a + a + a = b \Rightarrow 4a = b \Rightarrow$$
$$4b - a = 4(4a) - a = 16a - a = 15a$$

EXAMPLE 8: Which of the following is equal to $2\sin^2 3\theta + 2\cos^2 3\theta$?

(A) 1
(B) 2
(C) 3
(D) 6
(E) $2\tan^2 3\theta$

Solution (using TACTIC 3): First note that since you have one variable, θ, and no equations, θ is an extra variable, and so you can replace it by any number. Pick a really easy-to-use number, say $\theta = 0$. Then

$$3\theta = 0 \text{ and } 2\sin^2 3\theta + 2\cos^2 3\theta = 2(\sin 0)^2 + 2(\cos 0)^2$$
$$= 2(0)^2 + 2(1)^2 = 0 + 2 = 2$$

Immediately eliminate choices A, C, and D and keep B. Now check whether choice E equals 2 when $\theta = 0$: $2\tan^2 3(0) = 2(\tan 0)^2 = 2(0) = 0 \neq 2$, so E is not correct. The answer is B.

Mathematical Solution: Let $x = 3\theta$. Then

$$2\sin^2 3\theta + 2\cos^2 3\theta = 2\sin^2 x + 2\cos^2 x$$
$$= 2(\sin^2 x + \cos^2 x) = 2(1) = 2$$

EXAMPLE 9: If $2^a = 3^b$, what is the ratio of a to b?

(A) 0.63
(B) 0.67
(C) 1.5
(D) 1.58
(E) 1.66

The correct solution, using logarithms, is as follows:

$$\log 2^a = \log 3^b \Rightarrow a\log 2 = b\log 3 \Rightarrow \frac{a}{b} = \frac{\log 3}{\log 2} = 1.58$$

If you have no idea how to solve the given equation, or if you know that it can be done with logarithms but you do not remember how, use TACTIC 3. Since there are two variables and only one equation, there is one extra variable. Pick a number for either a or b, so let $b = 2$.

Then $2^a = 3^2 = 9$. Immediately, you should see that since $2^3 = 8$, a must be slightly greater than 3 and $\frac{a}{b}$ must be slightly more than $\frac{3}{2} = 1.5$. Certainly, the answer is D or E.

At this point you could guess, but you shouldn't. Instead, you should now use TACTIC 1 (use your calculator) and TACTIC 2 (backsolve) to be sure.

Since $b = 2$, if $\frac{a}{b} = 1.58$, then $a = (1.58)(2) = 3.16$ and if $\frac{a}{b} = 1.66$, then $a = (1.66)(2) = 3.32$. Finally, $2^{3.16}$ is much closer to 9 than is $2^{3.32}$.

Alternatively, you could have graphed $y = 2^x$ and traced to find where $2^x = 9$; or you could have graphed $y = 2^x$ and $y = 9$ and found the point where the two graphs intersect.

Tactic 4

Draw diagrams

On some geometry questions, diagrams are provided, sometimes drawn to scale, sometimes not. Frequently, however, a geometry question does not have a diagram. In those cases, you must draw one. The diagram can be a sketch, drawn quickly, but it should be reasonably accurate. Never answer a geometry question without having a diagram, either one provided by the test or one you have drawn.

Sometimes looking at the diagram will help you find the correct solution. Sometimes it will prevent you from making a careless error. Sometimes it will enable you to make an educated guess.

EXAMPLE 10: If the diagonal of a rectangle is twice as long as one of the shorter sides, what is the measure of each angle that the diagonal makes with the longer sides.

(A) 15°
(B) 30°
(C) 45°
(D) 60°
(E) 75°

Solution (using TACTIC 4): The first step is to sketch a rectangle quickly, but don't be sloppy. Don't draw a square, and don't draw a rectangle such as the one below in which the diagonal is 4 or 5 or 6 times as long as a short side.

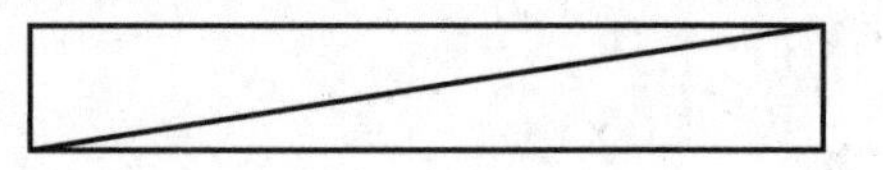

Draw a rectangle such as this:

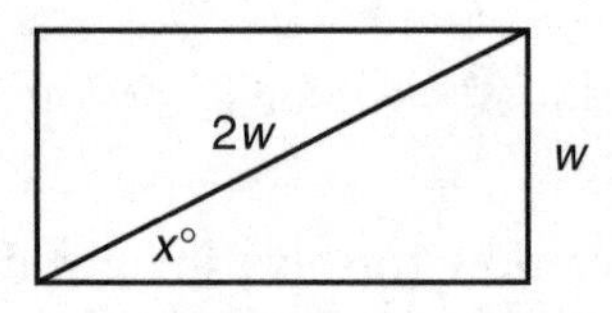

From the second sketch, it is clear that $x < 45$, and the angle is not nearly skinny enough for x to be 15. The answer must be 30°, choice B. In this case, you can be sure you have the right answer. If the answer choices had been

(A) 25°
(B) 30°
(C) 35°
(D) 45°
(E) 60°

you could have eliminated D and E but might have had to guess from among A, B, and C.

Mathematical Solution: Here are two correct solutions.

- If the length of one leg of a right triangle is half the length of the hypotenuse, the triangle is a 30-60-90 triangle, and the measure of the angle opposite the shorter leg is 30°.
- From the diagram drawn on page 24, you can see that

$$\sin x = \frac{w}{2w} = \frac{1}{2} \Rightarrow x = 30°.$$

EXAMPLE 11: $\overline{AB}$ is a diameter of a circle whose center is at (1, 1). If A is at (−3, 3), what are the coordinates of B?

(A) (−5, 1)
(B) (−1, 2)
(C) (5, −1)
(D) (5, 1)
(E) (5, 5)

Solution (using TACTIC 4): Even if you think you know exactly how to do this, first make a quick sketch.

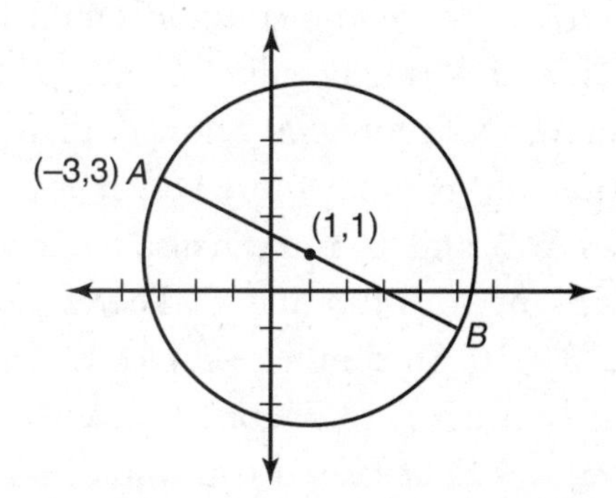

Even if your sketch wasn't drawn carefully enough, it would be clear that the x-coordinate of B is positive and the y-coordinate is near 0. So you could eliminate choices A, B, and E. If (5, −1) and (5, 1) are too close to tell from your sketch and if you don't know a correct way to proceed, just guess between C and D. If you drew your diagram carefully (as we did), you could definitely tell that the y-coordinate is negative, and so the answer must be C.

Mathematical Solution: Since the center of the circle is the midpoint of any diameter, (1, 1) is the midpoint of $\overline{AB}$ where A is (−3, 3) and B is (x, y). Use the midpoint formula: $(1, 1) = \left(\frac{-3+x}{2}, \frac{3+y}{2}\right)$. So

$$1 = \frac{-3+x}{2} \Rightarrow 2 = -3 + x \Rightarrow x = 5$$

$$1 = \frac{3+y}{2} \Rightarrow 2 = 3 + y \Rightarrow y = -1$$

Even if you know how to do this, you should sketch a diagram. If you make a careless error and get $y = 5$, for example, your diagram would alert you and prevent you from bubbling in E.

Tactic 5

Trust figures that are drawn to scale

On the Math 1 test, some diagrams have the following caption underneath them: "Note: Figure not drawn to scale." All other diagrams are absolutely accurate, and you may rely upon them in determining your answer.

EXAMPLE 12: In the diagram at the right, the radius of circle O is 4 and diameters $\overline{AB}$ and $\overline{CD}$ are perpendicular. What is the perimeter of ΔBOD?

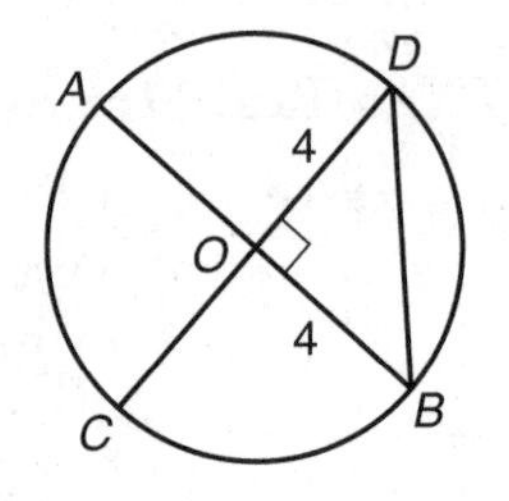

(A) 6
(B) 6.83
(C) 12
(D) 13.66
(E) 16

Solution (using TACTIC 5): Since the diagram is drawn to scale, you may trust it. The question states that radii $\overline{OD}$ and $\overline{OB}$ are each 4. Looking at the diagram, you can see that chord $\overline{BD}$ is longer than $\overline{OD}$ and is therefore greater than 4. Therefore, the perimeter of ΔBOD is greater than $4 + 4 + 4 = 12$. Eliminate choices A, B, and C. You can also see that chord $\overline{BD}$ is much shorter than diameter $\overline{AB}$ and so is less than 8. Therefore, the perimeter of ΔBOD is less than $4 + 4 + 8 = 16$. So eliminate choice E. The answer must be D.

If choice E had been 13.5, 13.75, or 14, you would not have known whether the correct answer was D or E, and you would have had to guess.

TIP

TACTIC 5 always allows you to eliminate choices, but you are not always able to eliminate all four incorrect choices.

Mathematical Solution: Since $OB = OD = 4$, ΔBOD is isosceles. Since $\overline{AB}$ and $\overline{CD}$ are perpendicular, ΔBOD is a right triangle. Therefore, $BD = 4\sqrt{2} = 5.66$, and the perimeter of ΔBOD is $4 + 4 + 5.66 = 13.66$.

Tactic 6

Redraw figures that are not drawn to scale

Recall that on the Math 1 test, the words "Note: Figure not drawn to scale" appear under some diagrams. When this occurs, you cannot trust anything in the figure to be accurate unless it is specifically stated in the question. When figures have not been drawn to scale, you can make *no* assumptions. Lines that look perpendicular may not be; an angle that appears to be acute may, in fact, be obtuse; two line segments may have the same length even though one looks twice as long as the other.

Often when you encounter a figure not drawn to scale, it is very easy to fix. You can redraw one or more of the line segments or angles so that the resulting figure will be accurate enough to trust. Of course, the first step in redrawing the figure is recognizing what is wrong with it.

When you take the Math 1 test, if you see a question such as the one in Example 13 below and if you are sure that you know exactly how to answer it, just do so. Don't be concerned that the figure isn't drawn to scale. Remember that most tactics should be used only when you are not sure of the correct solution. If, however, you are not sure what to do, quickly try to fix the diagram.

EXAMPLE 13: In ΔABC below, $m\angle A = 15°$ and $AC = 8$.

What is the value of x?

(A) 2.07
(B) 2.14
(C) 4
(D) 7.72
(E) 8.23

B
x
15°
A
C
8

Note: Figure not drawn to scale.

Solution: In the diagram, $\overline{AC}$ and $\overline{BC}$ appear to be about the same length. If the figure had been drawn to scale you would be pretty confident that the answer is D or E. However, the figure is not drawn to scale. Therefore, you can make no such assumptions.

You are told that the figure is not drawn to scale, and in fact, it isn't. The measure of $\angle A$ is 15°, but in the diagram it looks to be much more, perhaps 45°. To fix it, create a 45° angle by sketching a diagonal of a square, and then divide that angle into thirds. Now you have an accurate diagram, and $\overline{BC}$ is clearly much less than $\overline{AC}$, nowhere near 8.

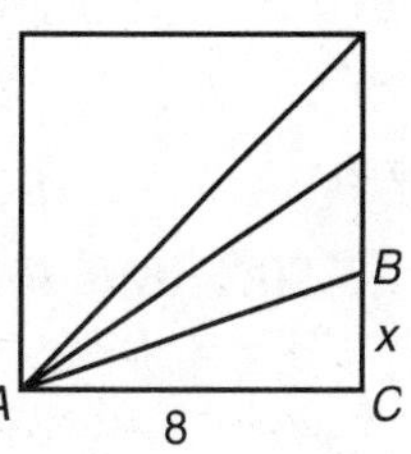

In fact, it is clearly less than 4. So the answer must be A or B. Unfortunately, no matter how carefully you draw the new diagram, you cannot distinguish between 2.07 and 2.14. Unless you know how to solve for x, you have to guess between A and B. If choice A had been 1.07 instead of 2.07, you would not have had to guess. From your redrawn diagram, you can tell that $\overline{AC}$ is about four times as long as $\overline{BC}$, not eight times as long, and you would know that the answer has to be B.

TIP

You *must* guess whenever you can eliminate choices.

Mathematical Solution:

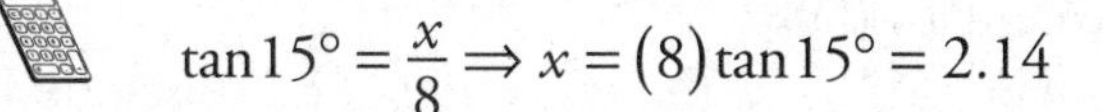

$$\tan 15° = \frac{x}{8} \Rightarrow x = (8)\tan 15° = 2.14$$

Tactic 7

Treat Roman numeral problems as three true-false questions

On the Math 1 test, some questions contain three statements labeled with the Roman numerals I, II, and III, and you must determine which of them are true. The five answer choices are phrases such as "None" or "I and II only," meaning that none of the three statements is true or that statements I and II are true and statement III is false, respectively. Sometimes what follows each of the three Roman numerals are only phrases or numbers. In such cases, those phrases or numbers are just abbreviations for statements that are either true or false. Do not attempt to analyze all three of them together. Treat each one separately. After determining whether or not it is true, eliminate the appropriate answer choices. Be sure to read those questions carefully. In particular, be aware of whether you are being asked what *must* be true or what *could* be true.

Now try using TACTIC 7 on the next two examples.

EXAMPLE 14: ΔABC is a right triangle with right angle C. If $m\angle A > m\angle B$, which of the following statements must be true?

I. $\sin A > \cos B$
II. $\cos A > \cos B$
III. $\tan A > \tan B$

(A) II only
(B) III only
(C) I and III only
(D) II and III only
(E) I, II, and III

Solution: First use TACTIC 4 and draw a diagram. Since you are told that $m\angle A > m\angle B$, make $\angle A$ much bigger than $\angle B$. Now test each statement.

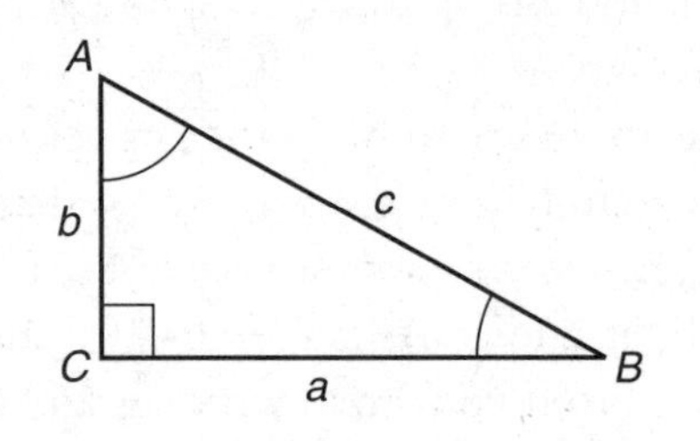

I. $\sin A > \cos B$ Is that true or false?

- $\sin A = \frac{a}{c}$ and $\cos B = \frac{a}{c}$.
- So $\sin A = \cos B$.
- Statement I is false.
- Eliminate C and E, the two choices that include I.

II. $\cos A > \cos B$ Is that true or false?

- $\cos A = \frac{b}{c}$ and $\cos B = \frac{a}{c}$. Clearly from the diagram $a > b$, and so

  $\frac{a}{c} > \frac{b}{c}$.
- Statement II is false.
- Eliminate A and D.

Having crossed out choices A, C, D, and E, you know the answer must be B. You do not have to verify that statement III is true. (Of course it is:

$\tan A = \frac{a}{b}$, which is greater than 1 since $a > b$ and $\tan B = \frac{b}{a}$, which is less than 1. So $\tan A > \tan B$.)

EXAMPLE 15: If the lengths of two sides of a triangle are 4 and 9, which of the following could be the area of the triangle?

I. 8
II. 18
III. 28

(A) II only
(B) III only
(C) I and II only
(D) II and III only
(E) I, II, and III

Solution: Think of Roman numeral I as the statement, "The area of the triangle could be 8," (and similarly for Roman numerals II and III). You are free to check the statements in any order.

Start by drawing a right triangle whose legs have lengths of 4 and 9. Then use the formula $A = \frac{1}{2}bh$ to calculate the area: $\frac{1}{2}(9)(4) = 18$. Therefore, statement II is true. Eliminate choice B, the only choice that does not include II.

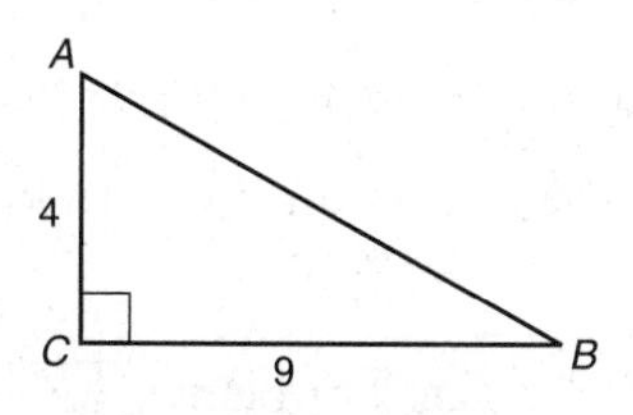

At this point, there are several ways to analyze the other choices. Knowing what the area is when C is a right angle, consider what would happen if angle C were acute or obtuse.

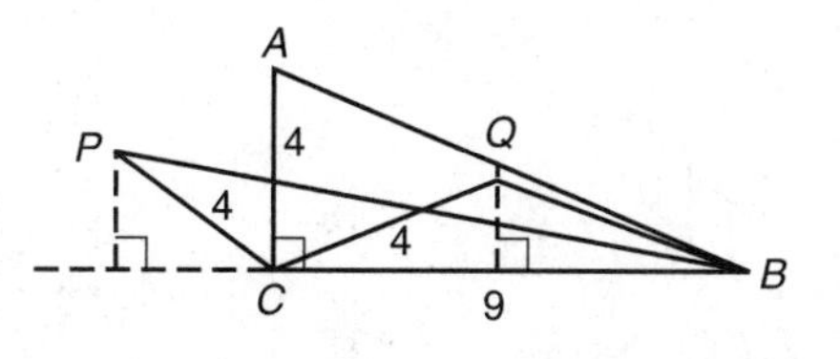

If you superimpose ΔPBC, in which $\angle C$ is obtuse, and ΔQBC, in which $\angle C$ is acute, onto ΔABC, you can see that in each case the height to base $\overline{BC}$ is less than 4. So in each case, the area is less than $\frac{1}{2}(9)(4) = 18$. The area of the triangle could *not* be 28. III is false. Eliminate D and E, the two remaining choices that include III. The areas of ΔPBC and ΔQBC are each less than 18, but it may not be clear whether either triangle, or any other triangle with sides 4 and 9, could have an area equal to 8. If you cannot determine whether I is true, guess between A (II only) and C (I and II only).

In fact, the area *could* be 8, or any other positive number less than 18. For the area to be 8, just let the height be $\frac{16}{9}$.

A particularly nice way to solve Example 15 is to use the formula for the area of a triangle that relies on trigonometry: $A = \frac{1}{2}ab\sin\theta$, where a and b are the lengths of two of the sides and θ is the measure of the angle between them. Since the maximum value of $\sin\theta$ is 1, the maximum possible area is $\frac{1}{2}(4)(9)(1) = 18$. Therefore, III is false. Eliminate choices B, D, and E. Finally, ask, "Could $\frac{1}{2}(4)(9)\sin\theta = 8$?"

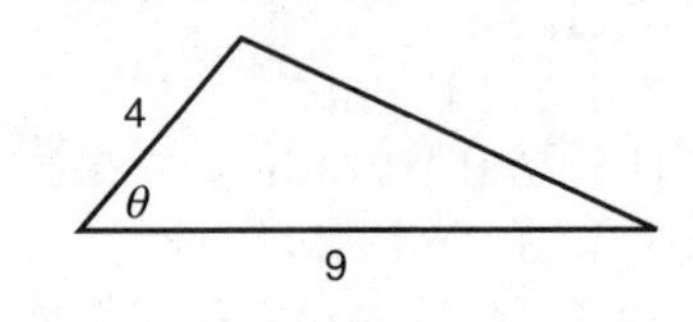

$$\frac{1}{2}(4)(9)\sin\theta = 8 \Rightarrow (18)\sin\theta = 8 \Rightarrow \sin\theta = \frac{8}{18}$$

and since $\frac{8}{18}$ is in the range of the sine function, the answer is yes. (In fact, $\theta = \sin^{-1}\left(\frac{8}{18}\right) \approx 26.4°$, which you should *not* take the time to evaluate.) I is true. Eliminate A. The answer is C.

Note that the formula $A = \frac{1}{2}ab\sin\theta$ is *not* part of the Math 1 syllabus, and no question on the test requires you to know it. If, however, you do know it, you are free to use it.

EXAMPLE 16: In the diagram below, $\overline{EF}$ is a diameter and chords $\overline{AB}$ and $\overline{CD}$ are parallel. Which of the following statements must be true?

I. $x = y$
II. $a = b$
III. $AB = CD$

(A) None
(B) I only
(C) II only
(D) III only
(E) I, II, and III

Note: Figure not drawn to scale.

Remember

Sometimes when you see "Note: Figure not drawn to scale," there is nothing wrong with the diagram. The diagram just did not have to be drawn the way it was.

Solution: Here, nothing is wrong with the diagram. You are told that chord $\overline{EF}$ is a diameter, and since it passes through the center, it is correctly drawn. You are told that $\overline{AB}$ and $\overline{CD}$ are parallel, and they are. However, there are many ways you could redraw the diagram consistent with the given conditions.

For example, you don't have to draw $\overline{EF} \perp \overline{AB}$, and you could draw $\overline{AB}$ closer to $\overline{CD}$. Both of the following diagrams satisfy all the given conditions.

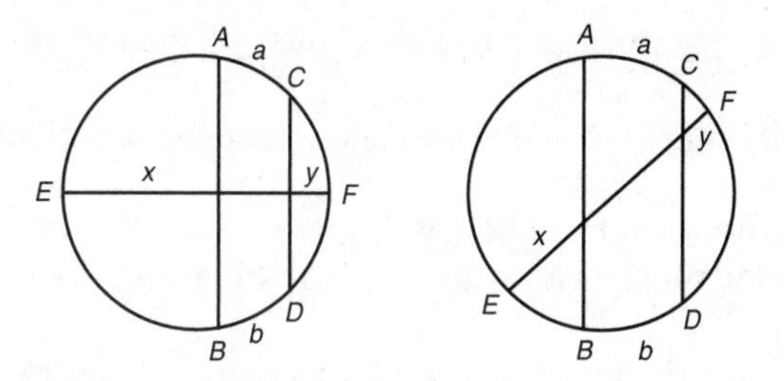

In each of the diagrams shown, $x \neq y$, and $AB \neq CD$, so I and III are false. Eliminate choices B, D, and E. Either II is true (in which case the answer is C) or II is false (in which case the answer is A). In both redrawn diagrams, a and b appear to be equal. Unless you can draw a diagram in which a and b are clearly unequal, you should guess that II is true.

In fact, in a circle parallel chords *always* cut off congruent arcs.

Tactic 8

Eliminate absurd choices

There will likely be some questions on the Math 1 test that you do not know how to answer. Before deciding to omit them, look at the answer choices. Very often, two or three choices are absurd. In that case, eliminate the absurd choices and guess among the remaining ones. Occasionally, four of the choices are absurd. In that case, your answer is not a guess, it is a certainty.

What makes an answer choice absurd? Lots of things. For example, you may know that the answer to a question must be positive, but two or three of the choices are negative. You may know that the measure of an angle must be acute, but three of the choices are numbers greater than or equal to 90°. You may know that a ratio must be greater than 1, but two or three of the choices are less than or equal to 1. Even if you know the correct mathematical method for answering a question, sometimes it is faster to start by eliminating answers that are clearly impossible.

EXAMPLE 17: If $a = 1 + \frac{1}{b}$, where $b > 1$, which of the following could be the value of a?

(A) $\frac{1}{15}$

(B) $\frac{11}{15}$

(C) $\frac{21}{15}$

(D) $\frac{31}{15}$

(E) $\frac{41}{15}$

Smart Strategy

If you have eliminated four answer choices, the remaining choice *must* be the correct answer. Don't waste time by verifying it.

Solution: Since you are asked for a possible value of *a*, you could use TACTIC 2 and backsolve. Before doing that, however, carefully look at the choices. Since b is positive, $\frac{1}{b}$ is positive, and so a must be greater than 1: eliminate choices A and B. Also, $b > 1 \Rightarrow \frac{1}{b} < 1$, and so a must be less than 2: eliminate choices D and E. The answer must be C.

You do not then have to prove that the answer is C by solving the equation $\frac{21}{15} = 1 + \frac{1}{b}$. You do not get any extra credit for determining that $a = \frac{21}{15}$ when $b = 2.5$.

EXAMPLE 18: In the figure below, a square is inscribed in a circle of radius 4. What is the area of the shaded region?

(A) 4.57
(B) 9.13
(C) 18.26
(D) 25.13
(E) 34.26

Solution (using TACTIC 8): In a question such as this some of the choices are always absurd. The area of a circle of radius 4 is $\pi(4)^2 = 16\pi$, which is approximately 50. Since the diagram is drawn to scale, you can trust it. Clearly, more of the circle is white than is shaded, so the area of the square is more than 25 and the area of the shaded region is less than 25. Eliminate choices D and E. If the area of the shaded region were 9.13, the area of the white square would be greater than 40, which is more than 4 times 9.13. That is surely wrong and 4.57 is an even worse answer. Eliminate A and B. The answer must be 18.26, choice C.

TOPICS IN ARITHMETIC

Basic Arithmetic

- The Number Line
- Absolute Value
- Addition, Subtraction, Multiplication, Division
- Integers
- Exponents and Roots
- Squares and Square Roots
- Logarithms
- PEMDAS
- Exercises
- Answers Explained

To do well on the Math Level 1 test, you need a good working knowledge of arithmetic—not because there are several questions on arithmetic (in fact, there are very few) but because arithmetic is the basis for much of the math that *is* on the test. In theory, you do not have to know how to subtract negative numbers, how to add fractions, or what fractional and negative exponents mean because you can use your calculator to evaluate $(-3) - (-5)$ and $\frac{3}{8} + \frac{4}{5}$ and $\left(4^{-\frac{1}{2}}\right)(4^{-2})$. However, you must know the rules so that you can subtract $(-3xy) - (-5xy)$ and add $\frac{3}{x} + \frac{4}{y}$ and simplify $a^{\frac{1}{2}}a^{-2}$.

THE NUMBER LINE

On the Math 1 test, one or two questions will involve i, the imaginary unit, which you will read about later in Chapter 17. Otherwise, the word ***number*** *always* means ***real number***, a number that can be represented by a point on the number line.

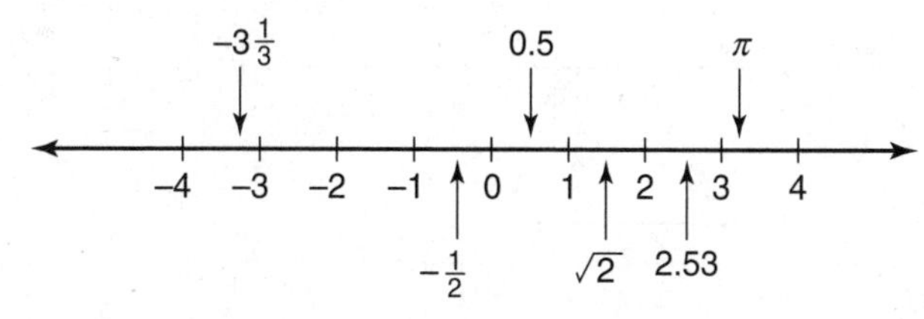

A ***positive number*** is a number that lies to the right of 0 on the number line. A ***negative number*** lies to the left of 0 on the number line.

Key Fact A1

TRICHOTOMY LAW

For any real number a, exactly one of the following statements is true:

- **a is negative**
- **$a = 0$**
- **a is positive**

ABSOLUTE VALUE

The ***absolute value*** of a number a, denoted $|a|$, is the distance between a and 0 on the number line. Since 4 is 4 units to the right of 0 on the number line and −4 is 4 units to the left of 0, both have an absolute value of 4:

Remember

The absolute value of a number is never negative.

- $|4| = 4$
- $|-4| = 4$

Since 4 and −4 are the only numbers that are 4 units from 0, if $|x| = 4$, then $x = 4$ or $x = -4$. If $|x| < 4$, then x is less than 4 units from 0, which means $-4 < x < 4$. If $|x| > 4$, then x is more than 4 units from 0, which means either that $x < -4$ or $x > 4$.

Key Fact A2

If b is a positive number, then

- $|x| = b \Rightarrow x = b$ **or** $x = -b$
- $|x| < b \Rightarrow -b < x < b$
- $|x| > b \Rightarrow x < -b$ **or** $x > b$

These results are displayed graphically on the three number lines below.

$|x| = b$

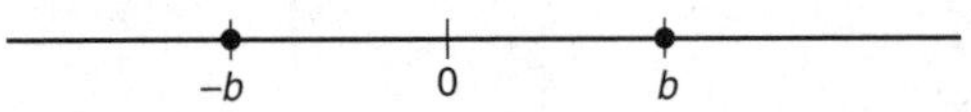

−b 0 b

$|x| < b$

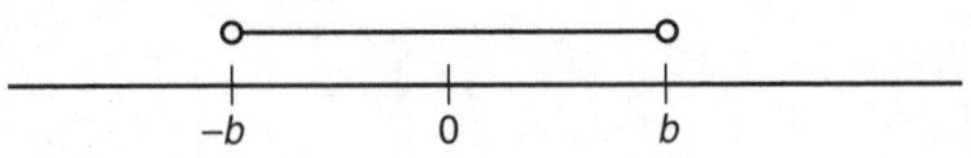

−b 0 b

$|x| > b$

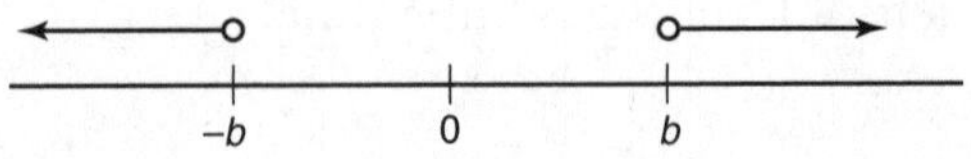

−b 0 b

Given any two numbers x and y, you can *always* find their sum $(x + y)$, difference $(x - y)$, product (xy), and quotient $(x \div y)$—with your calculator, whenever necessary—except that you may *never divide by zero*. For example, $4 \div 0$ is meaningless. If you attempt to divide by 0 on your calculator, you will get an error message.

ADDITION, SUBTRACTION, MULTIPLICATION, DIVISION

Even though you should know all the basic facts about the arithmetic of real numbers that are summarized in the next KEY FACT, it is important to remember that you can always use your calculator. Of course, you can evaluate $2 + 3$ and $2 - 3$ in your head. However, if you would hesitate for even a second when asked to evaluate $(-2) + (-3)$ or $-2 - (-3)$, then use your calculator.

TIP 

Whenever you enter a negative number into your calculator, be sure to put it in parentheses.

Key Fact A3

For any real numbers a and b:

- **If either a or b is 0, then $ab = 0$.**
- **If $ab = 0$, then $a = 0$ or $b = 0$.**
- **If a and b are both positive, ab and $\frac{a}{b}$ are positive.**
- **If a and b are both negative, ab and $\frac{a}{b}$ are positive.**
- **If either a or b is positive and the other is negative, ab and $\frac{a}{b}$ are negative.**
- **If both a and b are positive, $a + b$ is positive.**
- **If both a and b are negative, $a + b$ is negative.**
- **If either a or b is positive and the other is negative, $a + b$ has the same sign as the number whose absolute value is greater.**
- **To evaluate $a - b$, write it as $a + (-b)$ and use the above rules for addition.**

The following KEY FACT is a direct consequence of the third and fourth lines of KEY FACT A3.

Key Fact A4

- **The product of several numbers is positive if the number of negative factors is even.**
- **The product of several numbers is negative if the number of negative factors is odd.**

INTEGERS

Remember

Although 0 is neither positive nor negative, it *is* an integer.

The numbers in the set $\{\ldots, -4, -3, -2, -1, 0, 1, 2, 3, 4, \ldots\}$ are called ***integers***. Naturally, the integers that are greater than 0 are called ***positive integers***, and those that are less than 0 are called ***negative integers***.

Consecutive integers are two or more integers written in sequence in which each integer is 1 more than the preceding one. For example:

3, 4
15, 16, 17, 18,
−3, −2, −1, 0, 1, 2
$n, n+1, n+2, n+3, \ldots$

TIP 

Never assume that *number* means *integer*. If you are told that $1 < x < 4$, of course x could be 2 or 3, but it could also be 2.5, 3.33333, π, $\sqrt{13}$, or any of infinitely many other numbers.

The sum, difference, and product of two integers is *always* an integer. The quotient of two integers may or may not be an integer. The quotient $86 \div 10$ can be expressed as $\frac{86}{10}$ or $8\frac{6}{10}$ or 8.6. You can also say that the ***quotient*** is 8 and the ***remainder*** is 6. How you express this depends upon your point of view. For example, if you want to divide 86 dollars equally among 10 people, you can give each one $8.60 (8.6 dollars). However, if you divide 86 books into 10 piles, each pile will have 8 books, and 6 books will be left over (the remainder).

Whenever you have to answer a question involving remainders, think of piles of books. For example, suppose 3 is the remainder when an integer n is divided by 7. Then n could be 17 (2 piles of 7 with 3 left over), n could be 703 (100 piles of 7 with 3 left over), or n could be 3 more than *any* multiple of 7.

If m and n are integers, the following four terms are synonymous:

m is a ***divisor*** of n	m is a ***factor*** of n
n is ***divisible*** by m	n is a ***multiple*** of m

They all mean that when n is divided by m, there is no remainder (or, more precisely, the remainder is 0). For example:

4 is a divisor of 12	4 is a factor of 12
12 is divisible by 4	12 is a multiple of 4

Key Fact A5

If m and n are positive integers and if r is the remainder when n is divided by m, then n is r more than a multiple of m. That is, $n = mq + r$ where q is an integer and $0 \le r < m$.

Key Fact A6

Every integer has a finite set of factors (or divisors) and an infinite set of multiples.

The factors of 12:	−12, −6, −4, −3, −2, −1, 1, 2, 3, 4, 6, 12
The multiples of 12:	. . . , −48, −36, −24, −12, 0, 12, 24, 36, 48, . . .

The only positive divisor of 1 is 1. All other positive integers have at least two positive divisors—1 and itself—and possibly many more. For example, 12 is divisible by 1, 2, 3, 4, 6, and 12, whereas 7 is divisible by only 1 and 7. Positive integers, such as 7, which have exactly two positive divisors, are called ***prime numbers*** or ***primes***. The first few primes are

$$2, 3, 5, 7, 11, 13, 17, 19, 23$$

Memorize this list—it will come in handy.

Remember

1 is *not* a prime.

Key Fact A7

Every integer greater than 1 that is not a prime can be written as a product of primes.

EXAMPLE 1: To find the prime factorization of 180, write 180 as a product of any two factors; then write any of those factors that are not prime as a product of two factors. Continue this process until all the factors listed are prime.

$$180 = 18 \times 10 = (9 \times 2) \times (5 \times 2) = 3 \times 3 \times 2 \times 5 \times 2$$

This factorization is usually written with the primes in increasing order:

$$180 = 2 \times 2 \times 3 \times 3 \times 5 = 2^2 \times 3^2 \times 5$$

The ***least common multiple*** (**LCM**) of two or more integers is the smallest positive integer that is a multiple of each of them. For example, the LCM of 8 and 12 is 24. The ***greatest common factor*** (**GCF**) or ***greatest common divisor*** (**GCD**) of two or more integers is the largest integer that is a factor of each of them. For example, the GCF of 8 and 12 is 4. Note that the product of the GCF and LCM of 8 and 12 is $4 \times 24 = 96$, which is also the product of 8 and 12.

Key Fact A8

The product of the GCF and LCM of any two positive integers is equal to the product of the two integers.

Don't Get Confused

The terms odd and even apply only to integers

Key Fact A9

For any integers m and n:

- **If either m or n is even, then mn is even.**
- **If both m and n are odd, then mn is odd.**
- **If m and n are both even or both odd, then $m + n$ and $m - n$ are even.**
- **If either m or n is even and the other is odd, then $m + n$ and $m - n$ are odd.**

EXPONENTS AND ROOTS

In the expression b^n, which is read "b to the nth," "b to the nth power," or "b raised to the nth power," b is called the ***base*** and n is called the ***exponent***. Almost all of the exponents that appear in questions on the Math 1 test are integers. Integer exponents are defined in the next KEY FACT.

Key Fact A10

For any number b and positive integer n:

- **if $b \neq 0$, then $b^0 = 1$**
- **$b^1 = b$**
- **if $n > 1$, then $b^n = b \cdot b \cdot b \cdots b$, where b is a factor n times**
- **$b^{-n} = \frac{1}{b^n}$**

For example,

$4^0 = 1$ $\qquad$ $4^4 = 4 \times 4 \times 4 \times 4 = 256$

$4^1 = 4$ $\qquad$ $4^{-1} = \frac{1}{4^1} = \frac{1}{4}$

$4^2 = 4 \times 4 = 16$ $\qquad$ $4^{-2} = \frac{1}{4^2} = \frac{1}{16}$

Several rules govern the use of exponents in mathematics. The most important ones are listed in the next KEY FACT.

Key Fact A11

Memorize the laws of exponents. They come up quite often on the Math 1 test.

LAWS OF EXPONENTS

For any numbers b and c and integers m and n:

- $b^m b^n = b^{m+n}$
- $(b^m)^n = b^{mn}$
- $\frac{b^m}{b^n} = b^{m-n}$
- $b^m c^m = (bc)^m$

For example,

$$4^5 \times 4^3 = 4^8 \qquad (4^5)^3 = 4^{15}$$

$$\frac{4^5}{4^3} = 4^2 \qquad (4^2 \times 4^3)^4 = (4^5)^4 = 4^{20}$$

$$\frac{4^3}{4^5} = 4^{-2} = \frac{1}{4^2} \qquad 3^5 \times 4^5 = 12^5$$

EXAMPLE 2: If $4^m \times 4^n = 4^{10}$, what is the average (arithmetic mean) of m and n?

Since $4^m \times 4^n = 4^{m+n}$:

$$4^{10} = 4^{m+n} \Rightarrow m+n=10 \Rightarrow \frac{m+n}{2} = 5$$

SQUARES AND SQUARE ROOTS

The exponent that appears most often on the Math 1 test is 2. Although a^2 can be read "a to the second," it is usually read "a squared." You will see 2 as an exponent in many formulas. For example:

- $A = s^2$ (the area of a square)
- $A = \pi r^2$ (the area of a circle)
- $a^2 + b^2 = c^2$ (the Pythagorean theorem)
- $x^2 - y^2 = (x - y)(x + y)$ (factoring the difference of two squares)

Numbers that are the squares of integers are called ***perfect squares***. You should recognize at least the squares of the integers from 0 through 15.

x	0	1	2	3	4	5	6	7	8	9	10	11	12	13	14	15
x^2	0	1	4	9	16	25	36	49	64	81	100	121	144	169	196	225

Of course if you need to evaluate 13^2, you can use your calculator. However, it is often helpful to recognize these perfect squares. That way, if you see 169, you will immediately think "that is 13^2."

Two numbers, 5 and −5, satisfy the equation $x^2 = 25$. The positive one, 5, is called the ***square root*** of 25 and is denoted by the symbol $\sqrt{25}$. Clearly, each perfect square has a square root: $\sqrt{0} = 0$, $\sqrt{25} = 5$, $\sqrt{100} = 10$, and $\sqrt{169} = 13$. However, it is an important fact that *every positive number* has a square root.

Key Fact A12

For any *positive* number a, there is a positive number b that satisfies the equation $b^2 = a$. That number, b, is called the square root of a and is written $\sqrt{a}$. So, for any positive number a, $\sqrt{a} \times \sqrt{a} = \left(\sqrt{a}\right)^2 = a$.

Don't Get Confused

- $\sqrt{a+b} \neq \sqrt{a} + \sqrt{b}$

For example:

- $\sqrt{9} + \sqrt{16} = 3 + 4 = 7$
- $\sqrt{9+16} = \sqrt{25} = 5$
- $\sqrt{9} + \sqrt{16} \neq \sqrt{9+16}$

Key Fact A13

For any positive numbers a and b:

$$\sqrt{ab} = \sqrt{a} \times \sqrt{b} \text{ and } \sqrt{\frac{a}{b}} = \frac{\sqrt{a}}{\sqrt{b}}.$$

For example, $\sqrt{3600} = \sqrt{36} \times \sqrt{100} = 6 \times 10 = 60$ and $\sqrt{\frac{1}{9}} = \frac{\sqrt{1}}{\sqrt{9}} = \frac{1}{3}$.

The expression a^3 is often read "a cubed." Numbers that are the cubes of integers are called ***perfect cubes***. You should memorize the perfect cubes in the following table.

x	0	1	2	3	4	5	6	10
x^3	0	1	8	27	64	125	216	1000

The only other powers you should recognize immediately are the powers of 2 up to 2^{10}.

x	0	1	2	3	4	5
2^x	1	2	4	8	16	32

x	6	7	8	9	10
2^x	64	128	256	512	1024

In the same way that we write $b = \sqrt{a}$ to indicate that $b^2 = a$:

- We write $b = \sqrt[3]{a}$ to indicate that $b^3 = a$ and call b the *cube root* of a.
- We write $b = \sqrt[4]{a}$ to indicate that $b^4 = a$ and call b the fourth root of a.
- For any integer $n \geq 2$, we write $b = \sqrt[n]{a}$ to indicate that $b^n = a$ and call b the nth root of a.

For example:

- $\sqrt[3]{125} = 5$ because $5^3 = 125$.
- $\sqrt[3]{-8} = -2$ because $(-2)^3 = -8$.
- $\sqrt[10]{1024} = 2$ because $2^{10} = 1024$.

Note that $\sqrt{-64}$ is undefined because there is no real number x such that $x^2 = -64$. If you enter $\sqrt{-64}$ on your calculator, you will get an error message. (In Chapter 17, you will read about the imaginary unit i and will review the fact that $\sqrt{-64}$ is $8i$.)

Key Fact A14

For any real number a and integer $n \geq 2$:

- **If n is odd, then $\sqrt[n]{a}$ is the unique real number x that satisfies the equation $x^n = a$.**
- **If n is even and a is positive, then $\sqrt[n]{a}$ is the unique *positive* number x that satisfies the equation $x^n = a$.**

We can now expand our definition of exponents to include fractions.

Key Fact A15

For any positive number b and positive integers n and m with $n \geq 2$:

- $b^{\frac{1}{n}} = \sqrt[n]{b}$
- $b^{\frac{m}{n}} = \sqrt[n]{b^m} = \left(\sqrt[n]{b}\right)^m$

For example:

$$4^{\frac{1}{2}} = \sqrt{4} = 2$$

$$27^{\frac{1}{3}} = \sqrt[3]{27} = 3$$

$$8^{\frac{2}{3}} = \left(\sqrt[3]{8}\right)^2 = 2^2 = 4$$

$$8^{\frac{2}{3}} = \sqrt[3]{8^2} = \sqrt[3]{64} = 4$$

$$16^{-\frac{1}{4}} = \frac{1}{16^{\frac{1}{4}}} = \frac{1}{\sqrt[4]{16}} = \frac{1}{2}$$

The laws of exponents, listed in KEY FACT A11, are equally valid if any of the exponents are fractions. For example:

$$\left(2^{\frac{1}{2}}\right)\left(2^{\frac{1}{3}}\right) = 2^{\frac{1}{2}+\frac{1}{3}} = 2^{\frac{5}{6}}$$

$$\left(2^{\frac{3}{2}}\right)^6 = 2^{\left(\frac{3}{2}\right)(6)} = 2^9$$

LOGARITHMS

A logarithm is an exponent.

Recall that the statement $2^4 = 16$ is usually read as "2 to the 4th power equals 16." Another way to read this stresses the role of the exponent 4: "4 is the exponent to which the base 2 must be raised to equal 16." Mathematicians have a special word for this exponent—***logarithm***. The statement $2^4 = 16$ is equivalent to the statement $\log_2 16 = 4$, which is read, "the base 2 logarithm of 16 is 4."

Key Fact A16

If b is a positive number not equal to 1 and $x > 0$,

$\log_b x = y$ if and only if $b^y = x$.

For example:

- $\log_{10}100 = 2$ because $10^2 = 100$
- $\log_{10}\frac{1}{100} = -2$ because $10^{-2} = \frac{1}{10^2} = \frac{1}{100}$
- $\log_{100}10 = \frac{1}{2}$ because $100^{\frac{1}{2}} = \sqrt{100} = 10$

Although you can always estimate the value of a logarithm, in general there is no easy way to evaluate a logarithm exactly without a calculator. For example, $\log_{10} 5$ is the number x such that $10^x = 5$. Since $10^0 = 1$ and $10^1 = 10$, you know that $0 < \log_{10}5 < 1$. You could improve your estimate by noting that since $10^{\frac{1}{2}} = \sqrt{10}$, which is only slightly greater than 3, $\log_{10} 5 > \frac{1}{2}$, and so $0.5 < x < 1$. The LOG button on your calculator evaluates base 10 logarithms. So you can use your calculator to get that $\log_{10} 5 = \text{LOG } 5 = 0.699$.

TIP 

The LOG button on your calculator evaluates *only* base 10 logarithms. So LOG 5 = 0.699 means $\log_{10} 5 = 0.699$, which is often written log 5, without the base being indicated.

What if you want to know the value of $\log_2 5$? Clearly, you cannot just enter LOG 5 on your calculator since that means $\log_{10} 5$. Also, since $2^2 = 4$, $\log_2 4 = 2$ and so $\log_2 5 > 2$, certainly not 0.699. The following KEY FACT shows you how to use your calculator to evaluate $\log_2 5$ exactly.

Key Fact A17

CHANGE OF BASE FORMULA

$$\log_b x = \frac{\log_{10} x}{\log_{10} b}$$

$$\text{So, } \log_2 5 = \frac{\log_{10} 5}{\log_{10} 2} = \frac{0.699}{0.301} = 2.32.$$

$$\text{Similarly, } \log_3 5 = \frac{\log 5}{\log 3} = 1.46 \text{ and } \log_5 5 = \frac{\log 5}{\log 5} = 1.$$

Reminder: In the line above, log 5 and log 3 are abbreviations for $\log_{10}5$ and $\log_{10}3$, respectively.

You need to know a few important laws of logarithms. They are all direct consequences of the laws of exponents (KEY FACT A11).

Key Fact A18

LAWS OF LOGARITHMS

For any positive base $b \neq 1$ and any positive numbers x, y, and n:

- $\log_b (xy) = \log_b x + \log_b y$
- $\log_b \left(\frac{x}{y}\right) = \log_b x - \log_b y$
- $\log_b x^n = n \log_b x$
- $\log_b b^n = n$ **(In particular,** $\log_b 1 = \log_b b^0 = 0$ **and** $\log_b b = \log_b b^1 = 1$**)**

For example, if $\log_b 2 = x$ and $\log_b 3 = y$, then:

$$\log_b 6 = \log_b(2 \times 3) = \log_b 2 + \log_b 3 = x + y$$

$$\log_b \frac{2}{3} = \log_b 2 - \log_b 3 = x - y$$

$$\begin{aligned} \log_b 72 &= \log_b(8 \times 9) = \log_b 8 + \log_b 9 \\ &= \log_b 2^3 + \log_b 3^2 \\ &= 3 \log_b 2 + 2 \log_b 3 \\ &= 3x + 2y \end{aligned}$$

As you will see in Chapter 6, the third rule in KEY FACT A17 allows you to solve equations in which the variable is an exponent by bringing the variable down to the base line. To solve the equation $2^x = 512$, for example, take the logarithm of both sides: $\log_b 2^x = \log_b 512$. Then

$$\log_b 512 = \log_b 2^x = x \log_b 2 \Rightarrow x = \frac{\log_b 512}{\log_b 2}$$

Note that what you choose for the base b does not matter. As a consequence, you might as well let $b = 10$. Therefore, $x = \frac{\log 512}{\log 2}$. Now use your calculator:

$$x = (\log 512) \div (\log 2) = 9$$

PEMDAS

When a calculation requires performing more than one operation, those operations must be carried out in the correct order. For decades, students have memorized the sentence "Please Excuse My Dear Aunt Sally," or just the first letters, PEMDAS, to remember the proper order of operations.

PEMDAS stands for:

- Parentheses: first do whatever appears in parentheses, following PEMDAS within the parentheses if necessary.
- Exponents: next evaluate all terms with exponents.
- Multiplication and Division: next do all multiplications and divisions *in order from left to right*—do not multiply first and then divide.
- Addition and Subtraction: finally do all additions and subtractions *in order from left to right*—do not add first and then subtract.

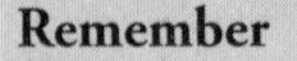

Remember

Your calculator automatically uses PEMDAS if you enter the numbers properly.

To be sure you get the right answer, when you enter numbers in your calculator, always use parentheses around negative numbers and fractions.

For example:

- $2 \times 3^2 = 2 \times 9 = 18$, whereas
  $(2 \times 3)^2 = 6^2 = 36$
- $12 \div 6 \times 2 = 2 \times 2 = 4$, whereas
  $12 \div (6 \times 2) = 12 \div 12 = 1$
- $3 \times 4 + 5 = 12 + 5 = 17$, whereas
  $3 \times (4 + 5) = 3 \times 9 = 27$
- $-5^2 = -25$, whereas
  $(-5)^2 = 25$

Key Fact A19

TIP

Sometimes when you do a calculation mentally, it is easier to use the distributive law than to use PEMDAS to evaluate what is in the parentheses.

DISTRIBUTIVE LAW

For any real numbers *a*, *b*, *c*, and *d* where $d \neq 0$:

- $a(b + c) = ab + ac$
- $a(b - c) = ab - ac$
- $\frac{b+c}{d} = \frac{b}{d} + \frac{c}{d}$
- $\frac{b-c}{d} = \frac{b}{d} - \frac{c}{d}$

Here are two ways to evaluate $7(100 - 1)$:

- Use PEMDAS: $7(100 - 1) = 7(99)$, which is not so easy.
- Use the distributive law: $7(100 - 1) = 7(100) - 7(1) = 700 - 7 = 693$.

Similarly, here are two ways to evaluate $(777 + 49) \div 7$:

- Use PEMDAS: $(777 + 49) \div 7 = 826 \div 7$, which is not so easy.
- Use the distributive law: $(777 + 49) \div 7 = \frac{777}{7} + \frac{49}{7} = 111 + 7 = 118$.

Of course, your calculator can multiply 7×99 as quickly as it can multiply 7×100, but you cannot.

Exercises

1. $||-3| - |-4| - |-5|| =$

   (A) −2
   (B) 2
   (C) 4
   (D) 6
   (E) 12

2. What is the sum of the product and quotient of −5 and 5?

   (A) −26
   (B) −24
   (C) −1
   (D) 0
   (E) 1

3. Which of the following statements are true?

   I. The product of the integers from −5 to 4 is equal to the product of the integers from −4 to 5.
   II. The sum of the integers from −5 to 4 is equal to the sum of the integers from −4 to 5.
   III. The absolute value of the sum of the integers from −4 to 5 is equal to the sum of the absolute values of the integers from −4 to 5.

   (A) I only
   (B) III only
   (C) I and II only
   (D) I and III only
   (E) I, II, and III

4. When the positive integer n is divided by 7, the remainder is 4. What is the remainder when n^2 is divided by 7?

   (A) 2
   (B) 4
   (C) 8
   (D) 16
   (E) It cannot be determined from the information given.

5. If $(7^a)(7^b) = \frac{7^c}{7^d}$, what is d in terms of a, b, and c?

   (A) $\frac{c}{ab}$
   (B) $c - a - b$
   (C) $a + b - c$
   (D) $c - ab$
   (E) $\frac{c}{a+b}$

6. What is the sum of the prime factors of 140?

   (A) 7
   (B) 9
   (C) 12
   (D) 14
   (E) 15

7. Which of the following is the smallest integer that has a remainder of 1 when it is divided by 2, 3, 4, 5, 6, and 7?

   (A) 61
   (B) 141
   (C) 211
   (D) 421
   (E) 841

8. What is the value of $\left(16^{\frac{1}{2}}\right)\left(16^{\frac{1}{3}}\right)\left(16^{\frac{1}{6}}\right)$?

   (A) 1
   (B) $16^{\frac{1}{36}}$
   (C) 2
   (D) 4
   (E) 16

9. If x and y are positive integers, and $(17^x)^y = 17^{17}$, what is the average (arithmetic mean) of x and y?
   (A) $\sqrt{17}$
   (B) 8.5
   (C) 9
   (D) 17
   (E) It cannot be determined from the information given.

10. Which of the following equations have exactly one solution?
    I. $|x| = 1$
    II. $x^2 = 1$
    III. $\sqrt{x} = 1$
    (A) I only
    (B) II only
    (C) III only
    (D) II and III only
    (E) I, II, and III

11. For any integer $n \geq 2$, $\langle n \rangle$ is defined to be the smallest prime factor of n. Which of the following is equal to $\langle 100 \rangle + \langle 45 \rangle$?
    (A) $\langle 10 \rangle$
    (B) $\langle 15 \rangle$
    (C) $\langle 35 \rangle$
    (D) $\langle 50 \rangle$
    (E) $\langle 135 \rangle$

12. Which of the following equations is equivalent to $y = \log_b x$?
    (A) $y = b^x$
    (B) $x = b^y$
    (C) $b = x^y$
    (D) $b = y^x$
    (E) $x = \log_b y$

13. $\log_7 1 + \log_7 7 - \log_7 \frac{1}{7} =$
    (A) -1
    (B) 0
    (C) 1
    (D) 2
    (E) 7

14. Which of the following is equal to $2^{\log_2 22}$?
    (A) 2
    (B) 11
    (C) 22
    (D) 2^{11}
    (E) 2^{22}

15. Which of the following are equal to $\sqrt{a^2}$ for any real number a?
    I. a
    II. $(\sqrt{a})^2$
    III. $|a|$
    (A) I only
    (B) II only
    (C) III only
    (D) I and III only
    (E) I, II, and III

ANSWERS EXPLAINED

Answer Key

1. **(D)**	4. **(A)**	7. **(D)**	10. **(C)**	13. **(D)**
2. **(A)**	5. **(B)**	8. **(E)**	11. **(C)**	14. **(C)**
3. **(A)**	6. **(D)**	9. **(C)**	12. **(B)**	15. **(C)**

Solutions

Each of the problems in this set of exercises is typical of a question you could see on a Math 1 test. When you take the model tests in this book and, in particular, when you take the actual Math 1 test, if you get stuck on questions such as these, you don't have to leave them out—you can almost always answer them by using one or more of the strategies discussed in the "Tactics" chapter. The solutions given here do *not* depend on those strategies; they are the correct mathematical ones.

1. **(D)** $||-3| - |-4| - |-5|| = |3 - 4 - 5| = |-6| = 6$

2. **(A)** The product of 5 and -5 is $(5)(-5) = -25$. The quotient of 5 and -5 is $(5) \div (-5) = -1$. Their sum is $(-25) + (-1) = -26$.

3. **(A)** Since 0 is one of the integers from -5 to 4 and from -4 to 5, both products equal 0 (I is true). The sum of the integers from -5 to 4 is -5, whereas the sum of the integers from -4 to 5 is 5 (II is false). The absolute value of the sum of the integers from -4 to 5 is equal to the absolute value of 5, which is 5; the sum of the absolute values of each of the integers from -4 to 5 is 25 (III is false). Only statement I is true.

4. **(A)** The easiest solution is to choose an integer n that has a remainder of 4 when divided by 7, say $n = 11$. Then $n^2 = 121 = 7 \times 17 + 2$. So the remainder is 2.

   The correct mathematical solution is to note that by using KEY FACT A5, there is a positive integer q for which $n = 7q + 4$, and so

   $$\begin{aligned} n^2 &= (7q+4)^2 = 49q^2 + 56q + 16 \\ &= (49q^2 + 56q + 14) + 2 \end{aligned}$$

   Since the expression in the final parentheses above is clearly divisible by 7, the remainder when n^2 is divided by 7 is 2.

5. **(B)** $(7^a)(7^b) = 7^{a+b}$, and $\dfrac{7^c}{7^d} = 7^{c-d}$.

   Therefore, $a + b = c - d \Rightarrow a + b + d = c \Rightarrow d = c - a - b$.

6. **(D)** $140 = 14 \times 10 = (2 \times 7) \times (2 \times 5) = 2 \times 2 \times 5 \times 7$. The prime factors of 140 are 2, 5, and 7. Their sum is $2 + 5 + 7 = 14$.

7. **(D)** The LCM of 2, 3, 4, 5, 6, 7 is 420, so 420 is divisible by each of the integers from 2 to 7. Therefore, 421 will leave a remainder of 1 when divided by any of them.

8. **(E)** $\left(16^{\frac{1}{2}}\right)\left(16^{\frac{1}{3}}\right)\left(16^{\frac{1}{6}}\right) = 16^{\frac{1}{2}+\frac{1}{3}+\frac{1}{6}} = 16^1 = 16$

9. **(C)** Since $17^{17} = (17^x)^y = 17^{xy}$, then $xy = 17$. The only positive integers whose product is 17 are 1 and 17.

   Their average is $\dfrac{1+17}{2} = \dfrac{18}{2} = 9$.

10. **(C)** $|x| = 1 \Rightarrow x = 1$ or $x = -1$ (I is false).
    $x^2 = 1 \Rightarrow x = 1$ or $x = -1$ (II is false).
    $\sqrt{x} = 1 \Rightarrow x = 1$ (III is true).
    Only statement III is true.

11. **(C)** The smallest prime factor of 100 is 2, and the smallest prime factor of 45 is 3. So $\langle 100 \rangle + \langle 45 \rangle = 2 + 3 = 5$. Of the five choices, only $\langle 35 \rangle = 5$.

12. **(B)** By KEY FACT A16, $y = \log_b x$ if and only if $b^y = x$.

13. **(D)** By KEY FACT A18, $\log_7 1 = 0$, $\log_7 7 = 1$, and $\log_7 \frac{1}{7} = \log_7 (7^{-1}) = -1$. So

$$\log_7 1 + \log_7 7 - \log_7 \frac{1}{7} = 0 + 1 - (-1) = 2$$

14. **(C)** By definition, $\log_2 22$ is the exponent to which 2 must be raised to equal 22. Therefore, $2^{\log_2 22} = 22$.

15. **(C)** If $a = -1$, then $\sqrt{a^2} = \sqrt{(-1)^2} = \sqrt{1} = 1$. Of the three choices, only $|-1| = 1$. Only III is true.

Fractions, Decimals, and Percents

Numbers such as $\frac{5}{11}$, $\frac{3}{3}$, and $\frac{17}{5}$ in which one integer is written over a second integer are called ***fractions***. The center line is called the fraction bar. The integer above the fraction bar is called the ***numerator***, and the integer below the fraction bar is called the ***denominator***.

FRACTIONS

Most of the numbers that we deal with in life are fractions—or can be expressed in fractions. The correct mathematical term for such a number is a ***rational number***. A rational number is any number that can be written in the form $\frac{a}{b}$, where a and b are integers. When a number is actually written as $\frac{a}{b}$, we call it a fraction.

For example, of 4, 0.7, $0.\overline{2222}\ldots$, 20%, and $\frac{3}{11}$, only $\frac{3}{11}$ is a fraction. However, all of them are rational numbers because they can be expressed as fractions: $4 = \frac{4}{1}$, $0.7 = \frac{7}{10}$, $0.\overline{2222}\ldots = \frac{2}{9}$, and $20\% = \frac{20}{100}$.

Numbers that cannot be expressed as fractions are called ***irrational numbers***. Any nonterminating, nonrepeating decimal is an irrational number. For the Math 1 test, it is sufficient to know that π is irrational, as is $\sqrt[n]{a}$, unless $a = b^n$ where b is rational. For example, $\sqrt[3]{8}$ is rational because $8 = 2^3$ and 2 is rational, but $\sqrt{8}$ is irrational because there is no rational number whose square is 8.

A fraction is in ***lowest terms*** if no single positive integer greater than 1 is a factor of both the numerator and denominator. For example, $\frac{8}{15}$ is in lowest terms since no integer greater than 1 is a factor of both 8 and 15; but $\frac{8}{18}$ is not in lowest terms since 2 is a factor of both 8 and 18.

Smart Strategy

You can always use your calculator to reduce a fraction to lowest terms.

Key Fact B1

Every fraction can be *reduced* to lowest terms by dividing the numerator and the denominator by their greatest common factor (GCF). If the GCF is 1, the fraction is already in lowest terms.

For example, by dividing 8 and 18 by 2, their GCF, we can reduce $\frac{8}{18}$ to $\frac{4}{9}$. Since the GCF of 4 and 9 is 1, $\frac{4}{9}$ is in lowest terms.

Key Fact B2

Every fraction can be expressed as a decimal (or a whole number) by dividing the numerator by the denominator.

- **If a fraction is written in lowest terms and if the only prime factors of the denominator are 2 or 5, the decimal terminates.**
- **If a fraction is written in lowest terms and if the denominator has any prime factor other than 2 or 5, the decimal repeats.**

EXAMPLE 1: Since 4, 5, 8, 10, 16, 20, 25, and 40 have no prime factors other than 2 and 5, the decimal equivalents of each of the following fractions terminate:

$\frac{7}{10} = 0.7$ $\quad \frac{3}{4} = 0.75$ $\quad \frac{3}{8} = 0.375$ $\quad \frac{5}{16} = 0.3125$

$\frac{13}{5} = 2.6$ $\quad \frac{7}{20} = 0.35$ $\quad \frac{26}{25} = 1.04$ $\quad \frac{75}{40} = 1.875$

EXAMPLE 2: Since 6, 7, 9, 12, and 22 all have prime factors other than 2 and 5 (6, 9, and 12 are multiples of 3; 7 is a multiple of 7, and 22 is a multiple of 11), the decimal equivalents of each of the following fractions repeat:

$\frac{5}{6} = 0.8\overline{3333}\ldots$

$\frac{3}{7} = 0.\overline{428571428571}\ldots$ $\quad \frac{7}{9} = 0.\overline{77777}\ldots$

$\frac{17}{12} = 1.41\overline{6666}\ldots$ $\quad \frac{25}{22} = 1.1\overline{363636}\ldots$

To determine if two fractions are ***equivalent*** (have the same value) or if one is greater than the other, cross multiply.

Key Fact B3

To compare $\frac{a}{b}$ and $\frac{c}{d}$, cross multiply.

- **If $ad = bc$, then $\frac{a}{b} = \frac{c}{d}$**
- **If $ad > bc$, then $\frac{a}{b} > \frac{c}{d}$**
- **If $ad < bc$, then $\frac{a}{b} < \frac{c}{d}$**

EXAMPLE 3: $\frac{2}{3} > \frac{5}{8}$ because $(2 \times 8) > (3 \times 5)$. Although KEY FACT B3 is useful, remember that you can always compare fractions by using your calculator to convert them to decimals.

EXAMPLE 4: $\frac{2}{3} > \frac{5}{8}$ because $\frac{2}{3} = 0.\overline{666}$ and $\frac{5}{8} = 0.625$, and $0.666 > 0.625$.

ARITHMETIC OPERATIONS WITH FRACTIONS

Key Fact B4

To multiply two fractions, multiply their numerators and multiply their denominators:

$$\frac{3}{5} \times \frac{4}{7} = \frac{3 \times 4}{5 \times 7} = \frac{12}{35}$$

$$\frac{a}{b} \times \frac{c}{d} = \frac{ac}{bd}$$

TIP

All calculations with fractions can (in fact, should) be done on your calculator. So you really don't need to use these rules with numbers. However, on the Math 1 test, you might have to multiply algebraic fractions such as $\frac{3}{x} \cdot \frac{x^2 + x}{x^2 - 1}$, which you cannot do on your calculator.

Key Fact B5

To multiply a fraction by any other number, write that number as a fraction whose denominator is 1:

$$\frac{3}{5} \times 7 = \frac{3}{5} \times \frac{7}{1} = \frac{21}{5} \qquad \frac{3}{4} \times \pi = \frac{3}{4} \times \frac{\pi}{1} = \frac{3\pi}{4}$$

Tactic B1

Before multiplying fractions, reduce. You may reduce by dividing any numerator and any denominator by a common factor.

EXAMPLE 5: $\frac{\overset{1}{\cancel{3}}}{\underset{1}{\cancel{4}}} \times \frac{\overset{2}{\cancel{8}}}{\underset{3}{\cancel{9}}} = \frac{2}{3}$, and as you will see in Chapter 5:

$$\frac{3}{x} \cdot \frac{x^2 + x}{x^2 - 1} = \frac{3}{\cancel{x}} \cdot \frac{\cancel{x}(\cancel{x+1})}{(\cancel{x+1})(x-1)} = \frac{3}{x-1}$$

Tactic B2

When a problem requires you to find a fraction of a number, multiply. Since a percent is just a fraction whose denominator is 100, you also multiply to find a percent of a number.

EXAMPLE 6: If $\frac{2}{7}$ of the 840 students at Monroe High School are freshmen and if 30% of the freshmen play musical instruments, how many freshmen play musical instruments?

There are $\frac{2}{7} \times 840 = 240$ freshmen. Of these, 30% play an instrument:

$$30\% \text{ of } 240 = \frac{30}{100} \times 240 = 72$$

The ***reciprocal*** of any nonzero number, x, is the number $\frac{1}{x}$. The reciprocal of the fraction $\frac{a}{b}$ is the fraction $\frac{b}{a}$.

Key Fact B6

To divide any number by a fraction, multiply that number by the reciprocal of the fraction.

$$20 \div \frac{2}{3} = \frac{20}{1} \times \frac{3}{2} = 30$$

$$\frac{3}{5} \div \frac{2}{3} = \frac{3}{5} \times \frac{3}{2} = \frac{9}{10}$$

$$\sqrt{2} \div \frac{2}{3} = \frac{\sqrt{2}}{1} \times \frac{3}{2} = \frac{3\sqrt{2}}{2}$$

$$\frac{a}{b} \div \frac{c}{d} = \left(\frac{a}{b}\right)\left(\frac{d}{c}\right) = \frac{ad}{bc}$$

Key Fact B7

To add or subtract fractions with the same denominator, add or subtract the numerators and keep the denominator.

$$\frac{4}{9}+\frac{1}{9}=\frac{5}{9} \text{ and } \frac{4}{9}-\frac{1}{9}=\frac{3}{9}=\frac{1}{3}$$

$$\frac{2x+3}{x^2}+\frac{x+1}{x^2}=\frac{3x+4}{x^2} \text{ and } \frac{2x+3}{x^2}-\frac{x+1}{x^2}=\frac{x+2}{x^2}$$

Key Fact B8

To add or subtract fractions with different denominators, first rewrite the fractions as equivalent fractions with the same denominators.

$$\frac{1}{6}+\frac{3}{4}=\frac{2}{12}+\frac{9}{12}=\frac{11}{12}$$

$$\frac{2}{3a}+\frac{3}{2a}=\frac{4}{6a}+\frac{9}{6a}=\frac{13}{6a}$$

NOTE: The *easiest* denominator to get is the product of the denominators ($6 \times 4 = 24$, in the example). However, the *best* denominator to use is the ***least common denominator***, which is the least common multiple (LCM) of the denominators (12 in this case).

Smart Strategy

Using the least common denominator minimizes the amount of reducing necessary to express an answer in lowest terms.

Key Fact B9

If $\frac{a}{b}$ is the fraction of the whole that satisfies some property, then $1-\frac{a}{b}$ is the fraction of the whole that does not satisfy it.

EXAMPLE 7: In a jar, $\frac{1}{2}$ of the marbles are red, $\frac{1}{4}$ are white, and $\frac{1}{5}$ are blue. What fraction of the marbles are neither red, white, nor blue?

The red, white, and blue marbles constitute $\frac{1}{2}+\frac{1}{4}+\frac{1}{5}=\frac{10}{20}+\frac{5}{20}+\frac{4}{20}=\frac{19}{20}$ of the total.

So, $1-\frac{19}{20}=\frac{20}{20}-\frac{19}{20}=\frac{1}{20}$ of the marbles are neither red, white, nor blue.

EXAMPLE 8: There are 800 students at Central High School, each of whom takes exactly one of four science courses. If $\frac{1}{5}$ take earth science, $\frac{2}{5}$ take biology, and $\frac{3}{10}$ take chemistry, how many students take physics?

Since $\frac{1}{5}+\frac{2}{5}+\frac{3}{10}=\frac{2}{10}+\frac{4}{10}+\frac{3}{10}=\frac{9}{10}$, $\frac{9}{10}$ of the students take a science other than physics and $1-\frac{9}{10}=\frac{10}{10}-\frac{9}{10}=\frac{1}{10}$ take physics. Finally, $\frac{1}{10}\times 800=80$ students take physics.

ARITHMETIC OPERATIONS WITH MIXED NUMBERS

A *mixed number* is a number such as $3\frac{1}{2}$ that consists of an integer followed by a fraction. It is an abbreviation for the *sum* of the integer and the fraction; so, $3\frac{1}{2}=3+\frac{1}{2}$.

Tactic B3

To do arithmetic with mixed numbers, just enter them into your calculator as sums written in parentheses.

EXAMPLE 9: How you evaluate $1\frac{2}{3}\times 3\frac{1}{4}$ depends on your calculator. If you use a scientific calculator with a fraction button, you can enter $\left(1+\frac{2}{3}\right)\left(3+\frac{1}{4}\right)$. If you use a graphing calculator, you would probably enter it as (1 + 2 ÷ 3)(3 + 1 ÷ 4). Either way, you should get $\frac{65}{12}$ or $5\frac{5}{12}$ or $5.41\overline{6666}$, which you can convert to the fraction $\frac{65}{12}$.

COMPLEX FRACTIONS

A ***complex fraction*** is a fraction, such as $\frac{1+\frac{1}{6}}{2-\frac{3}{4}}$ or $\frac{2+\frac{1}{x}}{3-\frac{1}{x}}$, that has a fraction in its numerator, denominator, or both.

Key Fact B10

A complex fraction can be simplified in two ways: (i) multiply *every* term in the numerator and denominator by the least common multiple of all the denominators that appear in the fraction or (ii) simplify the numerator and the denominator and then divide.

EXAMPLE 10: To simplify $\dfrac{1+\frac{1}{6}}{2-\frac{3}{4}}$, multiply each term by 12, the

LCM of 6 and 4:

$$\frac{12(1)+12\left(\frac{1}{6}\right)}{12(2)-12\left(\frac{3}{4}\right)}=\frac{12+2}{24-9}=\frac{14}{15}$$

or write

$$\frac{1+\frac{1}{6}}{2-\frac{3}{4}}=\frac{\frac{7}{6}}{\frac{5}{4}}=\frac{7}{6}\times\frac{4}{5}=\frac{14}{15}$$

PERCENTS

As mentioned previously, the word ***percent*** means hundredth. We use the symbol % to express the word "percent." For example, "23 percent" means "23 hundredths" and can be written with a % symbol, as a fraction, or as a decimal:

Remember

A percent is just a fraction whose denominator is 100.

$$23\%=\frac{23}{100}=0.23$$

Recall from TACTIC B2 that to take a percent of a number, you multiply. Since many calculators don't have a % key, you first have to convert the percent to a decimal or fraction.

Key Fact B11

- **To convert a percent to a decimal, drop the % symbol and move the decimal point two places to the left, adding 0's if necessary. (Remember that we consider a whole number to have a decimal point to the right of it.)**
- **To convert a percent to a fraction, drop the % symbol, write the number over 100, and reduce.**

$$20\%=0.20=\frac{20}{100}=\frac{1}{5}$$

$$100\%=1.00=\frac{100}{100}=1$$

$$37.5\%=0.375=\frac{37.5}{100}=\frac{375}{1,000}=\frac{3}{8}$$

$$3\%=0.03=\frac{3}{100}$$

$$\frac{1}{2}\% = 0.5\% = 0.005 = \frac{0.5}{100} = \frac{1}{200}$$

$$350\% = 3.50 = \frac{350}{100} = \frac{7}{2}$$

Consider the three problems in the following example.

EXAMPLE 11:

I. What is 35% of 160?

II. 54 is 45% of what number?

III. 77 is what percent of 140?

Remember

Although 35% can be written as a fraction $\left(\frac{35}{100}\right)$ or a decimal (0.35), x% can *only* be written as $\frac{x}{100}$.

You can easily solve each of these problems using your calculator *if you first set them up properly*. In each case, there is one unknown. Call the unknown x. Translate each sentence, replacing "is" by " = " and the unknown by x.

I. $x = 35\%$ of $160 \Rightarrow x = (0.35)(160) = 56$

II. $54 = 45\%$ of $x \Rightarrow 54 = 0.45x \Rightarrow x = 54 \div 0.45 = 120$

III. $77 = x\%$ of $140 \Rightarrow 77 = \frac{x}{100}(140) \Rightarrow$

$77(100) = 140x \Rightarrow x = \frac{7{,}700}{140} = 55$

Many students have been taught to solve problems such as these by writing the proportion $\frac{\text{is}}{\text{of}} = \frac{\%}{100}$. To use this method, think of "is," "of," and "percent" as variables. In each percent problem, you are given two of the variables and need to find the third, which you label x. Of course, you then solve such equations by cross multiplying.

In Example 12, each of the three problems in Example 11 is solved by using the $\frac{\text{is}}{\text{of}} = \frac{\%}{100}$ proportion.

EXAMPLE 12:

I. What is 35% of 160? (Let x = the "is" number.)

$$\frac{x}{160} = \frac{35}{100} \Rightarrow 100x = 35(160) = 5{,}600 \Rightarrow x = 56$$

II. 54 is 45% of what number? (Let x = the "of" number.)

$$\frac{54}{x} = \frac{45}{100} \Rightarrow 5{,}400 = 45x \Rightarrow x = 120$$

III. 77 is what % of 140? (Let x = the percent.)

$$\frac{77}{140} = \frac{x}{100} \Rightarrow 140x = 7{,}700 \Rightarrow x = 55$$

PERCENT INCREASE AND DECREASE

Key Fact B12

The *percent increase* of a quantity is

$$\frac{\textbf{the actual increase}}{\textbf{the original amount}} \times \mathbf{100\%}$$

The *percent decrease* of a quantity is

$$\frac{\textbf{the actual decrease}}{\textbf{the original amount}} \times \mathbf{100\%}$$

EXAMPLE 13: From 1980 to 1990, the population of a town increased from 12,000 to 15,000. Since the actual increase in the population was 3,000, the percent increase in the population was

$$\frac{3,000}{12,000} \times 100\% = 25\%$$

EXAMPLE 14: If from 1980 to 1990 the population of a town increased from 12,000 to 15,000 and from 1990 to 2000 the population increased by the same percent, then what was the percent increase in population from 1980 to 2000?

In Example 13, you calculated that from 1980 to 1990 the town's population increased by 25%. If from 1990 to 2000 the population again increased by 25%, it increased by (0.25)(15,000) = 3,750. So, the population in 2000 was 18,750. Therefore, from 1980 to 2000, the population increased by 18,750 – 12,000 = 6,750. The percent increase was

$$\frac{6,750}{12,000} \times 100\% = 56.25\%$$

Notice that a 25% increase followed by a second 25% increase is not a 50% increase.

Key Fact B13

An increase of a% followed by an increase of b% *always* results in a larger increase than a single increase of $(a + b)$%. Similarly, a decrease of a% followed by a decrease of b% *always* results in a smaller decrease than a single decrease of $(a + b)$%.

Key Fact B14

To increase a number by $r\%$, multiply it by $(1 + r\%)$. To decrease a number by $r\%$, multiply it by $(1 - r\%)$.

EXAMPLE 15: Starting on January 1, 1990, the population of Centerville increased by 5% every year. If Centerville's population was 10,000 on January 1, 1990, what was its population on January 1, 1992?

Since the population on January 1, 1990 was 10,000, to find the population one year later on January 1, 1991, multiply by 1.05:

$$(1 + 5\%)(10{,}000) = (1 + 0.05)(10{,}000) = (1.05)(10{,}000) = 10{,}500$$

To get the population on January 1, 1992, again multiply by 1.05.

$$(1.05)(10{,}500) = 11{,}025$$

Note that $(1.05)(10{,}500) = (1.05)[(1.05)(10{,}000)] = (1.05)^2(10{,}000)$.

This process can be continued for any number of years.

Key Fact B15

If an initial quantity A increases $r\%$ per year, then the amount at the end of t years is given by $A(t) = A(1 + r\%)^t$.

EXAMPLE 16: If the population of Centerville (which was 10,000 on January 1, 1990) grew at a rate of 5% per year for 10 years, what was its population on January 1, 2000?

The population after 10 years was

$$A(10) = 10{,}000(1 + 0.05)^{10} = 10{,}000(1.05)^{10} = 10{,}000(1.6289) = 16{,}289.$$

Note that 10 increases of 5% resulted not in a 50% increase, but in a total increase of 6,289 or nearly 63%. Also note that the population on January 1, 2000 was 163% of the population on January 1, 1990. Similarly, if the area of one square is 3 and the area of a second square is 18, the area of the larger square is 600% of the area of the smaller square:

$$600\% \text{ of } 3 = \frac{600}{100} \times 3 = 6 \times 3 = 18$$

However, the area of the large square is 500% greater than the area of the small square. The increase in area from the small square to the large one is 15, and

$$\frac{15}{3} \times 100\% = 500\%.$$

Exercises

1. Which of the following is *not* an irrational number?

   (A) π
   (B) $\sqrt{\pi^2}$
   (C) $\frac{\pi^3}{\pi^2}$
   (D) $\frac{\sqrt{\pi^2}}{\pi}$
   (E) $\frac{\pi^0}{\pi}$

2. What fraction of 72 is $\frac{3}{5}$ of 80?

   (A) $\frac{1}{3}$
   (B) $\frac{2}{5}$
   (C) $\frac{5}{9}$
   (D) $\frac{3}{5}$
   (E) $\frac{2}{3}$

3. Which of the following lists the numbers $\frac{3}{8}$, $\frac{4}{11}$, and $\frac{5}{13}$ in increasing order?

   (A) $\frac{3}{8}, \frac{4}{11}, \frac{5}{13}$
   (B) $\frac{4}{11}, \frac{3}{8}, \frac{5}{13}$
   (C) $\frac{5}{13}, \frac{4}{11}, \frac{3}{8}$
   (D) $\frac{4}{11}, \frac{5}{13}, \frac{3}{8}$
   (E) $\frac{3}{8}, \frac{5}{13}, \frac{4}{11}$

4. If $0 < x < 1$, which of the following *could* be less than x?

   I. x^2
   II. $10x\%$ of x
   III. $\frac{1}{x}$

   (A) I only
   (B) II only
   (C) III only
   (D) I and II only
   (E) I and III only

5. Which of the following numbers satisfies the inequality $\frac{2}{3} < \frac{1}{x} < \frac{3}{4}$?

   (A) $\frac{3}{5}$
   (B) $\frac{7}{10}$
   (C) $\frac{13}{10}$
   (D) $\frac{10}{7}$
   (E) $\frac{5}{3}$

6. If $5x = 3$ and $3y = 5$, what is the value of $\frac{x}{y}$?

   (A) $\frac{9}{25}$
   (B) $\frac{3}{5}$
   (C) 1
   (D) $\frac{5}{3}$
   (E) $\frac{25}{9}$

7. What percent of 50 is w?

(A) $2w$

(B) $\frac{2}{w}$

(C) $\frac{50}{w}$

(D) $\frac{w}{2}$

(E) $\frac{w}{50}$

8. At Sally's Sale Shop everything is sold for 20% less than the price marked. If Sally buys sweaters for \$80, what price should she mark them if she wants to make a 20% profit on her cost?

(A) \$80
(B) \$96
(C) \$100
(D) \$120
(E) \$125

9. From 1990 to 2000, the number of students attending McKinley High School increased by 3.5% each year. If the school had 800 students in 1990, how many did it have in 2000?

(A) 1,080
(B) 1,090
(C) 1,128
(D) 1,168
(E) 1,215

10. On a test consisting of 80 questions, Marie answered 75% of the first 60 questions correctly. What percent of the other 20 questions did she need to answer correctly for her grade on the entire exam to be 80%?

(A) 85%
(B) 87.5%
(C) 90%
(D) 95%
(E) 100%

ANSWERS EXPLAINED

Answer Key

1. **(D)**	4. **(D)**	7. **(A)**	10. **(D)**
2. **(E)**	5. **(D)**	8. **(D)**	
3. **(B)**	6. **(A)**	9. **(C)**	

Solutions

Each of the problems in this set of exercises is typical of a question you could see on a Math 1 test. When you take the model tests in this book and, in particular, when you take the actual Math 1 test, if you get stuck on questions such as these, you do not have to leave them out—you can almost always answer them by using one or more of the strategies discussed in the "Tactics" chapter. The solutions given here do *not* depend on those strategies; they are the correct mathematical ones.

1. **(D)** $\frac{\sqrt{\pi^2}}{\pi} = \frac{\pi}{\pi} = 1$, which is a rational number. It is *not* irrational. Choices A, B, and C are each equal to π, which is irrational. Choice E is also irrational since $\frac{\pi^0}{\pi} = \frac{1}{\pi}$, and the reciprocal of an irrational number is irrational.

2. **(E)** $\frac{3}{5}$ of $80 = \left(\frac{3}{5}\right)(80) = 48$, and 48 is $\frac{48}{72}$ of 72.

   Finally, $\frac{48}{72} = \frac{2}{3}$.

3. **(B)** Use your calculator to convert each fraction to a decimal.

   $\frac{4}{11} = 0.\overline{3636}\ldots$, $\frac{3}{8} = 0.375$, $\frac{5}{13} = 0.3846\ldots$ This is the correct order.

4. **(D)** $0 < x < 1 \Rightarrow x(x) < 1(x) \Rightarrow x^2 < x$ (I is true).

   $10x\%$ of $x = \frac{10x}{100}(x) = \frac{1}{10}x^2 < x^2 < x$ (II is true).

   $0 < x < 1 \Rightarrow \frac{x}{x} < \frac{1}{x} \Rightarrow 1 < \frac{1}{x}$. So $\frac{1}{x} > 1 > x$ (III is false).

   Only I and II are true.

5. **(D)** Convert the fraction to decimals. Then the inequality becomes $0.\overline{666}\ldots < \frac{1}{x} < 0.75$. There are infinitely many values of $\frac{1}{x}$ that satisfy this inequality, but $\frac{1}{x} = 0.7$ is one obvious choice. Then $\frac{1}{x} = \frac{7}{10} \Rightarrow x = \frac{10}{7}$.

6. **(A)** $5x = 3 \Rightarrow x = \frac{3}{5}$ and $3y = 5 \Rightarrow y = \frac{5}{3}$. So $\frac{x}{y} = \frac{3}{5} \div \frac{5}{3} = \frac{3}{5} \times \frac{3}{5} = \frac{9}{25}$.

7. **(A)** $w = \frac{x}{100}(50) = \frac{x}{2} \Rightarrow x = 2w$.

8. **(D)** Since 20% of 80 is 16, Sally wants to get $\$80 + \$16 = \$96$ for each sweater she sells. Then if x is the marked price,

$$\$96 = x - 0.20x = 0.8x \Rightarrow x = \frac{\$96}{0.8} = \$120$$

9. **(C)** The number of students in the school in 2000 was $(1.035)^{10}(800) =$ 1,128.

10. **(D)** To earn an 80% on the entire exam, Marie needs to answer a total of 64 questions correctly (80% of 80). So far, she has answered 45 questions correctly (75% of 60). Therefore, on the last 20 questions, she needs $64 - 45 = 19$ correct answers, and $\frac{19}{20} = 95\%$.

Ratios and Proportions

CHAPTER 4

- Ratios
- Proportions
- Exercises
- Answers Explained

A ***ratio*** is a fraction that compares two quantities that are measured in the same units. The first quantity is the numerator, and the second quantity is the denominator.

RATIOS

TIP

Ratios can *always* be written as a fraction.

For example, if in right ΔABC, the length of leg $\overline{AC}$ is 6 inches and the length of leg $\overline{BC}$ is 8 inches, we say that the ratio of AC to BC is 6 to 8, which is often written as 6 : 8 but is just the fraction $\frac{6}{8}$. Like any fraction, a ratio can be reduced and can be converted to a decimal or a percent.

$$AC \text{ to } BC = 6 \text{ to } 8 = 6 : 8 = \frac{6}{8}$$

$$AC \text{ to } BC = 3 \text{ to } 4 = 3 : 4 = \frac{3}{4} = 0.75 = 75\%$$

If you know that $AC = 6$ inches and $BC = 8$ inches, you know that the ratio of AC to BC is 6 to 8. However, if you know that the ratio of AC to BC is 6 to 8, you cannot determine how long either side is. They *may be* 6 and 8 inches long but *not necessarily*. Their lengths, in inches, may be 60 and 80 or 300 and 400 since $\frac{60}{80}$ and $\frac{300}{400}$ are both equivalent to the ratio $\frac{6}{8}$. In fact, there are infinitely many possibilities for the lengths.

AC	6	3	24	2.4	300	$\mathbf{3x}$
BC	8	4	32	3.2	400	$\mathbf{4x}$

The important thing to observe is that the length of $\overline{AC}$ can be *any* multiple of 3 as long as the length of $\overline{BC}$ is the *same* multiple of 4.

Key Fact C1

If two numbers are in the ratio of $a : b$, then for some number x, the first number is ax and the second number is bx.

Tactic C1

In any ratio problem, write x after each number and use some given information to solve for x.

EXAMPLE 1: In a right triangle, the ratio of the length of the shorter leg to the length of the longer leg is 5 to 12. If the length of the hypotenuse is 65, what is the perimeter of the triangle?

Draw a right triangle and label it with the given information; then use the Pythagorean theorem.

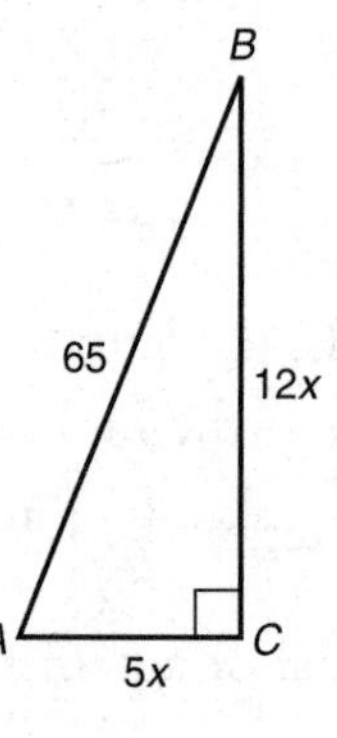

$$(5x)^2 + (12x)^2 = (65)^2 \Rightarrow 25x^2 + 144x^2 = 4{,}225$$
$$\Rightarrow 169x^2 = 4{,}225 \Rightarrow x^2 = 25 \Rightarrow x = 5$$

So $AC = 5(5) = 25$, $BC = 12(5) = 60$, and the perimeter equals $25 + 60 + 65 = 150$.

Ratios can be extended to 3 or 4 or more terms. For example, we can say that the ratio of freshmen to sophomores to juniors to seniors in a school band is $3 : 4 : 5 : 4$. This means that for every 3 freshmen in the band there are 4 sophomores, 5 juniors, and 4 seniors.

Note

TACTIC C1 applies to extended ratios as well.

EXAMPLE 2: What is the degree measure of the largest angle of a quadrilateral if the measures of the four angles are in the ratio of $2 : 3 : 3 : 4$?

Let the measures of the four angles be $2x$, $3x$, $3x$, and $4x$. Use the fact that the sum of the measures of the angles in any quadrilateral is 360°.

$$2x + 3x + 3x + 4x = 360 \Rightarrow 12x = 360 \Rightarrow x = 30$$

The measure of the largest angle is $4(30) = 120°$.

PROPORTIONS

A ***proportion*** is an equation that states that two ratios are equivalent. Since ratios are just fractions, any equation such as $\frac{6}{8}=\frac{3}{4}$, in which each side is a single fraction, is a proportion. The proportions you will see on the Math 1 test always involve one or more variables.

Tactic C2

Solve proportions by cross multiplying. If $\frac{a}{b}=\frac{c}{d}$, then $ad = bc$.

Often one of the first few questions on a Math 1 test will require you to solve a proportion. You should always do that by cross multiplying. Later, more difficult questions might require you to work out a word problem by first setting up a proportion and then solving it.

EXAMPLE 3: If $\frac{x+3}{19}=\frac{x+5}{20}$, then

$$20(x+3) = 19(x+5) \Rightarrow 20x + 60 = 19x + 95 \Rightarrow x = 35$$

EXAMPLE 4: What is the length of $\overline{EF}$ in the figure below?

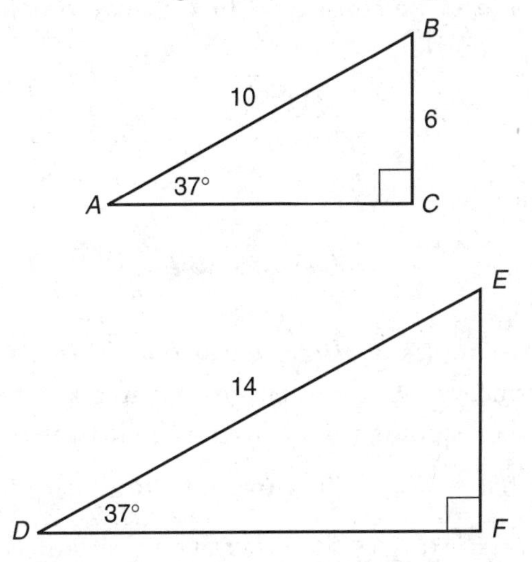

Since the measures of the angles in the two triangles are the same, the triangles are similar. Therefore their sides are in proportion. (See Chapter 9.)

$$\frac{10}{14}=\frac{6}{EF} \Rightarrow 10(EF) = 84 \Rightarrow EF = 8.4$$

NOTE: In Chapter 14, you will read about the trigonometric *ratios* and see that $\frac{BC}{AB}=\frac{EF}{DE}=\sin 37°$.

A ***rate*** is a fraction that compares two quantities measured in different units. Rates often use the word "per" as in miles per hour and dollars per week.

Tactic C3

Set up rate problems just like ratio problems. Solve the proportions by cross multiplying.

TIP 

A rate can *always* be written as a fraction.

EXAMPLE 5: Frank can type 600 words in 15 minutes. If Diane can type twice as fast, how many words can she type in 40 minutes?

Since Diane types twice as fast as Frank, she can type 1,200 words in 15 minutes. Now handle this rate problem exactly as you would a ratio problem. Set up a proportion and cross multiply:

$$\frac{\text{words}}{\text{minutes}} = \frac{1,200}{15} = \frac{x}{40} \Rightarrow$$

$$15x = (40)(1,200) = 48,000 \Rightarrow x = 3,200$$

Occasionally on the Math 1 test rate problems involve only variables. You handle them in exactly the same way.

EXAMPLE 6: If a apples cost c cents, find an expression that represents how many apples can be bought for d dollars.

Since the units must be consistent, first change d dollars to $100d$ cents, and then set up a proportion:

$$\frac{\text{apples}}{\text{cents}} = \frac{a}{c} = \frac{x}{100d}$$

Now cross multiply:

$$100ad = cx \Rightarrow x = \frac{100ad}{c}$$

TIP 

If two variables vary directly, their quotient is a constant.

Rate problems are examples of ***direct variation***. We say that one variable ***varies directly*** with a second variable, or that ***one*** variable is ***directly proportional*** to a second variable, if their quotient is a constant. So if y varies directly with x, there is a constant k such that $\frac{y}{x} = k$. When two quantities vary directly, as one quantity increases (or decreases), so does the other. The constant is the rate of increase or decrease. In Example 5, the number of words Diane types varies directly with the number of minutes she types. Diane's rate of typing is 80 words per minute. The quotient $\frac{\text{words}}{\text{minutes}}$ is constant:

$$\frac{1,200}{15} = 80 \text{ and } \frac{3,200}{40} = 80$$

In contrast to direct variation, in some problems, one variable increases as the other decreases. One example of this is ***inverse variation***. We say that one variable ***varies inversely*** with a second variable or one variable is ***inversely proportional*** to a second variable, if their product is a constant. So if y varies inversely with x, there is a constant k such that $xy = k$.

EXAMPLE 7: Boyle's law states that at a fixed temperature, the volume of a gas varies inversely as the pressure on the gas. If a certain gas occupies a volume of 1.2 liters at a pressure of 20 kilograms per square meter, what volume, in liters, will the gas occupy if the pressure is increased to 30 kilograms per square meter?

TIP 

If two variables vary indirectly, their product is a constant.

Since the volume, V, varies inversely with the pressure, P, there is a constant k such that $VP = k$. Then

$$k = (1.2)(20) = 24 \Rightarrow 24 = V(30) \Rightarrow V = \frac{24}{30} = 0.8$$

EXAMPLE 8: Assume x varies directly with y and inversely with z and that when x is 6, y and z are each 9. What is the value of $y + z$ when $x = 9$?

Since x varies directly with y, there is a constant k such that $\frac{x}{y} = k$. Then $k = \frac{6}{9} = \frac{2}{3}$ and when $x = 9$, $\frac{9}{y} = \frac{2}{3} \Rightarrow 2y = 27 \Rightarrow y = 13.5$.

Since x varies inversely with z, there is a constant c such that $xz = c$. Then $c = (6)(9) = 54$. When $x = 9$, $9z = 54 \Rightarrow z = 6$.

The value of $y + z$ is $13.5 + 6 = 19.5$.

Note: In Example 8, as x increased from 6 to 9, y increased (from 9 to 13.5) and z decreased (from 9 to 6).

Exercises

1. If the ratio of teachers to administrators on a committee is 3 : 5, what percent of the committee members are administrators?

   (A) 37.5%
   (B) 40%
   (C) 60%
   (D) 62.5%
   (E) It cannot be determined from the information given.

2. If the ratio of the measures of the two acute angles in a right triangle is 3 : 7, what is the degree measure of the smallest angle in the triangle?

   (A) 9
   (B) 18
   (C) 27
   (D) 36
   (E) 54

3. If the ratio of the diameter of circle I to the diameter of circle II is 2 : 3, what is the ratio of the area of circle I to the area of circle II?

   (A) 1 : 3
   (B) 2 : 3
   (C) 2 : 5
   (D) 3 : 5
   (E) 4 : 9

4. What is the ratio of the circumference of a circle to its radius?

   (A) 1
   (B) $\frac{\pi}{2}$
   (C) $\sqrt{\pi}$
   (D) π
   (E) 2π

5. If $x > 0$ and $\frac{x-3}{9} = \frac{8}{x+3}$, then $x =$

   (A) 3
   (B) 9
   (C) 27
   (D) 36
   (E) 72

6. If a varies inversely with b, and $a = 3m$ when $b = 5n$, what is b when $a = 5m$?

   (A) $\frac{3}{5}$
   (B) $\frac{5}{3}n$
   (C) $3m$
   (D) $3n$
   (E) $5n$

7. Jeremy can read 36 pages per hour. At this rate, how many pages can he read in 36 minutes?

   (A) 3.6
   (B) 21.6
   (C) 43.2
   (D) 72
   (E) 1,296

8. If x varies directly with y^2, and if $x = 50$ when $y = 5$, what is the value of x when $y = 10$?

   (A) 5
   (B) 25
   (C) 50
   (D) 100
   (E) 200

9. Meri can type p pages in $\frac{1}{m}$ minutes. At this rate, how many pages can she type in m minutes?

   (A) mp
   (B) pm^2
   (C) $\frac{1}{p}$
   (D) $\frac{m}{p}$
   (E) $\frac{1}{mp}$

ANSWERS EXPLAINED

Answer Key

1. **(D)**	4. **(E)**	7. **(B)**
2. **(C)**	5. **(B)**	8. **(E)**
3. **(E)**	6. **(D)**	9. **(B)**

Solutions

Each of the problems in this set of exercises is typical of a question you could see on a Math 1 test. When you take the model tests in this book and, in particular, when you take the actual Math 1 test, if you get stuck on questions such as these, you do not have to leave them out—you can almost always answer them by using one or more of the strategies discussed in the "Tactics" chapter. The solutions given here do *not* depend on those strategies; they are the correct mathematical ones.

1. **(D)** Of every 8 committee members, 3 are teachers and 5 are administrators. Therefore, administrators make up $\frac{5}{8} = 62.5\%$ of the committee.

2. **(C)** The sum of the measures of the two acute angles in a right triangle is 90°. So $3x + 7x = 90 \Rightarrow 10x = 90 \Rightarrow x = 9$. So the measure of the smaller acute angle is $3(9) = 27°$.

3. **(E)** If the diameters of the two circles are $2x$ and $3x$, their radii are x and $\frac{3}{2}x$.

   So their areas are πx^2 and $\pi\left(\frac{3}{2}x\right)^2 = \pi\left(\frac{9}{4}x^2\right)$. Therefore, the ratio of their areas is $1 : \frac{9}{4} = 4 : 9$.

4. **(E)** By definition, π is the ratio of the circumference to the diameter of a circle, so

$$\pi = \frac{C}{d} = \frac{C}{2r} \Rightarrow \frac{C}{r} = 2\pi$$

5. **(B)** Cross multiply: $(x-3)(x+3) = 72 \Rightarrow x^2 - 9 = 72 \Rightarrow x^2 = 81 \Rightarrow x = 9$.

6. **(D)** If a varies inversely as b, there is a constant k such that $ab = k$. So $(3m)(5n) = k \Rightarrow k = 15mn$. Then if $a = 5m$:

$$(5m)b = k = 15mn \Rightarrow b = 3n$$

7. **(B)** Set up a proportion:

$$\frac{36 \text{ pages}}{1 \text{ hour}} = \frac{36 \text{ pages}}{60 \text{ minutes}} = \frac{x \text{ pages}}{36 \text{ minutes}}$$

and cross multiply:

$$(36)(36) = 60x \Rightarrow 1{,}296 = 60x \Rightarrow x = 21.6$$

8. **(E)** Since x varies directly with y^2, there is a constant k such that $\frac{x}{y^2} = k$. Then

$$k = \frac{50}{5^2} = \frac{50}{25} = 2$$

So when $y = 10$:

$$2 = \frac{x}{10^2} = \frac{x}{100} \Rightarrow x = 200$$

9. **(B)** Meri types at the rate of $\dfrac{p \text{ pages}}{\frac{1}{m} \text{ minutes}} = \dfrac{p}{\frac{1}{m}}$ pages per minute $= mp$ pages per minute. Since she can type mp pages each minute, in m minutes she can type $m(mp) = pm^2$ pages.

ALGEBRA

Polynomials

- Polynomials
- Algebraic Fractions
- Exercises
- Answers Explained

More questions on the Math 1 test fall under the broad category of algebra than any other topic: 30% of the test (15 questions) are strictly algebraic. Several other questions that the College Board would assign to other categories, such as geometry, functions, and trigonometry, require the use of the algebra that you will review in this chapter.

It is very important to stress that many algebra questions can be answered using one of the tactics and strategies discussed in Chapter 1. If, while taking the Math 1 test or a model test in this book, you get stuck on an algebra question, you should use one of those tactics rather than skip the question. In this chapter, however, you will review the correct mathematical way to handle these questions.

POLYNOMIALS

The terms *monomial*, *binomial*, *trinomial*, and *polynomial* do not appear on the Math 1 test. However, it will be easier for you to understand this chapter if you first review these terms.

A ***monomial*** is a number or a variable or a product of numbers and variables. Each of the following is a monomial:

$$-5 \quad 7 \quad x \quad y \quad -xy \quad -5xyz \quad 7x^3 \quad \pi r^2 \quad \frac{1}{3}\pi r^2 h$$

Note

Although $x7y$ is a monomial, the coefficient always appears *before* the variables: $7xy$.

The number that appears in front of the variables in a monomial is called the ***coefficient***. The coefficient of $7x^3$ is 7. If there is no number in front of a variable, the coefficient is 1 or −1, because x means $1x$ and $-xy$ means $-1xy$.

A ***polynomial*** is a monomial or the sum of two or more monomials. Each monomial that makes up a polynomial is called a ***term*** of the polynomial. Each of the following is a polynomial:

$$7x^2 \qquad 2x^2+3 \qquad 3x^2-7 \qquad x^2+2x-1$$
$$a^2b^3+b^2a^3 \qquad x^2-y^2 \qquad a^2-2a+1$$

Note that $3x^2 - 7 = 3x^2 + (-7)$ and so is the sum of the monomials $3x^2$ and −7. Also, the sum, difference, and product of any polynomials is itself a polynomial.

The first polynomial in the preceding list $(7x^2)$ is a monomial because it has one term. The second $(2x^2 + 3)$, third $(3x^2 - 7)$, fifth $(a^2b^3 + b^2a^3)$, and sixth $(x^2 - y^2)$ polynomials are called ***binomials*** because they have two terms. The fourth

$(x^2 + 2x - 1)$ and seventh $(a^2 - 2a + 1)$ polynomials are called ***trinomials*** because they have three terms. On the Math 1 test, a question could require you to evaluate a polynomial for specific values of the variables.

EXAMPLE 1: To evaluate $-3x^2y - (x - 2y)$ when $x = -2$ and $y = \frac{1}{2}$, rewrite the polynomial, replacing each x by -2 and each y by $\frac{1}{2}$. Be sure to write each number in parentheses. Then evaluate mentally or with your calculator:

$$\begin{aligned} -3(-2)^2\left(\frac{1}{2}\right) - \left(-2 - 2\left(\frac{1}{2}\right)\right) &= -3(4)\left(\frac{1}{2}\right) - (-2 - 1) \\ &= -6 - (-3) \\ &= -6 + 3 = -3 \end{aligned}$$

Helpful hint: Be sure to follow PEMDAS. If you enter everything in parentheses, your calculator will do this automatically. In Example 1, you *cannot* multiply -3 by -2, get 6, and then square 6; you must first square -2.

Two terms are called ***like terms*** if they have exactly the same variables and exponents. They can differ only in their coefficients: $7a^2b$ and $-5a^2b$ are like terms, whereas a^2b and b^2a are not.

The polynomial $7x^2 + 4x - 2x + 3x^2 + x - 2$ has six terms, but some of them are like terms and can be combined:

$$7x^2 + 3x^2 = 10x^2 \text{ and } 4x - 2x + x = 3x$$

So, the original polynomial is equivalent to the trinomial $10x^2 + 3x - 2$.

Key Fact D1

The only terms of a polynomial that can be combined are like terms.

Key Fact D2

To add two polynomials, write each in parentheses and put a plus sign between. Then erase the parentheses and combine like terms.

EXAMPLE 2: To find the sum of $2x^2 + 3x - 7$ and $5x^2 - 4x + 12$, proceed as follows:

$$\begin{aligned} &(2x^2 + 3x - 7) + (5x^2 - 4x + 12) \\ &= 2x^2 + 3x - 7 + 5x^2 - 4x + 12 \\ &= (2x^2 + 5x^2) + (3x - 4x) + (-7 + 12) \\ &= 7x^2 - x + 5 \end{aligned}$$

Key Fact D3

To subtract two polynomials, write each one in parentheses and put a minus sign between them. Then change the minus sign between them to a plus sign, change the sign of every term in the second parentheses, and use KEY FACT D2 to add them.

EXAMPLE 3: To subtract $5x^2 - 4x + 12$ from $2x^2 + 3x - 7$, proceed as follows.

Start with the second polynomial and subtract the first:

$$\begin{aligned}&(2x^2 + 3x - 7) - (5x^2 - 4x + 12)\\ &= (2x^2 + 3x - 7) + (-5x^2 + 4x - 12)\\ &= -3x^2 + 7x - 19\end{aligned}$$

Key Fact D4

To multiply monomials, first multiply the coefficients, and then multiply their variables (one by one) by adding their exponents.

EXAMPLE 4: To find the product of $5xy^2z^3$ and $-2x^2y^2$, first multiply 5 by -2. Then multiply x by x^2 and y^2 by y^2, and keep the z^3:

$$(5xy^2z^3)(-2x^2y) = (5)(-2)(x)(x^2)(y^2)(y^2)(z^3) = -10x^3y^4z^3$$

All other polynomials are multiplied by using the distributive law.

Key Fact D5

To multiply a polynomial by a monomial, just multiply each term of the polynomial by the monomial.

EXAMPLE 5: The product of $2x$ and $3x^2 - 6xy + y^2$ is

$$2x(3x^2 - 6xy + y^2) = 2x(3x^2) - 2x(6xy) + 2x(y^2) = 6x^3 - 12x^2y + 2xy^2$$

Key Fact D6

To multiply two polynomials, multiply each term in the first polynomial by each term in the second polynomial and simplify by combining terms, if possible.

Note that if the two polynomials are binomials, KEY FACT D6 is just the FOIL (First terms, Outer terms, Inner terms, Last terms) method.

EXAMPLE 6:

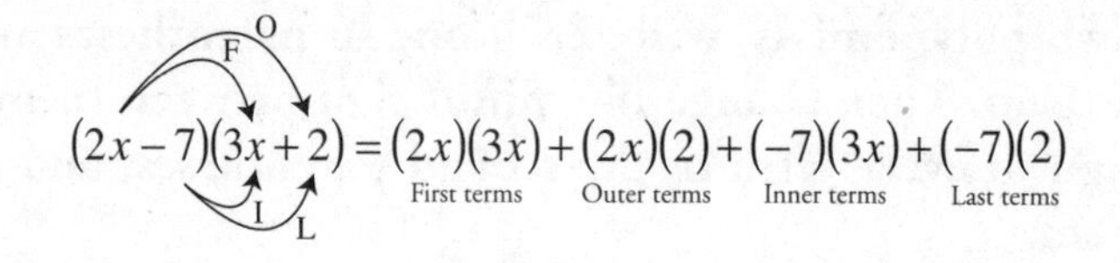

$$= 6x^2 + 4x - 21x - 14 = 6x^2 - 17x - 14$$

EXAMPLE 7:

$$(x+3)(x^2+2x-5) = x^3+2x^2-5x+3x^2+6x-15$$

$$= x^3 + 5x^2 + x - 15$$

Key Fact D7

For the Math 1 test, the three most important binomial products are

TIP

Be sure to memorize these three binomial products.

- $(x-y)(x+y) = x^2 - y^2$
- $(x-y)^2 = x^2 - 2xy + y^2$
- $(x+y)^2 = x^2 + 2xy + y^2$

EXAMPLE 8: If $a - b = 5$ and $a + b = 17$, what is the value of $a^2 - b^2$?

Later in this chapter, you will review how to solve this system of equations. However, here you do not have to solve for a and b, and so you shouldn't. Rather, you should see $a^2 - b^2$ and immediately think $(a - b)(a + b)$.

$$a^2 - b^2 = (a - b)(a + b) = (5)(17) = 85$$

Key Fact D8

To divide a polynomial by a monomial, divide each term by the monomial. Then simplify each term by taking the fractions formed by the coefficients, reducing them to lowest terms, and applying the laws of exponents to the variables.

For example, to divide $(12x^3y^2 - 3xyz)$ by $6x^2y$, write the problem as:

$$\frac{12x^3y^2}{6x^2y} - \frac{3xyz}{6x^2y} = 2xy - \frac{z}{2x}$$

On the Math 1 test you will probably not have to divide two polynomials. If you should have one such question, it is almost always faster to test the answer choices by multiplying than by actually performing long division.

Recall that if a and b are integers, a is a factor of b if there is an integer c such that $ac = b$. For example, 6 is a factor of 48 because there is an integer, namely 8, whose product with 6 is 48: $6 \times 8 = 48$.

Similarly, $x + 4$ is a factor of $x^2 + x - 12$ because there is a polynomial, namely $x - 3$, whose product with $x + 4$ is $x^2 + x - 12$:

$$(x + 4)(x - 3) = x^2 + x - 12$$

The process of finding the factors of a polynomial is called ***factoring***.

The Math 1 test always has at least one or two questions—usually solving a quadratic equation or simplifying an algebraic fraction—that requires you to factor a polynomial.

Key Fact D9

The first step in factoring a polynomial is to look for the greatest common factor of all the terms and, if there is one, to use the distributive property to remove it.

For example:

- To factor $12x^2y + 8xyz$, observe that the greatest common factor of $12x^2y$ and $8xyz$ is $4xy$, and so
$$12x^2y + 8xyz = 4xy(3x + 2z).$$
- To factor $x^3 + x^2 - 12x$, observe that the greatest common factor of x^3, x^2, and $12x$ is x, and so
$$x^3 + x^2 - 12x = x(x^2 + x - 12).$$
- We can't use KEY FACT D9 to factor $x^2 + 3x - 18$, because the greatest common factor of x^2, $3x$, and 18 is 1.

Note: In the second example $x^3 + x^2 - 12x$ has not been completely factored, and in the third example, $x^2 + 3x - 18$ can be factored (but *not* using Key Fact D9). See the examples that follow.

Besides removing common factors, most, if not all, of the factoring you will have to do on the Math 1 test will be limited to using the formula $x^2 - y^2 = (x - y)(x + y)$ for the difference of two squares and factoring trinomials.

Key Fact D10

To factor a trinomial, remove a common factor, if there is one, and then use trial and error to find the two binomials whose product is that trinomial.

For example:

- $x^2 + 12x + 36 = (x + 6)(x + 6) = (x + 6)^2$
- $x^2 + 3x - 18 = (x + 6)(x - 3)$
- $2x^2 + 6x - 36 = 2(x^2 + 3x - 18) = 2(x - 3)(x + 6)$
- $x^3 + x^2 - 12x = x(x^2 + x - 12) = x(x + 4)(x - 3)$
- $x^4 - 1 = (x^2 - 1)(x^2 + 1) = (x - 1)(x + 1)(x^2 + 1)$

ALGEBRAIC FRACTIONS

Although the coefficient of any term in a polynomial can be a fraction, such as $\frac{1}{2}x^2 - 2x + \frac{3}{5}$, the variable itself cannot be in the denominator. Expressions such as $\frac{3}{x}$ and $\frac{2x^2-18}{x^2+5x+6}$, which do have variables in their denominators, are called ***algebraic fractions***. You will have no trouble manipulating algebraic fractions if you treat them as regular fractions, using all the standard rules for adding, subtracting, multiplying, and dividing.

TIP

All the rules for numerical fractions apply to algebraic fractions.

Whenever you have to simplify an algebraic fraction, factor the numerator and denominator, and divide out any common factors.

EXAMPLE 9: Simplify $\frac{2x^2-18}{x^2+5x+6}$.

$$\frac{2x^2-18}{x^2+5x+6} = \frac{2(x^2-9)}{(x+2)(x+3)} = \frac{2(x-3)\cancel{(x+3)}}{(x+2)\cancel{(x+3)}} = \frac{2(x-3)}{x+2}$$

Note that $\frac{2x^2-18}{x^2+5x+6} = \frac{2(x-3)}{x+2}$ is an identity. For every real number (except −2 and −3, for which the original fraction is undefined), the expressions have the same value. For example, when $x = 8$:

$$\frac{2x^2-18}{x^2+5x+6} = \frac{2(8)^2-18}{8^2+5(8)+6} = \frac{110}{110} = 1 \quad \text{and}$$

$$\frac{2(8-3)}{8+2} = \frac{10}{10} = 1$$

Similarly, when $x = 3$, both expressions equal 0, and when $x = -4$, both expressions equal 7.

Exercises

1. What is the value of $\frac{a^2 - b^2}{a - b}$ when $a = -5.3$ and $b = 3.5$?
   (A) –8.8
   (B) –1.8
   (C) 1.8
   (D) 4.4
   (E) 8.8

2. If $a^2 + b^2 = 4$ and $(a - b)^2 = 2$, what is the value of ab?
   (A) 1
   (B) $\sqrt{2}$
   (C) 2
   (D) 3
   (E) 4

3. What is the result when $(x + 3)(x - 5)$ is subtracted from $(2x + 3)(2x - 5)$?
   (A) $x(3x - 2)$
   (B) $x(2 - 3x)$
   (C) $3x^2$
   (D) $5x^2 - 6x$
   (E) $5x^2 - 6x - 30$

4. If $x > 10$, then
   $$\frac{3x^2 - 12}{x^2 - 2x - 8} \div \frac{x^2 - 5x + 6}{2x^2 - 14x + 24} =$$
   (A) $\frac{3}{2}$
   (B) 6
   (C) $\frac{6(x-2)^2}{(x-4)^2}$
   (D) $\frac{3(x-2)^2}{2(x-4)^2}$
   (E) $\frac{3(x-2)}{2(x-4)}$

5. If $\left(\frac{1}{a} + a\right)^2 = 100$, what is the value of $\frac{1}{a^2} + a^2$?
   (A) 10
   (B) 98
   (C) 100
   (D) 102
   (E) 110

ANSWERS EXPLAINED

Answer Key

1. **(B)** 2. **(A)** 3. **(A)** 4. **(B)** 5. **(B)**

Solutions

Each of the problems in this set of exercises is typical of a question you could see on a Math 1 test. When you take the model tests in this book and, in particular, when you take the actual Math 1 test, if you get stuck on questions such as these, you do not have to leave them out—you can almost always answer them by using one or more of the strategies discussed in the "Tactics" chapter. The solutions given here do *not* depend on those strategies; they are the correct mathematical ones.

1. **(B)** $\frac{a^2-b^2}{a-b}=\frac{(a+b)\cancel{(a-b)}}{\cancel{a-b}}=a+b=-5.3+3.5=-1.8$

Of course you could replace a by -5.3 and b by -1.8 in the original expression and evaluate it with your calculator.

2. **(A)** $2=(a-b)^2=a^2-2ab+b^2=4-2ab \Rightarrow 2ab=2 \Rightarrow ab=1$

3. **(A)** $(2x+3)(2x-5)=4x^2-4x-15$ and $(x+3)(x-5)=x^2-2x-15$

$$\begin{gathered}(4x^2-4x-15)-(x^2-2x-15)= \\ 4x^2-4x-15-x^2+2x+15= \\ 3x^2-2x=x(3x-2)\end{gathered}$$

4. **(B)**

$$\frac{3x^2-12}{x^2-2x-8} \div \frac{x^2-5x+6}{2x^2-14x+24}=$$

$$\frac{3x^2-12}{x^2-2x-8} \bullet \frac{2x^2-14x+24}{x^2-5x+6}=$$

$$\frac{3\left(x^2-4\right)}{(x-4)(x+2)} \bullet \frac{2\left(x^2-7x+12\right)}{(x-3)(x-2)}=$$

$$\frac{3\cancel{(x-2)}\cancel{(x+2)}}{\cancel{(x-4)}\cancel{(x+2)}} \bullet \frac{2\cancel{(x-3)}\cancel{(x-4)}}{\cancel{(x-3)}\cancel{(x-2)}}=6$$

5. **(B)** $100=\left(\frac{1}{a}+a\right)^2=\left(\frac{1}{a}+a\right)\left(\frac{1}{a}+a\right)=\frac{1}{a^2}+2+a^2 \Rightarrow$

$\frac{1}{a^2}+a^2=98$

Equations and Inequalities

CHAPTER 6

On a typical Math 1 test, several questions require you to solve equations and inequalities. In this section, we will review the different types of equations and inequalities you will encounter and the methods for solving them.

The basic principle that you must adhere to in solving *any equation* is that you can manipulate it in any way as long as *you do the same thing to both sides.* For example, you may add the same number to each side, subtract the same number from each side, multiply or divide each side by the same number (except 0), square each side, take the square root of each side (if the quantities are positive), take the reciprocal of each side, take the logarithm of each side, and so on. These comments apply to inequalities as well. However, you must be very careful when working with inequalities because some procedures, such as multiplying or dividing by a negative number and taking reciprocals, reverse inequalities.

FIRST-DEGREE EQUATIONS AND INEQUALITIES

The simplest equations and inequalities that you will have to solve on the Math 1 test have only one variable and no exponents. They are called first-degree or linear equations and inequalities. You can always use the six-step method described below to solve them.

EXAMPLE 1: The following solution of the equation

$\frac{1}{2}x + 3(x-2) = 2(x+1) + 1$ illustrates each of the six steps.

Step	What You Should Do	$\frac{1}{2}x+3(x-2)=2(x+1)+1$
1	Get rid of fractions (and decimals) by multiplying both sides by a common denominator.	Multiply each term on both sides of the equation by 2: $x+6(x-2)=4(x+1)+2$
2	Get rid of all parentheses by using the distributive law.	$x+6x-12=4x+4+2$
3	Combine like terms on each side.	$7x-12=4x+6$
4	By adding or subtracting, get all the variables on one side.	Subtract $4x$ from each side: $3x-12=6$
5	By adding or subtracting, get all the plain numbers onto the other side.	Add 12 to each side: $3x=18$
6	Divide both sides by the coefficient of the variable.*	Divide both sides by 3: $x=6$

**NOTE:* If you use this method on an inequality and in Step 6 you divide by a negative number, remember to reverse the inequality.*

The example worked out above is unusual because it required all six steps. When you solve an equation on the Math 1 test, you will probably not need to do every step. Think of the six steps as a list of questions that must be answered *in that order*. Ask if each step is necessary. If it isn't, move on to the next one; if it is necessary, do it. Example 2, below, does not require all six steps.

EXAMPLE 2: For what value of x is $5(x-10)=x+10$?

Step	Question	Yes/No	$5(x-10)=x+10$
1	Are there any fractions or decimals?	No	
2	Are there any parentheses?	Yes	Distribute: $5x-50=x+10$
3	Are there any like terms to combine?	No	
4	Are there variables on both sides?	Yes	Subtract x from each side: $4x-50=10$
5	Is there a plain number on the same side as the variable?	Yes	Add 50 to each side: $4x=60$
6	Does the variable have a coefficient?	Yes	Divide both sides by 4: $x=15$

Tactic E1

Memorize these six steps *in order*, and use this method whenever you have to solve this type of equation or inequality.

When you use the six-step method, do not actually write out a table. Rather, use the method as a guideline and mentally go through each step, doing whichever ones are required. From now on, we will just do each step, using the symbol ⇒ to move from one step to the next.

Sometimes an equation on the Math 1 test contains several variables and you have to solve for one of them in terms of the others.

Tactic E2

When you have to solve for one variable in terms of others, treat all the others as if they were numbers, and apply the six-step method.

EXAMPLE 3: If $\frac{2}{3}x = 3y - 4z$, what is the value of y in terms of x and z?

$$\frac{2}{3}x = 3y - 4z$$

Multiply both sides by 3: $2x = 3(3y - 4z) = 9y - 12z$

Add $12z$ to each side: $2x + 12z = 9y$

Divide both sides by 9: $y = \frac{2x + 12z}{9}$

EXAMPLE 4: If v is an integer and the average (arithmetic mean) of 1, 2, 3, and v is less than 20, what is the greatest possible value of v?

Write the inequality $\frac{1+2+3+v}{4} < 20$ and use the six-step method:

$$\frac{1+2+3+v}{4} < 20 \Rightarrow \frac{6+v}{4} < 20 \Rightarrow$$

$$6 + v < 80 \Rightarrow v < 74$$

Since v is an integer, the greatest possible value of v is 73.

ABSOLUTE VALUE, RADICAL, AND FRACTIONAL EQUATIONS AND INEQUALITIES

The six-step method works in other situations as well. The next four examples illustrate the proper use of this method in equations and inequalities in which the variables are in denominators, between absolute value bars, or under radical signs.

EXAMPLE 5: For what value of x is $\frac{3}{4} - \frac{5}{x} = \frac{7}{x}$?

$$\frac{3}{4} - \frac{5}{x} = \frac{7}{x} \Rightarrow 4x\left(\frac{3}{4}\right) - 4x\left(\frac{5}{x}\right) = 4x\left(\frac{7}{x}\right) \Rightarrow$$
$$3x - 20 = 28 \Rightarrow 3x = 48 \Rightarrow x = 16$$

A second, and in this case slightly easier, solution to Example 5 is to start by adding $\frac{5}{x}$ to each side:

$$\left[\frac{3}{4} - \frac{5}{x} = \frac{7}{x}\right] \Rightarrow \left[\frac{3}{4} = \frac{7}{x} + \frac{5}{x}\right] \Rightarrow \frac{3}{4} = \frac{12}{x}$$

Therefore, $3x = 48$ and $x = 16$

EXAMPLE 6: For what values of x is $\frac{3}{4} - \frac{5}{x} > \frac{7}{x}$?

As in the second solution to Example 5:

$$\left[\frac{3}{4} - \frac{5}{x} > \frac{7}{x}\right] \Rightarrow \left[\frac{3}{4} > \frac{7}{x} + \frac{5}{x}\right] \Rightarrow \frac{3}{4} > \frac{12}{x}$$

Now be careful. Cross multiplying the inequality $\frac{3}{4} > \frac{12}{x}$ is essentially multiplying each side of the inequality by $4x$.

So now you must consider two cases. If x is positive, so is $4x$, and multiplying by $4x$ will preserve the inequality. However, if x is negative, so is $4x$, and multiplying by $4x$ will reverse the inequality. Therefore, you must treat the two cases separately.

Case I: $x > 0$.

$$\frac{3}{4} > \frac{12}{x} \Rightarrow 3x > 48 \Rightarrow x > 16$$

So the positive solutions to the inequality are all numbers greater than 16.

Case II: $x < 0$.

$$\frac{3}{4} > \frac{12}{x} \Rightarrow 3x < 48 \Rightarrow x < 16$$

So the negative solutions to the inequality are all negative numbers that are less than 16. Since every negative number is less than 16, every negative number is a solution.

Finally, by combining the two cases, we see that the solution set is $\{x \mid x < 0 \text{ or } x > 16\}$.

EXAMPLE 7: For what values of x is $3|x+5|-5=7$?

$$3|x+5|-5=7 \Rightarrow 3|x+5|=12 \Rightarrow |x+5|=4 \Rightarrow$$
$$x+5=4 \text{ or } x+5=-4 \Rightarrow x=-1 \text{ or } x=-9$$

EXAMPLE 8: For what values of x is $3|x+5|-5<7$?

$$3|x+5|-5<7 \Rightarrow 3|x+5|<12 \Rightarrow |x+5|<4$$

Now recall from KEY FACT A2 that this means:

$$-4<x+5<4 \Rightarrow -9<x<-1$$

If you have to solve an equation such as $4\sqrt{x}-5=7$, which involves a radical, use the six-step method, treating the radical as your variable. After step 6, when you have a radical equal to a number, perform one final step: raise each side to the same power. For example, if the radical is a square root, square both sides; if the radical is a cube root, cube both sides.

EXAMPLE 9: $4\sqrt{x}-5=7 \Rightarrow 4\sqrt{x}=12 \Rightarrow$
$$\sqrt{x}=3 \Rightarrow \left(\sqrt{x}\right)^2=3^2 \Rightarrow x=9$$

EXAMPLE 10: $4\left(\sqrt[3]{x}+1\right)=\sqrt[3]{x}+25 \Rightarrow 4\sqrt[3]{x}+4=\sqrt[3]{x}+25 \Rightarrow$
$$3\sqrt[3]{x}+4=25 \Rightarrow 3\sqrt[3]{x}=21 \Rightarrow \sqrt[3]{x}=7 \Rightarrow$$
$$\left(\sqrt[3]{x}\right)^3=7^3 \Rightarrow x=343$$

TIP

If you are solving an equation and at some point have a square root or any even root equal to a negative number, then the equation has no real solutions.

EXAMPLE 11: $4\sqrt{x}+15=7 \Rightarrow 4\sqrt{x}=-8 \Rightarrow \sqrt{x}=-2$

Since $\sqrt{x}$ cannot be negative, no real number x satisfies the equation $4\sqrt{x}+15=7$.

QUADRATIC EQUATIONS

A quadratic equation is an equation that can be written in the form $ax^2+bx+c=0$, where a, b, and c are any real numbers with $a \neq 0$. Any number, x, that satisfies the equation is called a ***solution*** or a ***root*** of the equation.

Although some quadratic equations can be solved by other, sometimes easier, methods—including factoring—that will be discussed later, *every* quadratic equation can be solved by using the quadratic formula given in KEY FACT E1.

Key Fact E1

QUADRATIC FORMULA

If a, b, and c are real numbers with $a \neq 0$ and if $ax^2 + bx + c = 0$, then

$$x = \frac{-b \pm \sqrt{b^2 - 4ac}}{2a}$$

Recall that the symbol $\pm$ is read "plus or minus" and that $x = \frac{-b \pm \sqrt{b^2 - 4ac}}{2a}$ is an abbreviation for $x = \frac{-b + \sqrt{b^2 - 4ac}}{2a}$ or $x = \frac{-b - \sqrt{b^2 - 4ac}}{2a}$.

As you can see, a quadratic equation has two roots, both of which are determined by the quadratic formula.

The expression $b^2 - 4ac$ that appears under the square root symbol is called the ***discriminant*** of the quadratic equation. As explained in KEY FACT E2, the discriminant provides valuable information about the nature of the roots of a quadratic equation. If we let D represent the discriminant, an alternative way to write the quadratic formula is $x = \frac{-b \pm \sqrt{D}}{2a}$. The following examples illustrate the proper use of the quadratic formula.

EXAMPLE 12: What are the roots of the equation $x^2 - 2x - 15 = 0$?

$$a = 1,\ b = -2,\ c = -15$$

and

$$D = b^2 - 4ac = (-2)^2 - 4(1)(-15) = 4 + 60 = 64$$

$$\text{So } x = \frac{-(-2) \pm \sqrt{64}}{2(1)} = \frac{2 \pm 8}{2} \Rightarrow$$

$$x = \frac{2+8}{2} = 5 \quad \text{or} \quad x = \frac{2-8}{2} = -3$$

EXAMPLE 13: What are the roots of the equation $x^2 = 10x - 25$?

First, rewrite the equation in the form $ax^2 + bx + c = 0$:

$$x^2 - 10x + 25 = 0$$

$$\text{Then } a = 1,\ b = -10,\ c = 25$$

and

$$D = b^2 - 4ac = (-10)^2 - 4(1)(25) = 100 - 100 = 0$$

$$\text{So } x = \frac{-(-10) \pm \sqrt{0}}{2(1)} = \frac{10 \pm 0}{2} \Rightarrow$$

$$x = \frac{10+0}{2} = 5 \quad \text{or} \quad x = \frac{10-0}{2} = 5$$

Notice that since $10 + 0 = 10$ and $10 - 0 = 10$, the two roots are each equal to 5. Although some people would say that the equation $x^2 - 10x + 25 = 0$ has only one root, you should say that the equation has two equal roots.

EXAMPLE 14: Solve the equation $2x^2 - 4x - 1 = 0$.

$$a = 2,\ b = -4,\ c = -1$$

and

$$D = b^2 - 4ac = (-4)^2 - 4(2)(-1) = 16 + 8 = 24$$

$$\text{So } x = \frac{-(-4) \pm \sqrt{24}}{2(2)} = \frac{4 \pm 2\sqrt{6}}{4} \Rightarrow$$

$$x = \frac{4 + 2\sqrt{6}}{4} = 1 + \frac{1}{2}\sqrt{6} \quad \text{or} \quad x = \frac{4 - 2\sqrt{6}}{4} = 1 - \frac{1}{2}\sqrt{6}.$$

EXAMPLE 15: Solve the equation $x^2 - 2x + 2 = 0$.

$$a = 1,\ b = -2,\ c = 2$$

and

$$D = b^2 - 4ac = (-2)^2 - 4(1)(2) = 4 - 8 = -4$$

$$\text{So } x = \frac{-(-2) \pm \sqrt{-4}}{2(1)} = \frac{2 \pm 2i}{2} = 1 \pm i \Rightarrow$$

$$x = 1 + i \quad \text{or} \quad x = 1 - i$$

See Chapter 17 for a discussion of the imaginary unit *i*.

Smart Strategy

Before using the quadratic formula, see if another method would be easier and quicker.

If in the equation $ax^2 + bx + c = 0$, either b or c is equal to 0, it is easier *not* to use the quadratic formula.

- If $b = 0$, you can just take a square root.

$$ax^2 + c = 0 \Rightarrow ax^2 = -c \Rightarrow x^2 = \frac{-c}{a} \Rightarrow x = \pm\sqrt{\frac{-c}{a}}$$

$$\text{So } x = \sqrt{\frac{-c}{a}} \quad \text{or} \quad x = -\sqrt{\frac{-c}{a}}$$

Note that if $c \le 0$, the two solutions will be real. If $c > 0$, the expression under the square root sign will be negative, and the two solutions will be imaginary.

EXAMPLE 16:

$$2x^2 - 18 = 0 \Rightarrow 2x^2 = 18 \Rightarrow x^2 = 9 \Rightarrow x = \pm\sqrt{9} = \pm 3$$

$$\text{So } x = 3 \quad \text{or} \quad x = -3$$

EXAMPLE 17:

$$2x^2 + 18 = 0 \Rightarrow 2x^2 = -18 \Rightarrow x^2 = -9 \Rightarrow x = \pm\sqrt{-9} = \pm 3i$$

$$\text{So } x = 3i \quad \text{or} \quad x = -3i$$

- If $c = 0$, you should just factor out an x.

$$ax^2 + bx = 0 \Rightarrow x(ax + b) = 0$$

Recall that if a product is equal to 0, one of the factors must be 0. So:

$$x(ax+b) = 0 \Rightarrow \quad x = 0 \quad \text{or} \quad ax + b = 0 \Rightarrow$$
$$x = 0 \quad \text{or} \quad x = \frac{-b}{a}$$

EXAMPLE 18:

$$2x^2 - 5x = 0 \Rightarrow x(2x-5) = 0 \Rightarrow$$
$$x = 0 \quad \text{or} \quad 2x - 5 = 0 \Rightarrow$$
$$x = 0 \quad \text{or} \quad x = \frac{5}{2}$$

EXAMPLE 19:

$$3x^2 = 4x \Rightarrow 3x^2 - 4x = 0 \Rightarrow x(3x-4) = 0 \Rightarrow$$
$$x = 0 \quad \text{or} \quad 3x - 4 = 0 \Rightarrow$$
$$x = 0 \quad \text{or} \quad x = \frac{4}{3}$$

- If you can easily factor the expression $ax^2 + bx + c$, then you can solve the equation $ax^2 + bx + c = 0$ by setting each factor equal to 0.

$$x^2 - x - 12 = 0 \Rightarrow (x-4)(x+3) = 0 \Rightarrow$$
$$x - 4 = 0 \quad \text{or} \quad x + 3 = 0 \Rightarrow$$
$$x = 4 \quad \text{or} \quad x = -3$$

Factoring is faster than using the quadratic formula *if you can immediately determine the factors.* However, factoring is a trial and error method. Even if an expression factors, it may take you a long time to do so. Worse, after spending a few minutes trying to factor, you may realize that the expression cannot be factored. Now you *have to* use the quadratic formula, and it would have taken less time to have done so right away.

Consider the equations $2x^2 + 11x - 24 = 0$ and $2x^2 + 13x - 24 = 0$. Only one of them can be solved by factoring. Do you know which one? Probably not. So if you were given one of these equations to solve and you attempted the factoring method, you might waste valuable time in a fruitless effort. Therefore, to solve a quadratic equation, use the quadratic formula unless $b = 0$ or $c = 0$ or unless you *immediately* see how to factor the quadratic expression.

Sometimes a question on a Math 1 test asks for information about the roots of a quadratic equation but does not specifically require you to solve the equation. Looking at the discriminant (D) will allow you to answer such a question.

- If $D = 0$, as in Example 13, the two roots are $x = \frac{-b+0}{2a} = -\frac{b}{2a}$ and $x = \frac{-b-0}{2a} = -\frac{b}{2a}$. So the two roots are equal. If a, b, and c are all rational numbers, then $x = -\frac{b}{2a}$ is also rational.
- If D is negative, as in Example 15, then the equation has no real roots. Both roots are complex (or imaginary) numbers. (See Chapter 17 for a discussion of complex numbers.) If one of the roots is $u + vi$, the other root is its conjugate: $u - vi$.
- If D is positive, as in Examples 12 and 14, then the equation has two unequal real roots. If a, b, and c are rational, then if D is a perfect square, as in Example 12, the two roots are rational; and if D is not a perfect square, as in Example 14, the two roots are irrational.

The preceding remarks are summarized in the following chart.

Key Fact E2

If a, b, and c are rational numbers with $a \neq 0$, if $ax^2 + bx + c = 0$, and if $D = b^2 - 4ac$, then

Value of Discriminant	Nature of the Roots
$D = 0$	2 equal rational roots
$D < 0$	2 unequal complex roots that are conjugates of each other
$D > 0$	
(i) D is a perfect square	2 unequal rational roots
(ii) D is not a perfect square	2 unequal irrational roots

Be sure to memorize the facts in this chart.

If $ax^2 + bx + c = 0$ and r_1 and r_2 are the two roots, then the sum of the roots is

$$r_1 + r_2 = \frac{-b + \sqrt{b^2 - 4ac}}{2a} + \frac{-b - \sqrt{b^2 - 4ac}}{2a} = \frac{-2b}{2a} = \frac{-b}{a}$$

and the product of the roots is

$$r_1 r_2 = \left(\frac{-b + \sqrt{b^2 - 4ac}}{2a}\right)\left(\frac{-b - \sqrt{b^2 - 4ac}}{2a}\right)$$

$$= \frac{b^2 - (b^2 - 4ac)}{4a^2} = \frac{4ac}{4a^2} = \frac{c}{a}$$

Key Fact E3

If $ax^2 + bx + c = 0$, then the sum of the two roots is $\frac{-b}{a}$ and the product of the two roots is $\frac{c}{a}$.

EXAMPLE 20: Find a quadratic equation for which the sum of the roots is 5 and the product of the roots is 5.

For simplicity, let $a = 1$. Then $\frac{-b}{a} = 5 \Rightarrow \frac{-b}{1} = 5 \Rightarrow b = -5$,

and $\frac{c}{a} = 5 \Rightarrow \frac{c}{1} = 5 \Rightarrow c = 5$. So the equation $x^2 - 5x + 5 = 0$

satisfies the given conditions.

EXPONENTIAL EQUATIONS

Sometimes on a Math 1 test you will have to solve an equation in which the variables are in the exponents. There are two ways to handle equations of this type: use the laws of exponents or use logarithms.

There is a big difference between the equations $2^{x-3} = 16$ and $2^{x-3} = 15$. The first equation is much easier to solve than the second *if you recognize that 16 is a power of 2.*

EXAMPLE 21: For what value of x is $2^{x-3} = 16$?

$$2^{x-3} = 16 \Rightarrow 2^{x-3} = 2^4 \Rightarrow x - 3 = 4 \Rightarrow x = 7$$

EXAMPLE 22: For what value of x is $2^{x-3} = 15$?

Since 15 is not a power of 2, you must use logarithms, which you evaluate on your calculator, to solve this equation. (See Chapter 2 to review the laws of logarithms.)

$$2^{x-3} = 15 \Rightarrow \log 2^{x-3} = \log 15 \Rightarrow$$

$$(x-3)\log 2 = \log 15 \Rightarrow x - 3 = \frac{\log 15}{\log 2} \Rightarrow$$

$$x = 3 + \frac{\log 15}{\log 2} = 3 + 3.91 = 6.91$$

Smart Strategy

Only use logarithms if you have to.

Of course, you *could have* used logarithms to solve the equation in Example 21. In that case, you would have

$$x = 3 + \frac{\log 16}{\log 2} = 3 + 4 = 7$$

Calculator Hint: Another option for solving exponential equations is to use a graphing calculator. For example, to solve the equation in Example 22, graph $y = 2^{x-3}$, and trace along the curve until $y = 15$, zooming in to get any desired degree of accuracy.

Tactic E3

Use logarithms to solve equations in which variables appear in the exponents.

Sometimes the exponents are on both sides of the equation. Again, if both sides can be written as powers of the same integer, you can use the laws of exponents. Otherwise, you must use logarithms.

EXAMPLE 23:

$$4^{x+3} = 8^{x-1} \Rightarrow (2^2)^{x+3} = (2^3)^{x-1} \Rightarrow 2^{2(x+3)} = 2^{3(x-1)} \Rightarrow$$
$$2^{2x+6} = 2^{3x-3} \Rightarrow 2x + 6 = 3x - 3 \Rightarrow x = 9$$

EXAMPLE 24:

$$4^{x+3} = 7^{x-1} \Rightarrow \log 4^{x+3} = \log 7^{x-1} \Rightarrow$$
$$(x+3)\log 4 = (x-1)\log 7 \Rightarrow$$
$$(x+3)(0.602) = (x-1)(0.845) \Rightarrow$$
$$0.602x + 1.806 = 0.845x - 0.845 \Rightarrow$$
$$0.243x = 2.651 \Rightarrow x = 10.9$$

Note:

- In Example 23, 4 and 8 are both powers of 2, so you don't need logarithms.
- In Example 24, 4 and 7 *cannot* be written as powers of the same integer, so you *must* use logarithms.

Occasionally, two different variables are in the exponents and you have to determine the relationship between the variables.

EXAMPLE 25: If $2^x = 4^y$, what is the ratio of x to y?

$$2^x = 4^y \Rightarrow 2^x = \left(2^2\right)^y \Rightarrow 2^x = 2^{2y} \Rightarrow x = 2y \Rightarrow \frac{x}{y} = 2$$

EXAMPLE 26: If $2^x = 5^y$, what is the ratio of x to y?

$$2^x = 5^y \Rightarrow \log 2^x = \log 5^y \Rightarrow$$
$$x \log 2 = y \log 5 \Rightarrow \frac{x}{y} = \frac{\log 5}{\log 2} = 2.32$$

SYSTEMS OF LINEAR EQUATIONS

A ***system of equations*** is a set of two or more equations involving two or more variables. A solution consists of a value for each variable that will simultaneously satisfy each equation.

The equations $2x + y = 13$ and $3x - y = 12$ each have infinitely many solutions. A few of them are given in the tables that follow.

$2x + y = 13$

x	0	2	5	–5	10	0.5	π
y	13	9	3	23	–7	12	$13 - 2\pi$
$2x + y$	13	13	13	13	13	13	13

$3x - y = 12$

x	0	2	5	–2	10	6.2	π
y	–12	–6	3	–18	18	6.6	$3\pi - 12$
$3x - y$	12	12	12	12	12	12	12

However, only one pair of numbers, $x = 5$ and $y = 3$, satisfies both equations simultaneously: $2(5) + 3 = 13$ and $3(5) - 3 = 12$. This then is the only solution of the system of equations: $\begin{Bmatrix} 2x + y = 13 \\ 3x - y = 12 \end{Bmatrix}$

Since the graphs of $2x + y = 13$ and $3x - y = 12$ are lines (see Chapter 13), these equations are called linear equations. You can use three basic methods to solve systems of linear equations, two algebraic ones—the addition method and the substitution method—and one graphic one.

THE ADDITION METHOD

EXAMPLE 27: The easiest way to solve the system of equations discussed above is to add the two equations:

$$\begin{aligned} 2x + y &= 13 \\ \underline{3x - y} &\underline{= 12} \\ 5x \quad &= 25 \Rightarrow x = 5 \end{aligned}$$

Now solve for y by replacing x with 5 in either of the two original equations. For example:

$$2(5) + y = 13 \Rightarrow 10 + y = 13 \Rightarrow y = 3$$

So the unique solution is $x = 5$ and $y = 3$.

The addition method worked so nicely in Example 27 because adding the two equations eliminated the variable y. That does not always happen, however, as the following example illustrates.

EXAMPLE 28: To solve the system $\begin{Bmatrix} 2x + y = 13 \\ 5x - 2y = 10 \end{Bmatrix}$, you cannot just add the two equations.

Fortunately, there is an easy remedy. If you multiply the first equation by 2, you will get an equivalent equation: $4x + 2y = 26$. Now you can use the addition method to solve the new system:

$$\begin{array}{l} 4x + 2y = 26 \\ \underline{5x - 2y = 10} \\ 9x \qquad = 36 \Rightarrow x = 4 \end{array}$$

Then, substitute 4 for x in one of the original equations:

$$2(4) + y = 13 \Rightarrow 8 + y = 13 \Rightarrow y = 5$$

So the solution is $x = 4$ and $y = 5$.

In Example 29 below, you have to multiply each of the equations in a system by a different number to yield a new system in which adding the equations will eliminate a variable.

EXAMPLE 29: To solve the system $\begin{Bmatrix} 2x + 3y = 31 \\ 3x + 2y = 29 \end{Bmatrix}$, multiply the first equation by 3 and the second equation by −2, and then add the new equations:

$$\begin{array}{rcr} 3(2x + 3y = 31) & \rightarrow & 6x + 9y = 93 \\ \underline{-2(3x + 2y = 29)} & \underline{\rightarrow} & \underline{-6x - 4y = -58} \\ & & 5y = 35 \Rightarrow y = 7 \end{array}$$

Now replace y by 7 in either of the *original* equations:

$$3x + 2(7) = 29 \Rightarrow 3x + 14 = 29 \Rightarrow 3x = 15 \Rightarrow x = 5$$

The solution is $x = 5$ and $y = 7$.

THE SUBSTITUTION METHOD

If in a system of equations either variable has a coefficient of 1 or −1, solving the system by the substitution method may be just as easy or even easier than solving it by the addition method. Look at the system of equations you solved in Example 27: $\begin{Bmatrix} 2x + y = 13 \\ 3x - y = 12 \end{Bmatrix}$. By subtracting $2x$ from each side of the first equation, you can rewrite that equation as $y = 13 - 2x$. Now in the second equation you can replace y by $13 - 2x$:

$$3x - (13 - 2x) = 12$$

This is now a simple equation in one variable that you can solve using the six-step method.

$$3x - (13 - 2x) = 12 \Rightarrow 3x - 13 + 2x = 12 \Rightarrow$$
$$5x - 13 = 12 \Rightarrow 5x = 25 \Rightarrow x = 5$$

Then substitute 5 for x:

$$y = 13 - 2(5) = 13 - 10 = 3$$

The solution is $x = 5$ and $y = 3$.

THE GRAPHING METHOD

Systems of linear equations can also be solved graphically. To solve $\begin{Bmatrix} 2x + y = 3 \\ 3x - y = 7 \end{Bmatrix}$, graph each of the lines and find the point where the two lines intersect. The x- and y-coordinates of the point of intersection are the x and y values of the solution. If you have a graphing calculator, you can rewrite the equations as $y = 3 - 2x$ and $y = 3x - 7$ and graph them. You can then trace along one of the lines to the point of intersection or use the calculator's capability to locate the point of intersection.

Smart Strategy

As a rule, you should consider using the graphing method only if your algebra skills are very weak.

Just because you have a graphing calculator, you do not have to use it. Many students can solve the preceding system of equations algebraically in less time than it takes to graph it on a calculator.

If you do not have a graphing calculator, you can use the methods discussed in Chapter 13 to make a rough sketch.

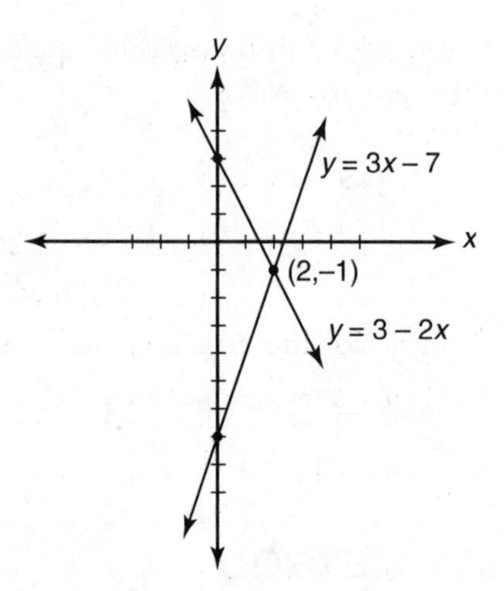

Even if your sketch is not perfect, you will probably be able to eliminate at least three, and maybe even four, choices.

SOLVING LINEAR-QUADRATIC SYSTEMS

A question on the Math 1 test could ask you to solve a system of equations in which one, or even both, of the equations are quadratic. The next example illustrates this.

EXAMPLE 30: To solve the system $\begin{Bmatrix} y = 2x - 1 \\ y = x^2 - 2x + 2 \end{Bmatrix}$, use the substitution method. Replace the y in the second equation by $2x - 1$.

$2x - 1 = x^2 - 2x + 2 \Rightarrow x^2 - 4x + 3 = 0 \Rightarrow$

$(x - 3)(x - 1) = 0 \Rightarrow x = 3$ or $x = 1$

If $x = 3$, then $y = 2(3) - 1 = 5$; and if $x = 1$, then $y = 2(1) - 1 = 1$.

So there are two solutions: $x = 3$, $y = 5$ and $x = 1$, $y = 1$.

As you will see in Chapter 13, solving the system of equations in Example 30 is equivalent to determining the points of intersection of the line $y = 2x - 1$ and the parabola $y = x^2 - 2x + 2$. Those points are (1, 1) and (3, 5).

It follows from the discussion in the previous paragraph that an alternative method of solving the system of equations in Example 30 is to graph them. If you have a graphing calculator, you can graph the given line and parabola and then determine their points of intersection. Which solution is preferable? This is a personal decision. If your algebra skills are strong, solving the system graphically is no faster and offers no advantage. If, on the other hand, your algebra skills are weak and your facility with the calculator is good, you should avoid the algebra and use your calculator.

Exercises

1. If $4x + 13 = 7 - 2x$, what is the value of x?

   (A) $-\frac{10}{3}$
   (B) -3
   (C) -1
   (D) 1
   (E) $\frac{10}{3}$

2. If $\frac{x+3}{2} + 3x = 5(x-3) + \frac{x+23}{5}$, what is the value of x?

   (A) 3
   (B) 5
   (C) 7
   (D) 9
   (E) 11

3. For what values of x is $|2x - 3| - 4 < 7$?

   (A) $x < 7$
   (B) $0 < x < 3$
   (C) $-4 < x < 7$
   (D) $x < 0$ or $x > 3$
   (E) $x < -4$ or $x > 7$

4. Which of the following is a solution of the equation $\sqrt{2x} + 5 = 3$?

   (A) 2
   (B) 4
   (C) 16
   (D) 32
   (E) This equation has no solution.

5. If $ax - b = c - dx$, what is the value of x in terms of a, b, c, and d?

   (A) $\frac{b+c}{a+d}$
   (B) $\frac{c-b}{a-d}$
   (C) $\frac{b+c-d}{a}$
   (D) $\frac{c-b}{a+d}$
   (E) $\frac{c}{b} - \frac{d}{a}$

6. If the sum of two numbers is 13 and the difference of the numbers is 23, what is the product of the numbers?

   (A) −90
   (B) −23
   (C) 0
   (D) 18
   (E) 36

7. If $4^a = 8^b$, what is the ratio of a to b?

   (A) $\frac{1}{2}$
   (B) $\frac{2}{3}$
   (C) $\frac{3}{2}$
   (D) $\frac{2}{1}$
   (E) $\frac{4}{1}$

8. If $4^a = 5^b$, what is the ratio of a to b?

   (A) 0.8
   (B) 0.86
   (C) 1.08
   (D) 1.16
   (E) 1.25

9. If r_1 and r_2 are the two roots of the equation $x^2 - 2x - 1 = 0$ and if $r_1 > r_2$, what is $r_1 - r_2$?

   (A) $-2\sqrt{2}$
   (B) -2
   (C) 0
   (D) $2\sqrt{2}$
   (E) 2

10. If $x^2 + 5x + c = 0$ has exactly one solution, what is the value of c?

   (A) -6.25
   (B) -2.25
   (C) 0
   (D) 3.75
   (E) 6.25

ANSWERS EXPLAINED

Answer Key

1. **(C)**	3. **(C)**	5. **(A)**	7. **(C)**	9. **(D)**
2. **(C)**	4. **(E)**	6. **(A)**	8. **(D)**	10. **(E)**

Solutions

Each of the problems in this set of exercises is typical of a question you could see on a Math 1 test. When you take the model tests in this book and, in particular, when you take the actual Math 1 test, if you get stuck on questions such as these, you do not have to leave them out—you can almost always answer them by using one or more of the strategies discussed in the "Tactics" chapter. The solutions given here do *not* depend on those strategies; they are the correct mathematical ones.

1. **(C)** Add $2x$ to each side: $6x + 13 = 7$. Subtract 13 from each side: $6x = -6$. Divide each side by 6: $x = -1$.

2. **(C)** Use the six-step method:
   (1) Multiply both sides of the equation by 10 to get rid of fractions:

   $$5(x + 3) + 30x = 50(x - 3) + 2(x + 23)$$

   (2) Use the distributive law to get rid of parentheses:

   $$5x + 15 + 30x = 50x - 150 + 2x + 46$$

   (3) Combine like terms:

   $$35x + 15 = 52x - 104$$

   (4) Get the variable onto one side:

   $$15 = 17x - 104$$

(5) Add 104 to each side:

$$119 = 17x$$

(6) Divide by 17:

$$x = 7$$

3. **(C)**

$$|2x - 3| - 4 < 7 \Rightarrow |2x - 3| < 11 \Rightarrow$$
$$-11 < 2x - 3 < 11 \Rightarrow -8 < 2x < 14 \Rightarrow -4 < x < 7$$

4. **(E)** $\sqrt{2x} + 5 = 3 \Rightarrow \sqrt{2x} = -2$. Since by definition a square root cannot be negative, this equation has no solution.

5. **(A)** Treat *a, b, c,* and *d* as constants, and use the six-step method to solve for *x*:

$$ax - b = c - dx \Rightarrow ax - b + dx = c \Rightarrow$$
$$ax + dx = c + b \Rightarrow x(a + d) = b + c \Rightarrow$$
$$x = \frac{b + c}{a + d}$$

6. **(A)** Represent the numbers by *x* and *y*. Then $x + y = 13$ and $x - y = 23$. Now add the equations:

$$\begin{aligned} x + y &= 13 \\ x - y &= 23 \\ \hline 2x \quad &= 36 \Rightarrow x = 18 \end{aligned}$$

So $18 + y = 13 \Rightarrow y = -5$. Finally, the product $xy = (18)(-5) = -90$.

7. **(C)**

$$4^a = (2^2)^a = 2^{2a} \text{ and } 8^b = (2^3)^b = 2^{3b}$$

Therefore, $4^a = 8^b \Rightarrow 2a = 3b \Rightarrow \frac{a}{b} = \frac{3}{2}$

Alternatively, you could use logarithms as in Solution 8 below.

8. **(D)**

$$4^a = 5^b \Rightarrow \log 4^a = \log 5^b \Rightarrow$$
$$a \log 4 = b \log 5 \Rightarrow \frac{a}{b} = \frac{\log 5}{\log 4} = 1.16$$

9. **(D)** Use the quadratic formula to find r_1 and r_2. Since $a = 1$, $b = -2$, and $c = -1$:

$$x = \frac{2 \pm \sqrt{4-(-4)}}{2} = \frac{2 \pm \sqrt{8}}{2} = \frac{2 \pm 2\sqrt{2}}{2} = 1 \pm \sqrt{2}$$

Then $r_1 = 1 + \sqrt{2}$ and $r_2 = 1 - \sqrt{2}$, and so:

$$r_1 - r_2 = \left(1+\sqrt{2}\right) - \left(1-\sqrt{2}\right) = 2\sqrt{2}$$

10. **(E)** If a quadratic equation has exactly one solution, then its discriminant is 0. So:

$$b^2 - 4ac = 25 - 4(1)c = 0 \Rightarrow 25 - 4c = 0 \Rightarrow 4c = 25 \Rightarrow c = 6.25$$

Word Problems

- Rate Problems
- Age Problems
- Percent Problems
- Integer Problems
- A Few Miscellaneous Problems
- Exercises
- Answers Explained

Every Math 1 test has at least a few word problems. You can encounter word problems on any math topic in the Math 1 syllabus. In this section, you will see a variety of problems involving several areas of mathematics. Read each example carefully, and try to follow the algebraic solution.

To solve word problems algebraically, you must treat algebra as a foreign language and translate "word for word" from English into algebra, just as you would from English into any foreign language. When translating into algebra, you should use some letter (often x) to represent the unknown quantity you are trying to determine. Review all of the examples in this section so that you master this translation process. Once the translation is complete, you will be able to use the techniques presented in Chapter 6 to solve the equation.

If you get stuck trying to set up or solve a word problem algebraically, you can use one of the tactics explained in Chapter 1 to get a correct answer.

RATE PROBLEMS

The basic formula used in all rate problems is

$$\text{distance} = \text{rate} \times \text{time} \quad \text{or} \quad d = rt$$

Of course, you can solve for one of the other variables:

$$\text{rate} = \frac{\text{distance}}{\text{time}} \quad \text{or} \quad r = \frac{d}{t}$$

$$\text{time} = \frac{\text{distance}}{\text{rate}} \quad \text{or} \quad t = \frac{d}{r}$$

Sometimes in rate problems, the word *speed* is used instead of *rate.* When you solve word problems, be sure you use consistent units.

EXAMPLE 1: John drove from his house to his office at an average speed of 30 miles per hour. If the trip took 40 minutes, how far, in miles, is it from John's house to his office?

You know the rate (30 mph) and the time (40 minutes), and you want to find the distance. Of course, you are going to use the formula $d = rt$. However, if you write $d = (30)(40) = 1{,}200$, you know something is wrong. Clearly, John didn't drive 1,200 miles in less than one hour. The problem is that the units are wrong. The formula is really:

$$d \text{ miles} = r\frac{\text{miles}}{\cancel{\text{hour}}}\cdot t\ \cancel{\text{hours}} = rt \text{ miles}$$

So you have to convert 40 minutes to hours:

$$40 \text{ minutes} = \left(x\ \cancel{\text{hours}}\right)\left(\frac{60 \text{ minutes}}{\cancel{\text{hour}}}\right) = 60x \text{ minutes}$$

Then $60x = 40 \Rightarrow x = \frac{40}{60} = \frac{2}{3}$. So it took John $\frac{2}{3}$ hours to get to his office.

Now you can use the formula $d = rt$, with $r = 30$ and $t = \frac{2}{3}$.

$$d = \left(30\frac{\text{miles}}{\cancel{\text{hour}}}\right)\left(\frac{2}{3}\ \cancel{\text{hours}}\right) = \left(\frac{2}{3}\right)(30) \text{ miles} = 20 \text{ miles}$$

Be sure you understand the explanation in Example 1. It is very important.

The d in the formula $d = rt$ does not have to represent *distance*; it could represent any type of work that is performed at a constant rate, r, for a certain time, t. In Example 1, instead of driving 20 miles at 30 miles per hour for 40 minutes, in 40 minutes John could have

- read 20 pages at a rate of 30 pages per hour.
- drawn 20 sketches at a rate of 30 sketches per hour.
- planted 20 flowers at a rate of 30 flowers per hour.

EXAMPLE 2: If Brian can paint a fence in 4 hours and Scott can paint the same fence in 6 hours, when working together, how long will it take the two of them to paint the fence?

Call painting the fence "the job." Then Brian works at the rate of $\frac{1 \text{ job}}{4 \text{ hours}} = \frac{1}{4}\frac{\text{jobs}}{\text{hour}}$. Similarly, Scott's rate of work is $\frac{1}{6}\frac{\text{jobs}}{\text{hour}}$. Together they can complete $\left(\frac{1}{4}+\frac{1}{6}\right) = \frac{5}{12}$ jobs per hour. Finally,

$$1 \text{ job} = \left(\frac{5 \text{ jobs}}{12 \text{ hours}}\right)(t \text{ hours}) \Rightarrow 1 = \frac{5}{12}t \Rightarrow t = \frac{12}{5}$$

So it will take Brian and Scott $\frac{12}{5} = 2\frac{2}{5} = 2.4$ hours = 2 hours and 24 minutes to paint the fence.

AGE PROBLEMS

In age problems, it often helps to organize the given data in a table.

EXAMPLE 3: In 2001, Lior was four times as old as Ezra, and in 2003, Lior was twice as old as Ezra. How many years older than Ezra is Lior?

Let x represent Ezra's age in 2001, and make the following table:

Year	Ezra	Lior
2001	x	$4x$
2003	$x + 2$	$4x + 2$

In 2003, Lior was twice as old as Ezra, so

$$4x + 2 = 2(x + 2) \Rightarrow 4x + 2 = 2x + 4 \Rightarrow 2x = 2 \Rightarrow x = 1$$

So in 2001, Ezra was 1 and Lior was 4. Lior is 3 years older than Ezra.

PERCENT PROBLEMS

If you have any difficulty with problems involving percents, including percent increase and percent decrease, review the material on percents in Chapter 3.

EXAMPLE 4: There are twice as many girls as boys in a biology class. If 30% of the girls and 45% of the boys have completed a lab, what percent of the students in the class have not yet completed the lab?

If x represents the number of boys in the class, then $2x$ represents the number of girls in the class. Then of the $3x$ students in the class, the number of students who have completed the lab is

$$0.30(2x) + 0.45x = 0.60x + 0.45x = 1.05x$$

The fraction of students who have completed the lab is

$$\frac{1.05x}{3x} = 0.35 = 35\%$$

So 65% of the students have not yet completed the lab.

INTEGER PROBLEMS

Since the difference between consecutive integers is 1, if n represents an integer, the next three consecutive integers are $n + 1$, $n + 2$, and $n + 3$.

Since the difference between consecutive even integers (or consecutive odd integers) is 2, if n represents an even integer, the next three consecutive even integers are $n + 2$, $n + 4$, and $n + 6$.

EXAMPLE 5: Ruth wrote down three consecutive odd integers. She then multiplied the first one by 2, the second one by 3, and the third one by 4. If the sum of all six numbers is 400, what is the first number she wrote?

Since the first three numbers were consecutive odd integers, represent them by n, $n + 2$, and $n + 4$. The other three numbers were $2n$, $3(n + 2)$, and $4(n + 4)$, so

$$n + (n + 2) + (n + 4) + 2n + 3(n + 2) + 4(n + 4) = 400$$
$$n + n + 2 + n + 4 + 2n + 3n + 6 + 4n + 16 = 400$$
$$12n + 28 = 400 \Rightarrow 12n = 372 \Rightarrow n = 31$$

A FEW MISCELLANEOUS PROBLEMS

EXAMPLE 6: Adam has five times as many baseball cards as Noah. If Adam gives Noah 200 cards, Noah will have five times as many cards as Adam. How many cards does each of them have?

If Noah now has x cards, Adam has $5x$. After Adam gives Noah 200 cards, he will have $5x - 200$ and Noah will have $x + 200$. So:

$$x + 200 = 5(5x - 200) \Rightarrow x + 200 = 25x - 1{,}000 \Rightarrow$$
$$1{,}200 = 24x \Rightarrow x = 50$$

Noah now has 50 cards and Adam has 250 cards.

EXAMPLE 7: An amusement park has two payment options. By using Plan A, you pay a \$10 admission fee plus \$3 for every ride you go on. By using Plan B, you pay a \$20 admission fee and \$1 per ride. What is the least number of rides you must go on for Plan B to be less expensive?

Let x = number of rides you go on. Then the cost using Plan A is $10 + 3x$, and the cost using Plan B is $20 + x$. So

$$20 + x < 10 + 3x \Rightarrow 10 < 2x \Rightarrow 5 < x$$

So you have to go on at least 6 rides for Plan B to be less expensive.

EXAMPLE 8: In 2000, the ratio of boys to girls performing community service at Central High School was 3 to 5. In 2001, the number of boys increased by 6 and the number of girls increased by 2. If the ratio of boys to girls performing community service in 2001 was 2 to 3, how many students performed community service in 2000?

Let $3x$ and $5x$ represent the number of boys and girls doing community service in 2000, respectively. Then in 2001, the numbers were $3x + 6$ and $5x + 2$. So

$$\frac{3x+6}{5x+2} = \frac{2}{3} \Rightarrow 9x + 18 = 10x + 4 \Rightarrow x = 14$$

So in 2000, there were $3(14) = 42$ boys and $5(14) = 70$ girls. The total number of students was $42 + 70 = 112$.

EXAMPLE 9: A concession stand sells soda in two sizes: small and large. A small soda costs $1.50, and a large soda costs $2.50. One day 200 sodas were sold for a total of $360. How many small sodas were sold?

If $x =$ the number of small sodas, and $y =$ the number of large sodas, then

$$x + y = 200 \text{ and } 1.50x + 2.50y = 360$$

$x + y = 200 \Rightarrow y = 200 - x$, so replace y by $200 - x$ in the second equation:

$$360 = 1.50x + 2.50(200 - x) = 1.50x + 500 - 2.50x = 500 - x \Rightarrow x = 140$$

Exercises

1. How many seconds longer does it take to drive 1 mile at 40 miles per hour than at 60 miles per hour?

   (A) 15
   (B) 30
   (C) 40
   (D) 60
   (E) 90

2. On Monday morning, Meri read part of her library book at a rate of 60 pages per hour. In the evening when she was tired, she read the same number of pages at a rate of 40 pages per hour. If Meri read for a total of two hours on Monday, how many minutes did she spend reading in the morning?

   (A) 36
   (B) 45
   (C) 48
   (D) 60
   (E) 72

3. The product of two consecutive positive integers is 12 more than 12 times the sum of those two integers. What is the smaller of the two integers?

   (A) 1
   (B) 23
   (C) 24
   (D) 25
   (E) 49

4. Charlie was 40 years old when his son Adam was born. How old was Charlie when he was 5 times as old as Adam?

   (A) 40
   (B) 45
   (C) 50
   (D) 60
   (E) 80

5. If Mary can address a box of envelopes in 5 hours and Jane can address the same box of envelopes in 10 hours, how many minutes will it take Mary and Jane working together to address all the envelopes in the box?

   (A) 80
   (B) 160
   (C) 200
   (D) 450
   (E) 900

6. Elaine has 50 coins, all nickels and dimes, that have a total value of $3.40. How many of her coins are nickels?

   (A) 18
   (B) 24
   (C) 25
   (D) 28
   (E) 32

7. The length of each side of a square is 3 more than the length of each side of a regular pentagon. If the perimeters of the square and pentagon are equal, how long is each side of the pentagon?

   (A) 6
   (B) 12
   (C) 10
   (D) 15
   (E) 18

8. At an amusement park, Lance bought 3 hamburgers and 4 sodas for a total of \$15. While paying the same prices, Karen bought 2 hamburgers and 3 sodas for \$10.50. What is the total cost of 1 hamburger and 1 soda?

   (A) \$1.50
   (B) \$3.00
   (C) \$4.00
   (D) \$4.50
   (E) \$5.00

ANSWERS EXPLAINED

Answer Key

1. **(B)**	3. **(C)**	5. **(C)**	7. **(B)**
2. **(C)**	4. **(C)**	6. **(E)**	8. **(D)**

Solutions

Each of the problems in this set of exercises is typical of a question you could see on a Math 1 test. When you take the model tests in this book and, in particular, when you take the actual Math 1 test, if you get stuck on questions such as these, you do not have to leave them out—you can almost always answer them by using one or more of the strategies discussed in the "Tactics" chapter. The solutions given here do *not* depend on those strategies; they are the correct mathematical ones.

1. **(B)** Since $d = rt$, $t = \frac{d}{r}$. The time required to drive 1 mile at 40 miles per hour is $t = \frac{1 \text{ mile}}{40 \text{ miles per hour}} = \frac{1}{40} \text{hour} = \left(\frac{1}{40}\right)(60) \text{minutes} = 1.5 \text{ minutes} =$ $(1.5)(60)$ seconds $= 90$ seconds.

   The time required to drive 1 mile at 60 miles per hour is $t = \frac{1 \text{ mile}}{60 \text{ miles per hour}} = \frac{1}{60} \text{hour} = \left(\frac{1}{60}\right)(60) \text{minutes} = 1 \text{ minute} = 60$ seconds. So it takes $90 - 60 = 30$ seconds more to drive 1 mile at 40 miles per hour than at 60 miles per hour.

2. **(C)** Let d = the number of pages Meri read in the morning (and in the evening). Then let t_1 and t_2 be the times Meri spent reading in the morning and evening, respectively:

$$\text{in the morning}: d = 60t_1 \Rightarrow t_1 = \frac{d}{60}$$

$$\text{in the evening}: d = 40t_2 \Rightarrow t_2 = \frac{d}{40}$$

Since $t_1 + t_2 = 2$, we have:

$$\frac{d}{60}+\frac{d}{40}=2 \Rightarrow 120\left(\frac{d}{60}\right)+120\left(\frac{d}{40}\right)=120(2) \Rightarrow$$
$$2d + 3d = 240 \Rightarrow 5d = 240 \Rightarrow d = 48$$

3. **(C)** Let n and $n + 1$ represent the two consecutive positive integers. Then:

$$n(n+1) = 12 + 12(n + n + 1) \Rightarrow$$
$$n^2 + n = 12 + 24n + 12 \Rightarrow n^2 - 23n - 24 = 0 \Rightarrow$$
$$(n+1)(n-24) = 0 \Rightarrow n = -1 \quad \text{or} \quad n = 24$$

Since it is given that n is positive, $n = 24$.

4. **(C)** x years after Adam was born, Charlie was $40 + x$ years old and Adam was x years old.

$$40 + x = 5x \Rightarrow 40 = 4x \Rightarrow x = 10$$

So Charlie was 50 years old (and Adam was 10).

5. **(C)** Mary addresses envelopes at the rate of $\frac{1 \text{ box}}{5 \text{ hours}} = \frac{1}{5}\frac{\text{boxes}}{\text{hour}}$ and Jane addresses envelopes at the rate of $\frac{1}{10}\frac{\text{boxes}}{\text{hour}}$. Together they can address $\left(\frac{1}{5}+\frac{1}{10}\right)=\frac{3}{10}$ boxes per hour. So the whole job will take $1 \div \frac{3}{10} = \frac{10}{3} = 3\frac{1}{3}$ hours = 200 minutes.

6. **(E)** Let x represent the number of nickels Elaine has, and let y represent the number of dimes she has. Then $x + y = 50$. Since each nickel is worth 5 cents and each dime is worth 10 cents and since \$3.40 is 340 cents: $5x + 10y = 340$. Also,

$$x + y = 50 \Rightarrow y = 50 - x. \text{ Therefore:}$$
$$5x + 10(50 - x) = 340 \Rightarrow 5x + 500 - 10x = 340 \Rightarrow$$
$$500 - 5x = 340 \Rightarrow 5x = 160 \Rightarrow x = 32$$

7. **(B)** Let $x =$ the length of each side of the pentagon. Then $x + 3 =$ the length of each side of the square. The perimeter of the pentagon is $5x$ and the perimeter of the square is $4(x + 3)$. Therefore:

$$5x = 4(x + 3) = 4x + 12 \Rightarrow x = 12$$

8. **(D)** Let $x =$ the cost of a hamburger, and let $y =$ the cost of a soda. Then:

$$3x + 4y = 15$$
$$2x + 3y = 10.50$$

After multiplying the top equation by 3 and the bottom equation by 4, we get

$$9x + 12y = 45$$
$$8x + 12y = 42$$

Subtracting the second equation from the first gives us $x = 3$. Then substitute 3 for x:

$$3x + 4y = 15 \Rightarrow 3(3) + 4y = 15 \Rightarrow$$
$$9 + 4y = 15 \Rightarrow 4y = 6 \Rightarrow y = 1.50$$

So 1 hamburger and 1 soda cost 3 + 1.50 = $4.50.

PLANE GEOMETRY

Lines and Angles

CHAPTER 8

- Angles
- Perpendicular and Parallel Lines
- Exercises
- Answers Explained

On the Math 1 test, approximately 20 of the 50 questions deal with geometry. Therefore, mastering geometry is crucial for anyone who is taking this test. Of the 20 geometry questions, about 10 of them are on plane geometry; the other 10 are split between solid geometry and coordinate geometry, which will be discussed in Chapters 12 and 13. Much of the geometry you have learned in your math classes is not covered on the Math 1 test. In this chapter, we will review all of the important topics in plane geometry that you need to know for the Math 1 test and only those.

Lines are usually referred to by lowercase letters, such as ℓ, m, and n. We can also name a line using two of the points on the line. If A and B are points on line ℓ, we can refer to line ℓ as line $\overleftrightarrow{AB}$.

$\overrightarrow{AB}$ represents the ***ray*** that consists of point A and all the points on $\overleftrightarrow{AB}$ that are on the same side of A as B.

$\overline{AB}$ represents the ***line segment*** that consists of points A and B and all the points on $\overleftrightarrow{AB}$ that are between them.

Finally, AB represents the ***length*** of segment $\overline{AB}$.

TIP

$\overline{AB} \cong \overline{XY}$ means the same thing as $AB = XY$.

If two line segments have the same length, we say they are ***congruent***. The symbol $\cong$ is used to indicate congruence, so in the figure below we have $\overline{AB} \cong \overline{XY}$.

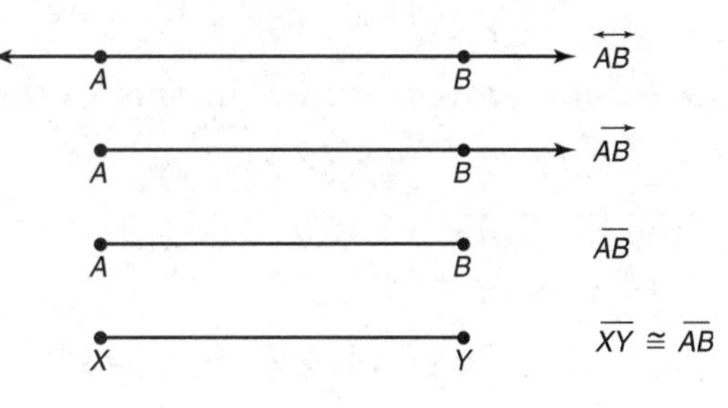

ANGLES

An ***angle*** is formed by the intersection of two line segments, rays, or lines. The point of intersection is called the ***vertex*** of the angle. On the Math 1 test, angles are always measured in degrees and are classified according to their degree measures.

Key Fact G1

- **An *acute* angle measures less than 90°.**
- **A right angle measures 90°.**
- **An obtuse angle measures more than 90° but less than 180°.**
- **A straight angle measures 180°.**

Note

A small square drawn at the vertex of an angle is used to indicate that the angle is a right angle. On the Math Level 1 test, if an angle has a square in it, it must be a 90° angle, *even if the figure has not been drawn to scale.*

An angle can be named by three points: a point on one side, the vertex, and a point on the other side, in that order. When there is no possible ambiguity, we can name the angle just by its vertex.

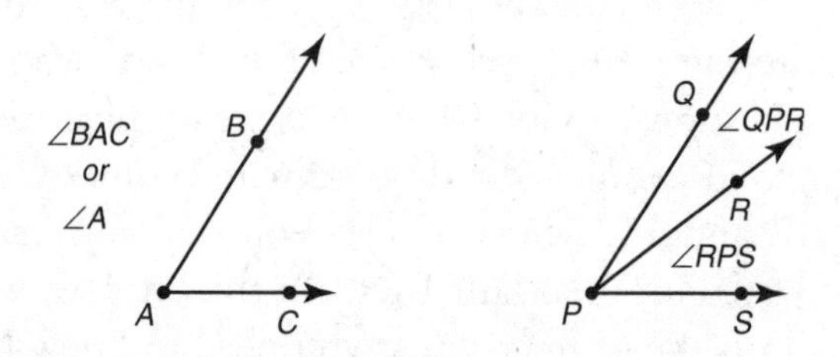

To indicate that the measure of angle A is 60°, say, we write $m\angle A = 60°$.

Two angles that have the same measure are said to be ***congruent***, denoted by the symbol ≅. So if $m\angle A = m\angle B$, we can write $\angle A \cong \angle B$.

TIP

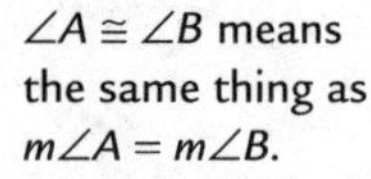

$\angle A \cong \angle B$ means the same thing as $m\angle A = m\angle B$.

If the sum of the measures of two angles is 180°, we say that the angles are ***supplementary***. If the sum of their measures is 90°, we say that the angles are ***complementary***. In the diagram below, $\angle APB$ is supplementary to $\angle BPC$ and $\angle DQE$ is complementary to $\angle EQF$.

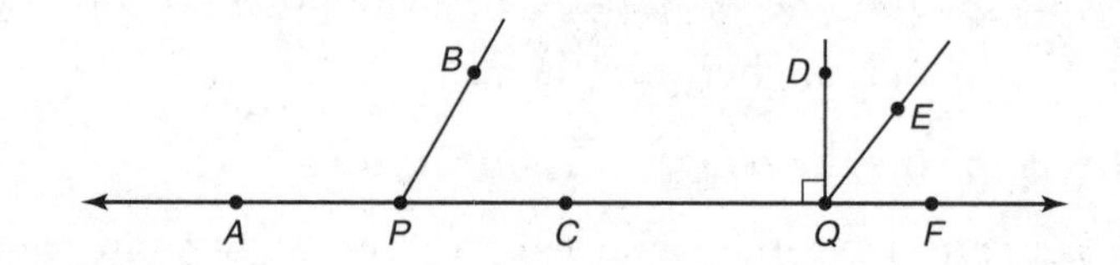

Key Fact G2

If two or more angles form a straight angle, the sum of their measures is 180°.

EXAMPLE 1: If in the figure below $a : b : c = 2 : 3 : 4$, then $a = 2x$, $b = 3x$, and $c = 4x$. So

$$2x + 3x + 4x = 180 \Rightarrow 9x = 180 \Rightarrow$$

$$x = 20° \Rightarrow a = 40°,\ b = 60°, \text{ and } c = 80°.$$

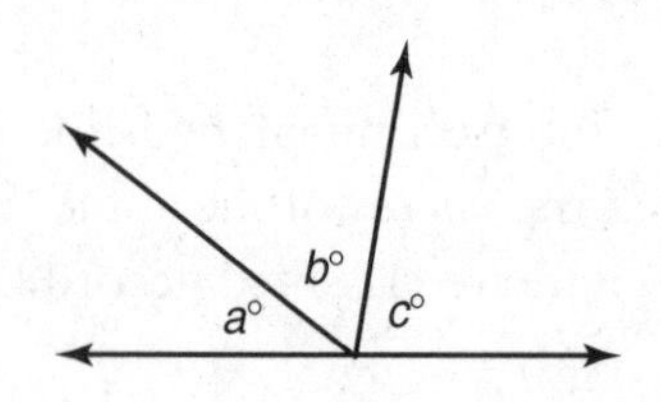

Key Fact G3

The sum of the measures of all the nonoverlapping angles around a point is 360°.

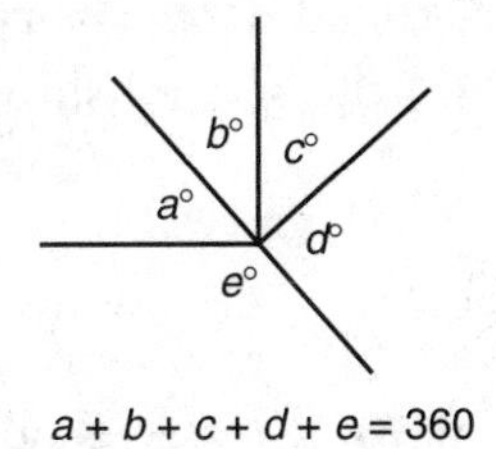

$a + b + c + d + e = 360$

When two lines intersect, four angles are formed. The two angles in each pair of opposite angles are called ***vertical angles***.

Key Fact G4

Vertical angles have equal measures.

TIP 

Key Fact G4 means that vertical angles are congruent.

EXAMPLE 2: To find the value of y in the figure below, note that $5x + 6 = 7(x - 2) \Rightarrow 5x + 6 = 7x - 14 \Rightarrow 2x = 20$. Therefore, $x = 10$. So the measure of each acute angle is $(5 \times 10 + 6)° = 56°$ and $y = 180 - 56 = 124$.

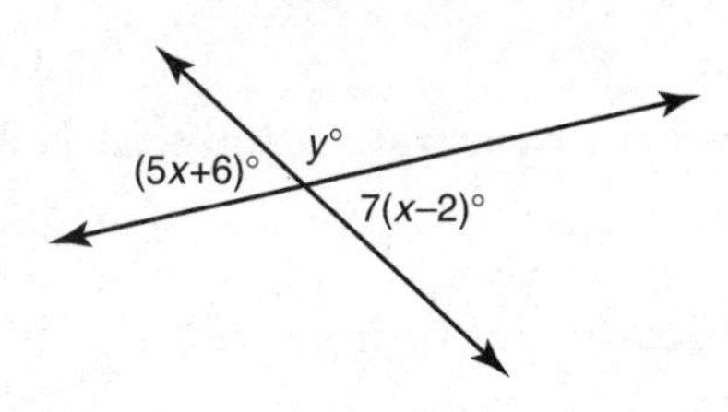

In the figure below, line ℓ divides $\angle PQR$ into two equal parts, and line k divides line segment $\overline{ST}$ into two equal parts. Line ℓ is said to ***bisect*** the angle, and line k ***bisects*** the line segment. Point M is called the ***midpoint*** of segment $\overline{ST}$.

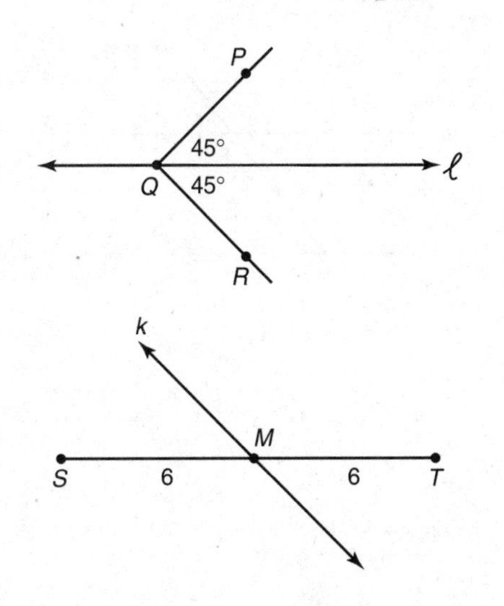

PERPENDICULAR AND PARALLEL LINES

Two lines that intersect to form right angles are called ***perpendicular***. Two lines that never intersect are said to be ***parallel***. Consequently, parallel lines form no angles. However, if a third line, called a ***transversal***, intersects a pair of parallel lines, eight angles are formed, and the relationships between these angles are very important.

Key Fact G5

If a pair of parallel lines is cut by a transversal that is perpendicular to the parallel lines, all eight angles are right angles.

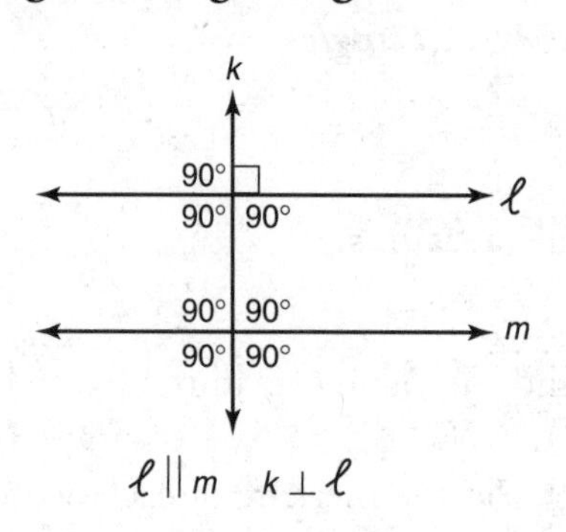

Key Fact G6

If a pair of parallel lines is cut by a transversal that is not perpendicular to the parallel lines:

TIP

The four acute angles all have the same measure, and the four obtuse angles all have the same measure.

- **Four of the angles are acute, and four are obtuse.**
- **All four acute angles are congruent.**
- **All four obtuse angles are congruent.**
- **The sum of the measures of any acute angle and any obtuse angle is 180°.**

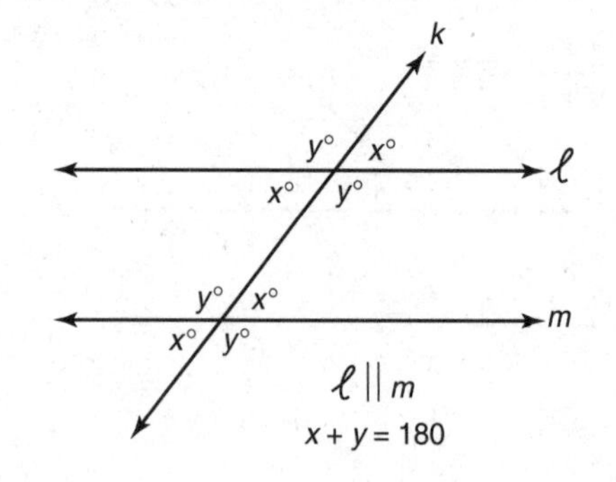

EXAMPLE 3: In the figure below, what is the value of $a + b + c + d$?

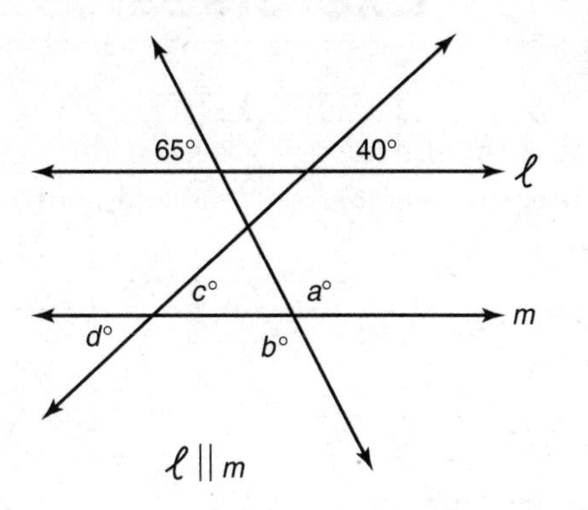

By KEY FACT G6, c and d each measure 40° and $65 + a = 180$. Therefore, $a = 115$. Since vertical angles have equal measures, b is also 115, and so

$$a + b + c + d = 115 + 115 + 40 + 40 = 310.$$

The converse of KEY FACT G6 is also true. If in the figure below, $x = y$, then lines ℓ and m are parallel.

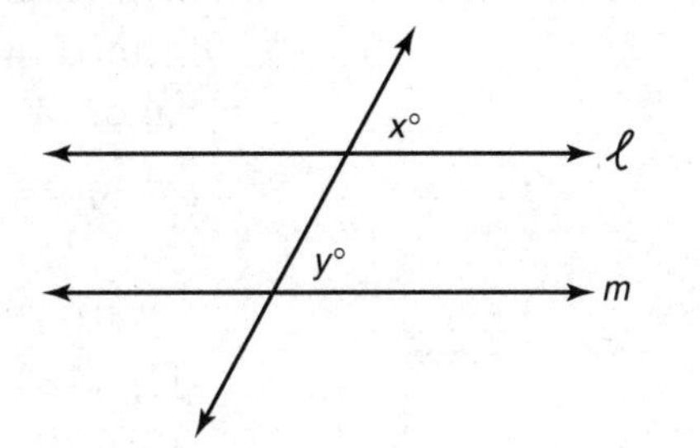

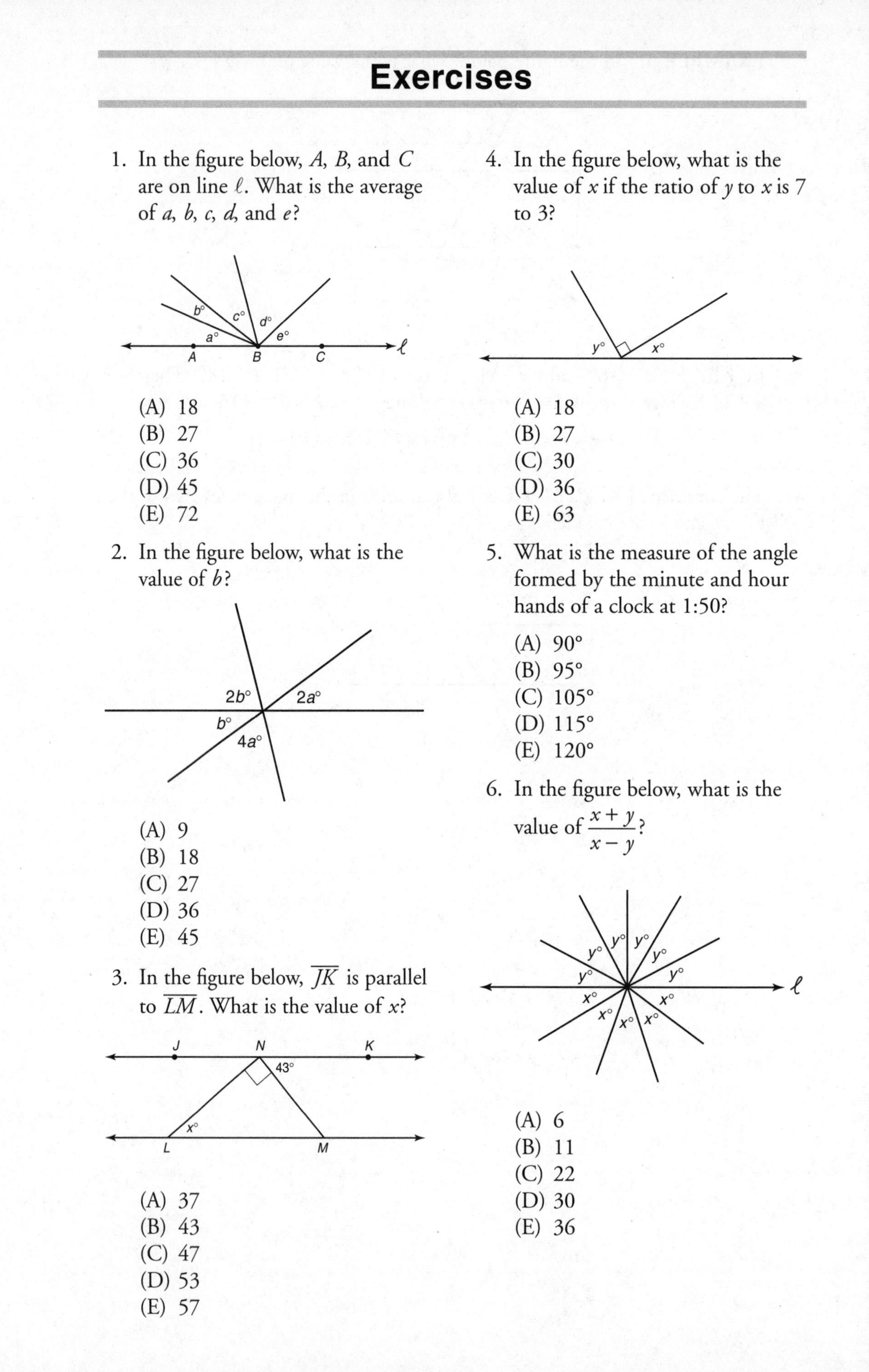

Exercises

1. In the figure below, A, B, and C are on line ℓ. What is the average of a, b, c, d, and e?

   (A) 18
   (B) 27
   (C) 36
   (D) 45
   (E) 72

2. In the figure below, what is the value of b?

   (A) 9
   (B) 18
   (C) 27
   (D) 36
   (E) 45

3. In the figure below, $\overline{JK}$ is parallel to $\overline{LM}$. What is the value of x?

   (A) 37
   (B) 43
   (C) 47
   (D) 53
   (E) 57

4. In the figure below, what is the value of x if the ratio of y to x is 7 to 3?

   (A) 18
   (B) 27
   (C) 30
   (D) 36
   (E) 63

5. What is the measure of the angle formed by the minute and hour hands of a clock at 1:50?

   (A) 90°
   (B) 95°
   (C) 105°
   (D) 115°
   (E) 120°

6. In the figure below, what is the value of $\frac{x+y}{x-y}$?

   (A) 6
   (B) 11
   (C) 22
   (D) 30
   (E) 36

7. In the figure below, $\overline{PQ}$ is parallel to $\overline{RS}$. What is the value of $a + b$?

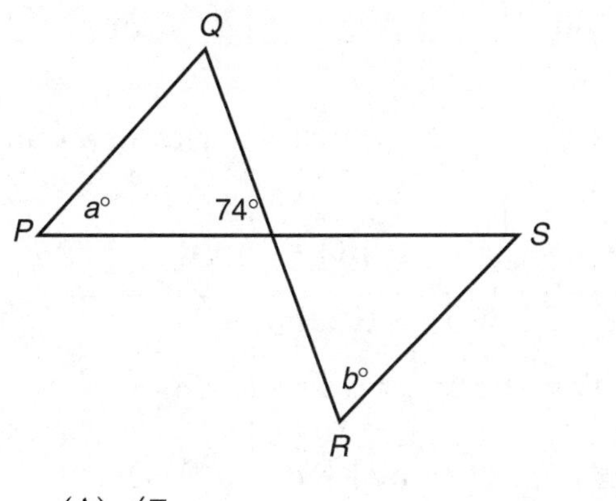

(A) 47
(B) 74
(C) 90
(D) 106
(E) It cannot be determined from the information given

8. In the figure below, $a : b = 3 : 5$ and $c : b = 2 : 1$. What is the measure of the largest angle?

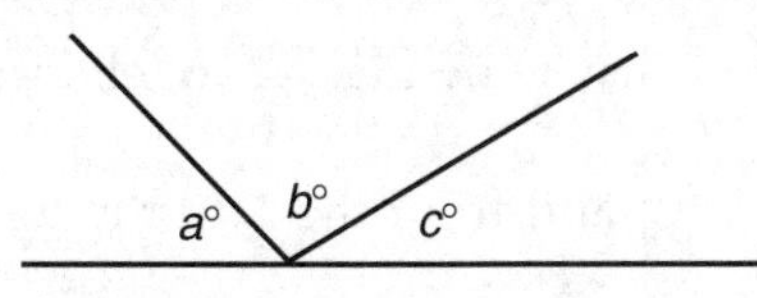

Note: Figure not drawn to scale.

(A) 50°
(B) 90°
(C) 100°
(D) 120°
(E) 150°

9. *A*, *B*, and *C* are points on a line with *B* between *A* and *C*. Let *M* and *N* be the midpoints of $\overline{AB}$ and $\overline{BC}$, respectively. If $AB : BC = 3 : 1$, what is $MN : BC$?

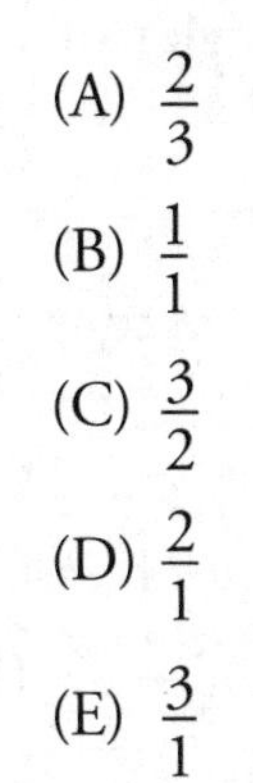

(A) $\frac{2}{3}$

(B) $\frac{1}{1}$

(C) $\frac{3}{2}$

(D) $\frac{2}{1}$

(E) $\frac{3}{1}$

10. In the figure below, what is the average (arithmetic mean) of the measures of the five angles?

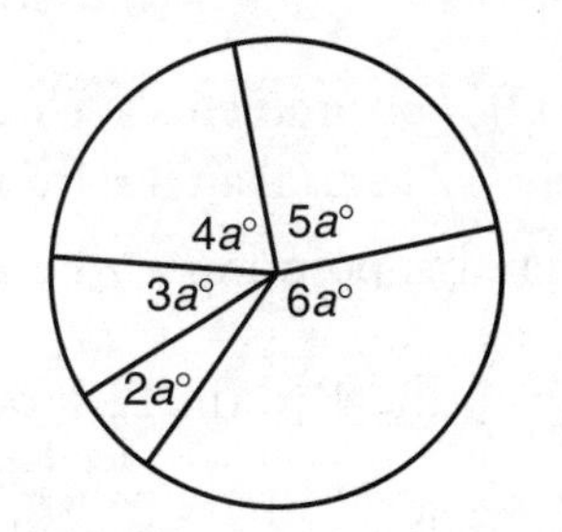

(A) 18°
(B) 36°
(C) 45°
(D) 72°
(E) It cannot be determined from the given information

ANSWERS EXPLAINED

Answer Key

1. **(C)**	3. **(C)**	5. **(D)**	7. **(D)**	9. **(D)**
2. **(D)**	4. **(B)**	6. **(B)**	8. **(C)**	10. **(D)**

Solutions

Each of the problems in this set of exercises is typical of a question you could see on a Math 1 test. When you take the model tests in this book and, in particular, when you take the actual Math 1 test, if you get stuck on questions such as these, you do not have to leave them out—you can almost always answer them by using one or more of the strategies discussed in the "Tactics" chapter. The solutions given here do *not* depend on those strategies; they are the correct mathematical ones.

1. **(C)** Since $\angle ABC$ is a straight angle, by KEY FACT G2, the sum of a, b, c, d, and e is 180, and so their average is $\frac{180}{5} = 36$.

2. **(D)** Since vertical angles are congruent, the two unmarked angles are $2b$ and $4a$. Since the sum of all six angles is 360°:

$$360 = 4a + 2b + 2a + 4a + 2b + b = 10a + 5b$$

Since vertical angles are congruent, $b = 2a \Rightarrow 5b = 10a$. Hence:

$$360 = 10a + 5b = 10a + 10a = 20a \Rightarrow a = 18 \Rightarrow b = 36$$

3. **(C)** Let y be the degree measure of $\angle JNL$. Then by KEY FACT G2:

$$43 + 90 + y = 180 \Rightarrow 133 + y = 180 \Rightarrow y = 47$$

Since $\overline{LM}$ is parallel to $\overline{JK}$, by KEY FACT G6, $x = y \Rightarrow x = 47$.

4. **(B)** $x + y + 90 = 180 \Rightarrow x + y = 90$

Since the ratio of y to x is 7 to 3, $y = 7t$ and $x = 3t$. So:

$$7t + 3t = 90 \Rightarrow 10t = 90 \Rightarrow t = 9 \Rightarrow x = 3(9) = 27$$

5. **(D)** For problems such as this, always draw a picture. The measure of each of the 12 central angles from one number to the next is 30°. At 1:50, the minute hand is pointing at the 10, and the hour hand has gone $\frac{50}{60} = \frac{5}{6}$ of the way from the 1 to the 2. From the 10 to the 1 is 90° and from the 1 to the hour hand is $\frac{5}{6}(30°) = 25°$, for a total of $90° + 25° = 115°$.

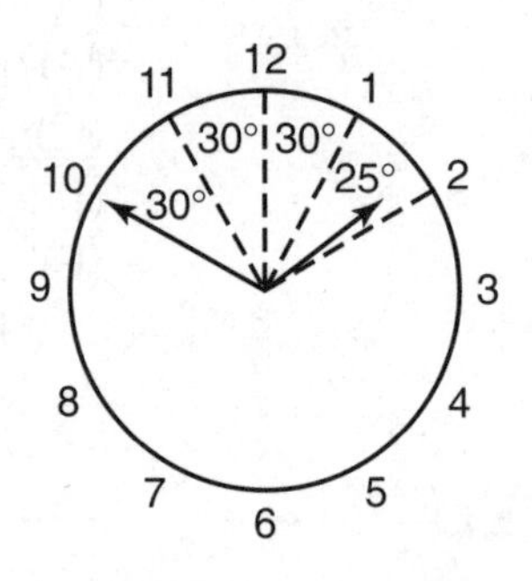

6. **(B)** From the diagram, we see that $6y = 180$, which implies that $y = 30$ and that $5x = 180$, which implies that $x = 36$. So:

$$\frac{x + y}{x - y} = \frac{36 + 30}{36 - 30} = \frac{66}{6} = 11$$

7. **(D)** Since $\overline{PQ}$ is parallel to $\overline{RS}$, $\angle R \cong \angle Q$ and so $b = m\angle Q$.

$$m\angle Q + a + 74 = 180 \Rightarrow a + b + 74 = 180 \Rightarrow a + b = 106$$

8. **(C)** Since $a : b = 3 : 5$, $a = 3x$ and $b = 5x$. Since $c : b = 2 : 1$, $c = 2b = 10x$. Then:

$$3x + 5x + 10x = 180 \Rightarrow 18x = 180 \Rightarrow x = 10 \Rightarrow c = 10x = 100$$

9. **(D)** If a diagram is not provided on a geometry question, draw one. From the figure below, you can see that $MN : BC = 2 : 1$.

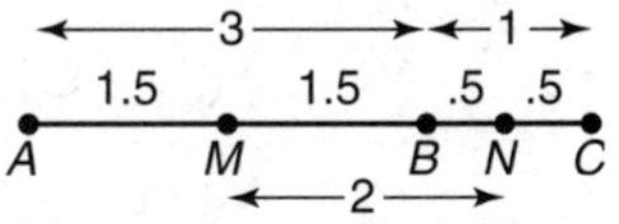

10. **(D)** The markings in the five angles are irrelevant. The sum of the measures of the five angles is 360°, and $360° \div 5 = 72°$.

Triangles

- Sides and Angles of a Triangle
- Right Triangles
- Pythagorean Theorem
- Special Right Triangles
- Perimeter and Area
- Similar Triangles
- Exercises
- Answers Explained

Since more geometry questions on the Math 1 test concern triangles than any other topic, it is imperative that you learn all the KEY FACTS in this section.

SIDES AND ANGLES OF A TRIANGLE

Key Fact H1

In any triangle, the sum of the measures of the three angles is 180°.

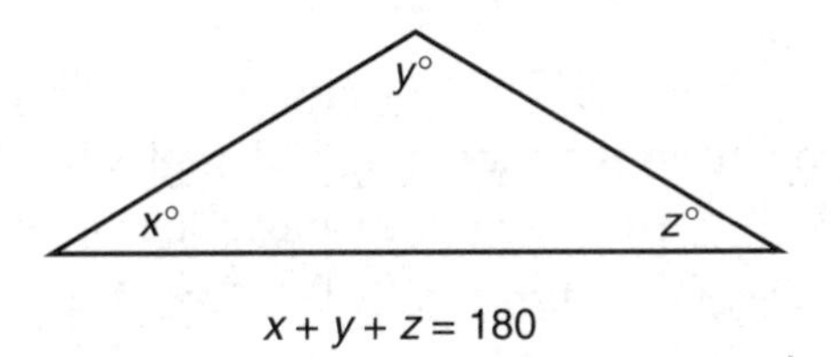

$x + y + z = 180$

EXAMPLE 1: To find the value of x in the figure below, use KEY FACT H1 twice. $130 + 30 + z = 180 \Rightarrow z = 20 \Rightarrow y = 20$, then $90 + x + 20 = 180 \Rightarrow x = 70$.

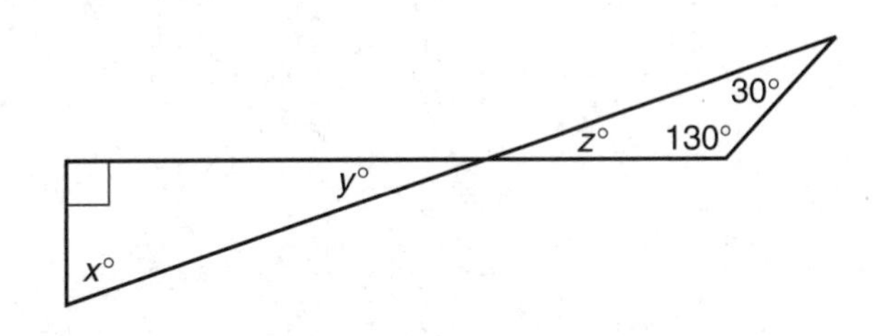

In the figure below, $\angle BCD$, which is formed by extending side $\overline{AC}$ of ΔABC, is called an ***exterior angle***.

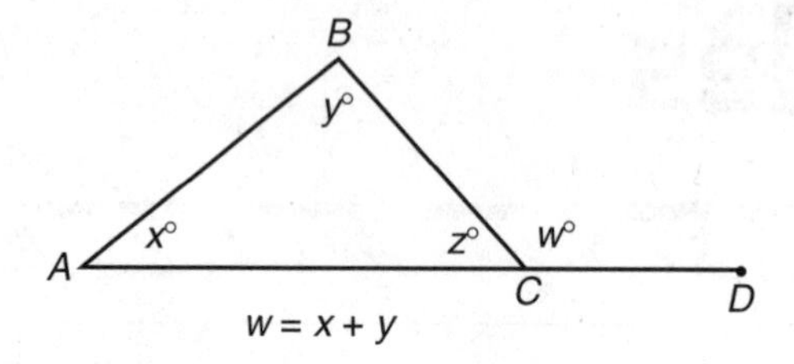

Key Fact H2

The measure of an exterior angle of a triangle is equal to the sum of the measures of the two opposite interior angles.

In the diagram above, $w = x + y$.

Key Fact H3

In any triangle:

- **The longest side is opposite the largest angle.**
- **The shortest side is opposite the smallest angle.**
- **Sides with equal lengths are opposite angles with equal measures (the angles opposite congruent sides are congruent).**

EXAMPLE 2: By using KEY FACT H3, we can draw the following conclusions concerning the angles in the figure below. In ΔCDE, since angles D and E are opposite congruent sides ($\overline{CE}$ and $\overline{CD}$, respectively), $m\angle D = m\angle E$. In ΔABC, angle C is the largest angle, so $\overline{AB}$ is the longest side. Specifically, $AB > 6$. Also, since $BC > AC$, $m\angle A > m\angle B$.

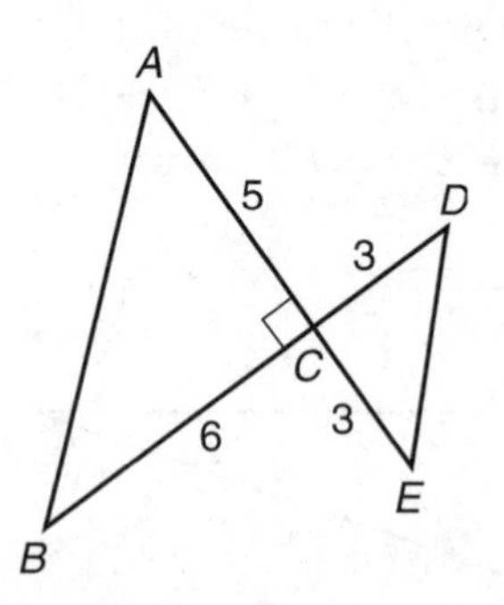

- A triangle is called ***scalene*** if the three sides all have different lengths. Then by KEY FACT H3, the three angles all have different measures.
- A triangle is called ***isosceles*** if two sides are congruent. By KEY FACT H3, the two angles opposite the congruent sides are congruent.

- A triangle is called ***equilateral*** if all three sides are congruent. By KEY FACT H3, all three angles are congruent. Since the sum of the measures of the three angles is 180°, each angle measures 60°.
- ***Acute triangles*** are triangles in which all three angles are acute. An acute triangle could be scalene, isosceles, or equilateral.
- ***Obtuse triangles*** are triangles in which one angle is obtuse and two are acute. An obtuse triangle could be scalene or isosceles.
- ***Right triangles*** are triangles that have one right angle and two acute angles. A right triangle could be scalene or isosceles. The side opposite the 90° angle is called the ***hypotenuse***, and by KEY FACT H3, it is the longest side. The other two sides are called the ***legs***.

RIGHT TRIANGLES

If a and b are the measures in degrees, of the acute angles of a right triangle, then by KEY FACT H1, $90 + a + b = 180 \Rightarrow a + b = 90$.

Key Fact H4

In any right triangle, the sum of the measures of the two acute angles is 90°.

EXAMPLE 3: To find the average of a and b in ΔABC below, note that by KEY FACT H4, $a + b = 90$, so $\frac{a+b}{2} = \frac{90}{2} = 45$.

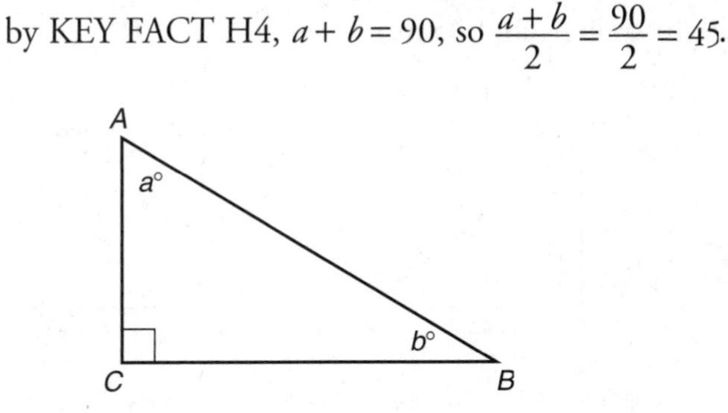

The most important fact concerning right triangles is the ***Pythagorean theorem***, which is given in the first line of KEY FACT H5. The second and third lines of KEY FACT H5 are important corollaries of the Pythagorean theorem.

Key Fact H5

PYTHAGOREAN THEOREM AND COROLLARIES

Let a, b, and c be the lengths of the sides of ΔABC, with $a \le b \le c$.

- **$a^2 + b^2 = c^2$ if and only if angle C is a right angle.**
- **$a^2 + b^2 < c^2$ if and only if angle C is obtuse.**
- **$a^2 + b^2 > c^2$, if and only if angle C is acute.**

TIP

The Pythagorean theorem is probably the most important theorem you need to know. Be sure to review all of its uses.

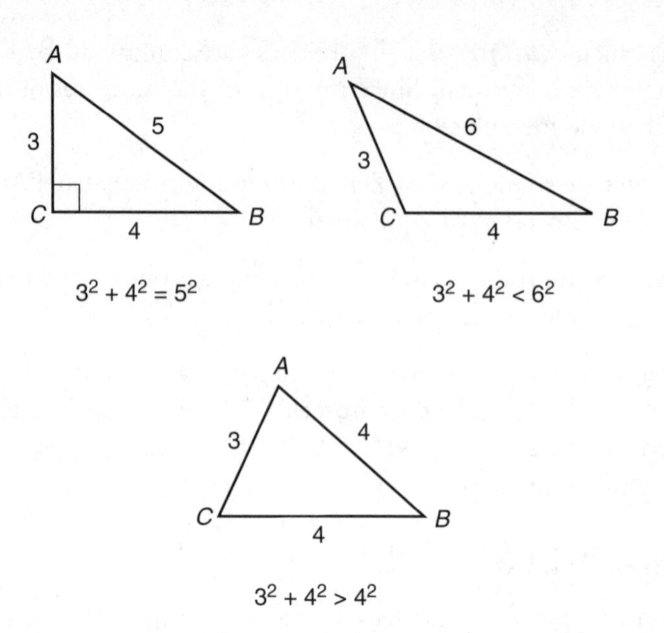

EXAMPLE 4: Since $8^2 + 15^2 = 64 + 225 = 289 = 17^2$, a triangle whose sides have lengths 8, 15, and 17 is a right triangle. Since $10^2 + 15^2 = 100 + 225 = 325 < 19^2$, a triangle whose sides have lengths 10, 15, and 19 is not a right triangle—it is obtuse.

On the Math 1 test, the most common right triangles whose sides are integers are the 3-4-5 right triangle and its multiples.

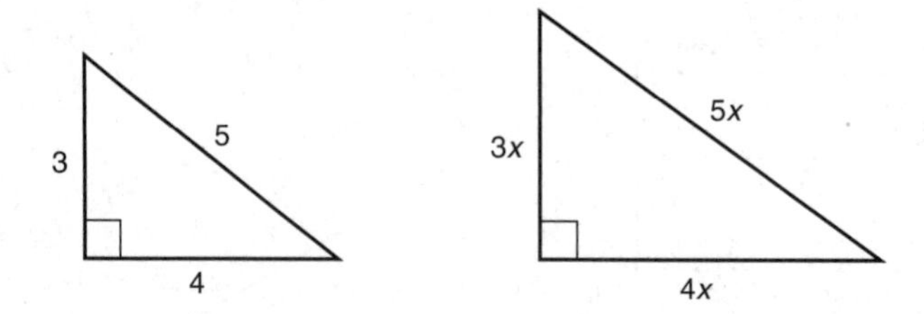

NOTE

KEY FACT H6 applies even if x is not an integer. For example:

$x = 0.5$ 1.5, 2, 2.5
$x = \pi$ $3\pi, 4\pi, 5\pi$

Key Fact H6

For any positive number x, there is a right triangle whose sides are $3x$, $4x$, $5x$.

For example:

$x = 1$	3, 4, 5	$x = 5$	15, 20, 25
$x = 2$	6, 8, 10	$x = 10$	30, 40, 50
$x = 3$	9, 12, 15	$x = 50$	150, 200, 250
$x = 4$	12, 16, 20	$x = 100$	300, 400, 500

Other right triangles with integer sides that you should recognize immediately are the ones whose sides are 5, 12, 13 and 8, 15, 17.

SPECIAL RIGHT TRIANGLES

Let x be the length of each leg and let h be the length of the hypotenuse of an isosceles right triangle. By the Pythagorean theorem,

$$x^2 + x^2 = h^2 \Rightarrow 2x^2 = h^2 \Rightarrow h = \sqrt{2x^2} = x\sqrt{2}.$$

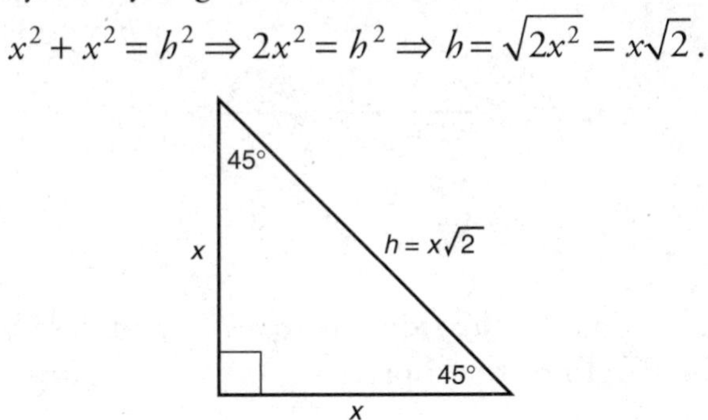

Key Fact H7

In a 45-45-90 right triangle, the sides are x, x, and $x\sqrt{2}$.

- **If you are given the length of a leg, multiply it by $\sqrt{2}$ to get the length of the hypotenuse.**
- **If you are given the length of the hypotenuse, divide it by $\sqrt{2}$ to get the length of each leg.**

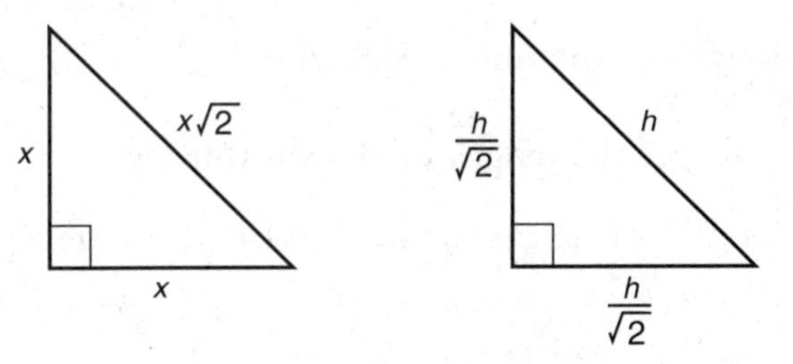

EXAMPLE 5: To find the area of a square whose diagonal is 8, note that the diagonal divides the square into two isosceles right triangles. So $s = \frac{8}{\sqrt{2}}$ and $A = s^2 = \left(\frac{8}{\sqrt{2}}\right)^2 = \frac{64}{2} = 32$.

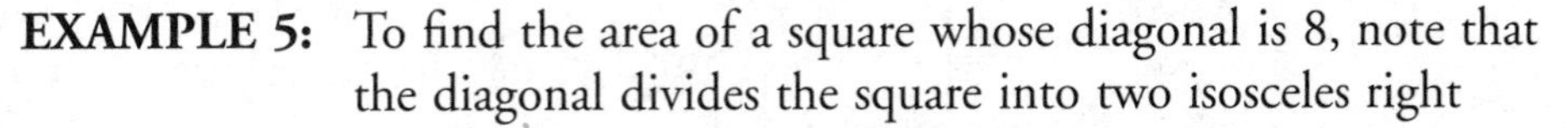

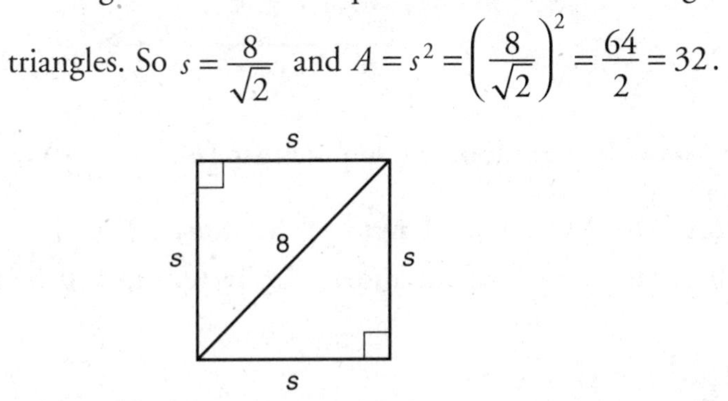

Let $2x$ be the length of each side of equilateral ΔPQR, shown at the top of page 128, in which altitude $\overline{PS}$ has been drawn. Then ΔPQS is a 30-60-90 right triangle, and its sides are x, $2x$, and h. By the Pythagorean theorem,

$$x^2 + h^2 = (2x)^2 = 4x^2 \Rightarrow h^2 = 3x^2 \Rightarrow h = \sqrt{3x^2} = x\sqrt{3}.$$

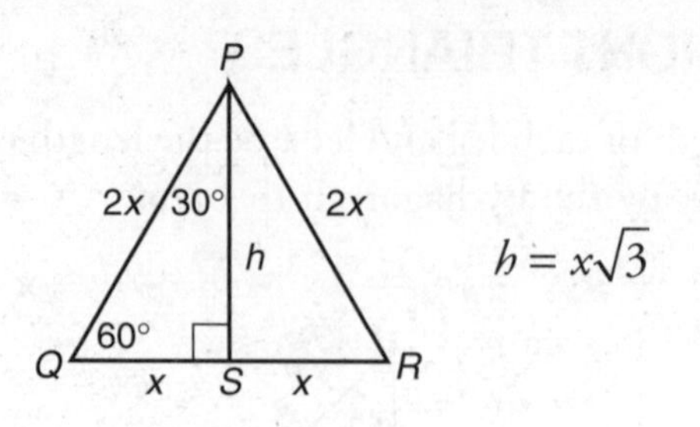

Key Fact H8

In a 30-60-90 right triangle, the sides are x, $x\sqrt{3}$, and $2x$.

If you know the length of the shorter leg (x):

- **Multiply it by $\sqrt{3}$ to get the length of the longer leg.**
- **Multiply it by 2 to get the length of the hypotenuse.**

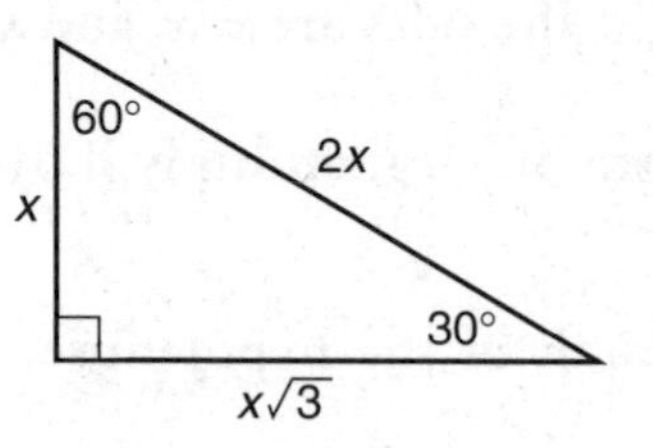

If you know the length of the longer leg (a):

- **Divide it by $\sqrt{3}$ to get the length of the shorter leg.**
- **Multiply the length of the shorter leg by 2 to get the length of the hypotenuse.**

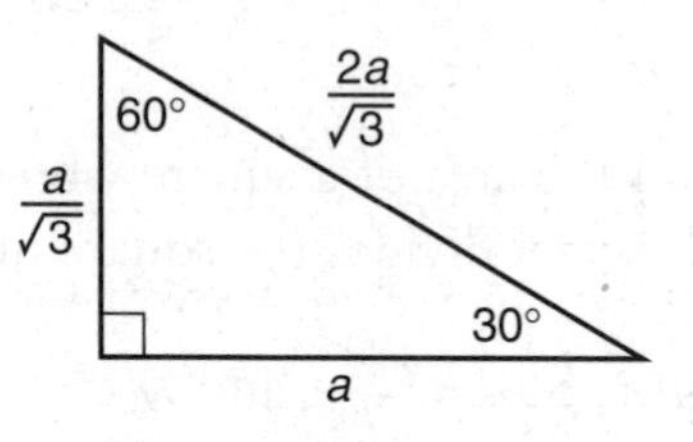

If you know the length of the hypotenuse (h):

- **Divide it by 2 to get the length of the shorter leg.**
- **Multiply the length of the shorter leg by $\sqrt{3}$ to get the length of the longer leg.**

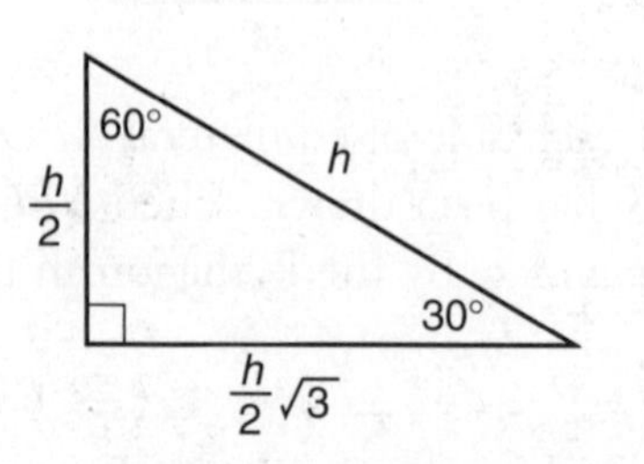

PERIMETER AND AREA

The ***perimeter*** of a triangle is the sum of the lengths of the three sides.

EXAMPLE 6: To find the perimeter of an equilateral triangle whose height is 12, note that the height divides the triangle into two 30-60-90 right triangles. In the figure below, by KEY FACT H8,

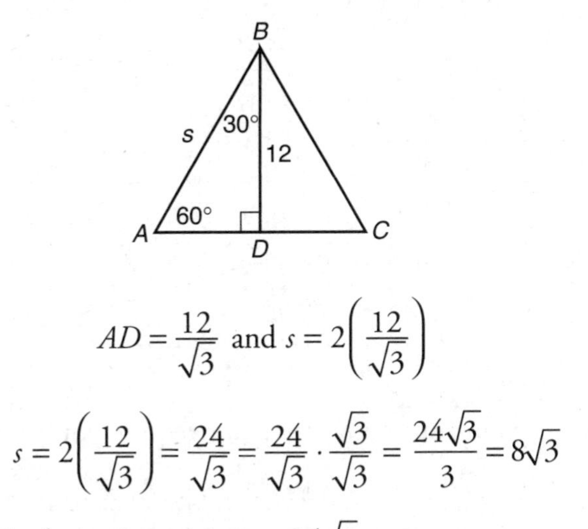

$$AD = \frac{12}{\sqrt{3}} \text{ and } s = 2\left(\frac{12}{\sqrt{3}}\right)$$

$$s = 2\left(\frac{12}{\sqrt{3}}\right) = \frac{24}{\sqrt{3}} = \frac{24}{\sqrt{3}} \cdot \frac{\sqrt{3}}{\sqrt{3}} = \frac{24\sqrt{3}}{3} = 8\sqrt{3}$$

So the perimeter is $3s = 24\sqrt{3}$.

Key Fact H9

TRIANGLE INEQUALITY

- **The sum of the lengths of any two sides of a triangle is greater than the length of the third side.**
- **The difference of the lengths of any two sides of a triangle is less than the length of the third side.**

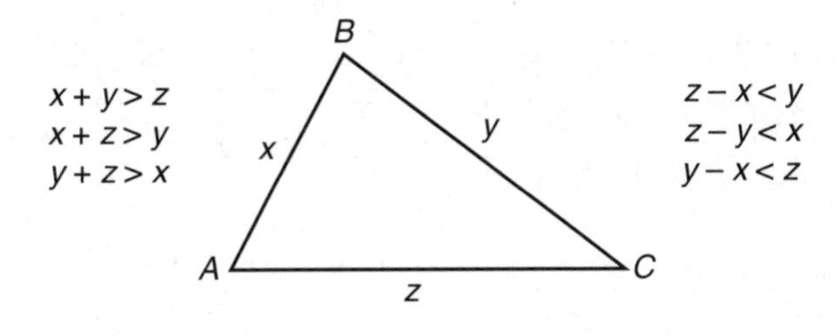

EXAMPLE 7: A teacher asked her class to draw triangles in which the lengths of two of the sides were 6 inches and 7 inches and the length of the third side was also a whole number of inches. To determine how many different triangles the class could draw, note that if x represents the length of the third side, then by KEY FACT H9, $6 + 7 > x$, so $x < 13$. Also,

$7 - 6 < x$, so $x > 1$. So x could have 11 different values: 2, 3, 4, 5, 6, 7, 8, 9, 10, 11, and 12. The perimeters of the triangles are the integers from 15 to 25. The following diagram illustrates four of the triangles the class could have drawn.

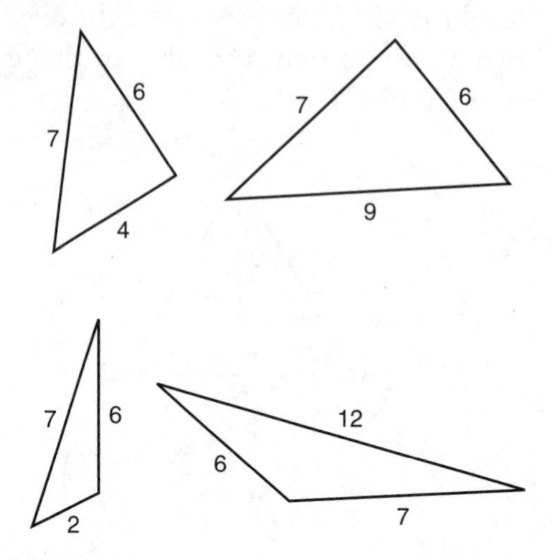

Frequently, questions on the Math 1 test require you to calculate the ***area*** of a triangle.

Key Fact H10

The area of a triangle is given by $A = \frac{1}{2}bh$, where b and h are the lengths of the base and height, respectively.

TIP

The height can be outside the triangle. See $\overline{AF}$ in the diagram shown below on the left .

(1) *Any* side of the triangle can be taken as the base.
(2) The ***height*** (which is also called an ***altitude***) is a line segment drawn perpendicular to the base from the opposite vertex.
(3) In a right triangle, either leg can be the base and the other the height.
(4) If one endpoint of the base is the vertex of an obtuse angle, then the height will be outside the triangle.
(5) In each figure below:

- If $\overline{AC}$ is the base, $\overline{BD}$ is the height.
- If $\overline{AB}$ is the base, $\overline{CE}$ is the height.
- If $\overline{BC}$ is the base, $\overline{AF}$ is the height.

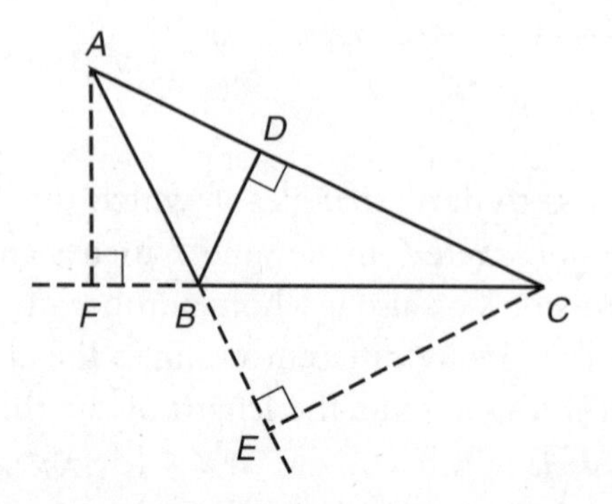

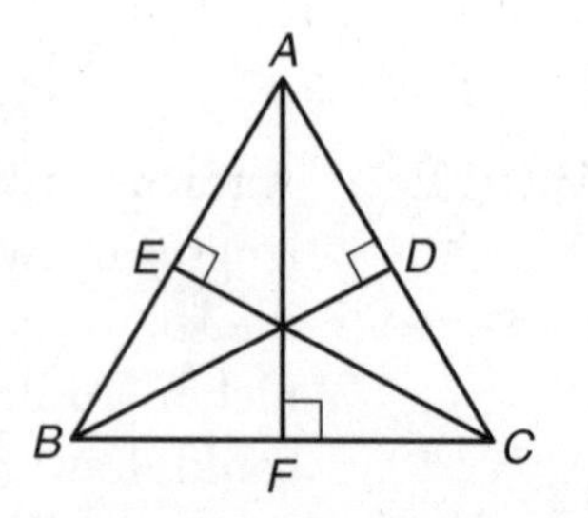

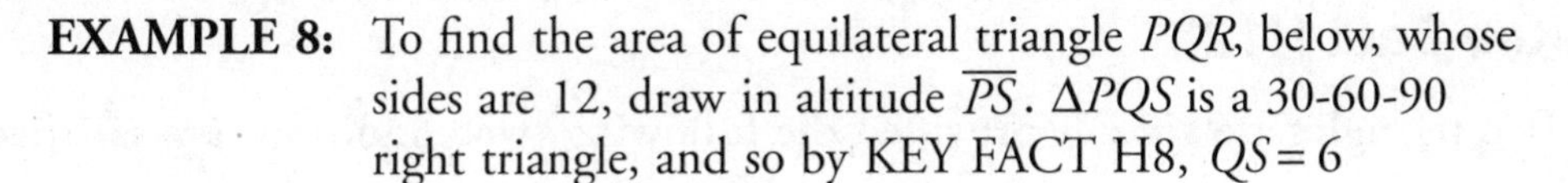

EXAMPLE 8: To find the area of equilateral triangle PQR, below, whose sides are 12, draw in altitude $\overline{PS}$. ΔPQS is a 30-60-90 right triangle, and so by KEY FACT H8, $QS = 6$ and $PS = 6\sqrt{3}$. Finally, the area of $\Delta PQS = \frac{1}{2}(12)(6\sqrt{3}) = 36\sqrt{3}$.

Replacing 12 by s in Example 8 yields a useful formula.

Key Fact H11

If A represents the area of an equilateral triangle with side s, then $A = \frac{s^2\sqrt{3}}{4}$.

Smart Strategy

Learning this formula for the area of an equilateral triangle can save you time.

SIMILAR TRIANGLES

Two triangles, such as triangle I and triangle II in the figure below, that have the same shape but not necessarily the same size are said to be ***similar.***

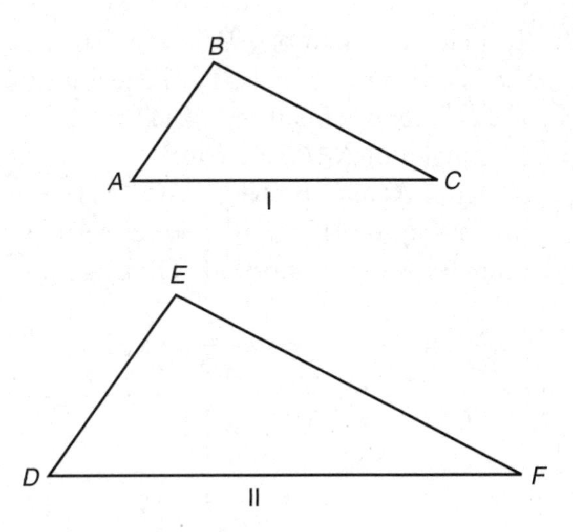

The following KEY FACT makes this intuitive definition mathematically precise.

Remember

Corresponding sides are sides that are opposite angles of the same measure.

Key Fact H12

Two triangles are *similar* provided the following two conditions are satisfied:

- **The three angles in the first triangle have the same measures as the three angles in the second triangle.**

  $m\angle A = m\angle D \qquad m\angle B = m\angle E \qquad m\angle C = m\angle F$

- **The lengths of the corresponding sides of the two triangles are in proportion.**

$$\frac{AB}{DE} = \frac{BC}{EF} = \frac{AC}{DF}$$

An important theorem in geometry states that if the first condition in KEY FACT H12 is satisfied, then the second condition is automatically satisfied. Therefore, to show that two triangles are similar, it is sufficient to show that their angles have the same measures. Furthermore, if the measures of two angles of one triangle are equal to the measures of two angles of a second triangle, then the measures of the third angles are also equal. KEY FACT H13 is an immediate consequence of the results in this paragraph.

Key Fact H13

If the measures of two angles of one triangle are equal to the measures of two angles of a second triangle, then the triangles are similar.

EXAMPLE 9: To find the length of $\overline{ED}$ in the diagram below, first note that since vertical angles have equal measures, $m\angle AEB = m\angle DEC$ and, of course, $m\angle A = m\angle C$. So the measures of two angles of ΔABE are equal to the measures of two angles of ΔCDE. By KEY FACT H13, the two triangles are similar. So, by KEY FACT H12, corresponding sides of the triangles are in proportion. Therefore:

$$\frac{AB}{DC} = \frac{BE}{ED} \Rightarrow \frac{6}{2.7} = \frac{10}{ED} \Rightarrow 6(ED) = 27 \Rightarrow$$

$$ED = \frac{27}{6} = 4.5$$

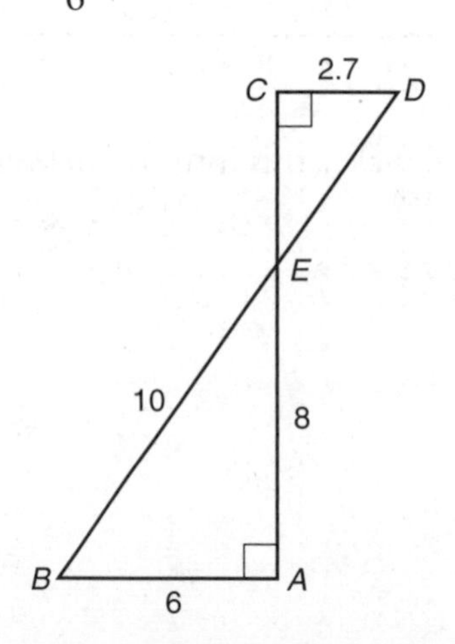

In ΔABC below, line ℓ is parallel to side $\overline{BC}$ and intersects sides $\overline{AB}$ and $\overline{AC}$ at D and E, respectively. Then by KEY FACT G6, $m\angle AED = m\angle C$ and $m\angle ADE = m\angle B$. So by KEY FACT H13, ΔAED is similar to ΔACB.

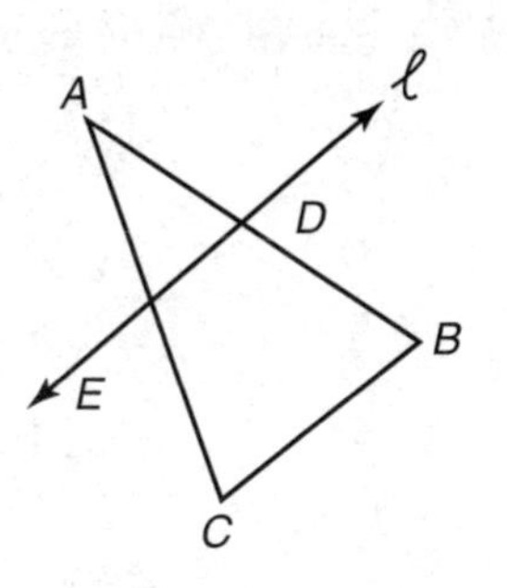

Key Fact H14

A line that intersects two sides of a triangle and is parallel to the third side creates a smaller triangle that is similar to the original one.

If two triangles are similar, the common ratio of their corresponding sides is called the ***ratio of similitude***.

Key Fact H15

If two triangles are similar, and if k is the ratio of similitude:

- **The ratio of all their linear measurements is k.**
- **The ratio of their areas is k^2.**

In the figure below, $\overline{DE} \parallel \overline{AC}$. So by KEY FACT H14, ΔABC and ΔDBE are similar.

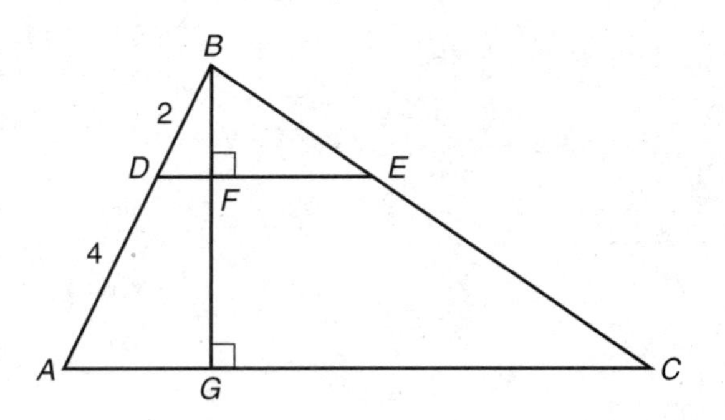

Then $\overline{AB}$ and $\overline{DB}$ are corresponding sides. Since $AB = 6$ and $DB = 2$, the ratio of similitude is $\frac{6}{2} = \frac{3}{1}$.

Therefore:

- All the sides are in the ratio of $\frac{3}{1}$: $BC = \mathbf{3} \times BE$
  $AC = \mathbf{3} \times DE$
- The altitudes are in the ratio of $\frac{3}{1}$: $BG = \mathbf{3} \times BF$

- The perimeters are in the ratio of $\frac{3}{1}$: perimeter of $\Delta ABC = \mathbf{3} \times$ (perimeter of ΔDBE)
- The areas are in the ratio of $\frac{9}{1}$: area of $\Delta ABC = \mathbf{9} \times$ (area of ΔDBE)

Exercises

1. In ΔABC, the measures of the exterior angles at vertices A and B are 145° and 125°, respectively. Which of the following statements about ΔABC must be true?

   I. The triangle is a right triangle.
   II. The triangle is an isosceles triangle.
   III. The triangle is a scalene triangle.

   (A) None
   (B) I only
   (C) III only
   (D) I and II only
   (E) I and III only

2. In the triangle below, what is the value of x?

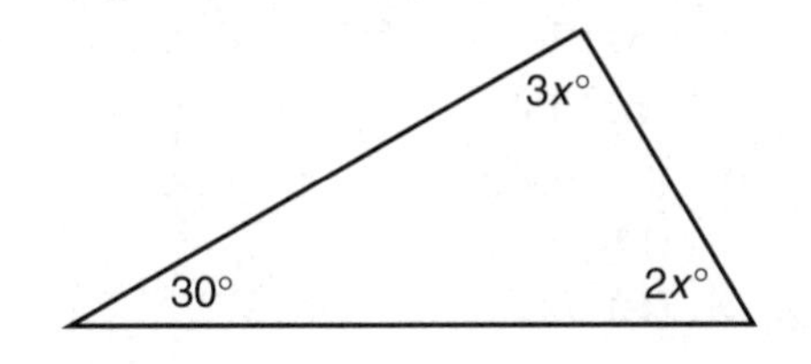

   (A) 20
   (B) 30
   (C) 40
   (D) 50
   (E) 60

3. In the figure below, what is the value of c?

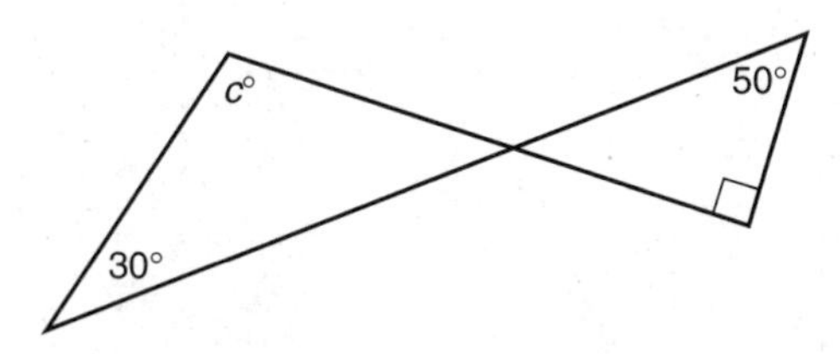

   (A) 100
   (B) 110
   (C) 120
   (D) 130
   (E) 140

4. The lengths of two sides of a right triangle are 15 and 17. Which of the following *could* be the length of the third side?

   I. 8
   II. 19
   III. $\sqrt{514}$

   (A) I only
   (B) II only
   (C) I and II
   (D) I and III
   (E) I, II, and III

5. What is the value of w in the figure below?

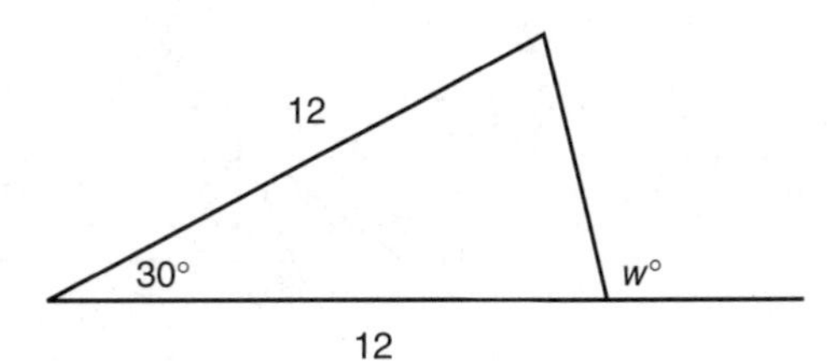

   (A) 60
   (B) 90
   (C) 105
   (D) 120
   (E) 150

6. What is the area of an equilateral triangle whose altitude is 6?

   (A) 18
   (B) $12\sqrt{3}$
   (C) $18\sqrt{3}$
   (D) 36
   (E) $24\sqrt{3}$

7. Which of the following expresses a true relationship between x and y in the figure below?

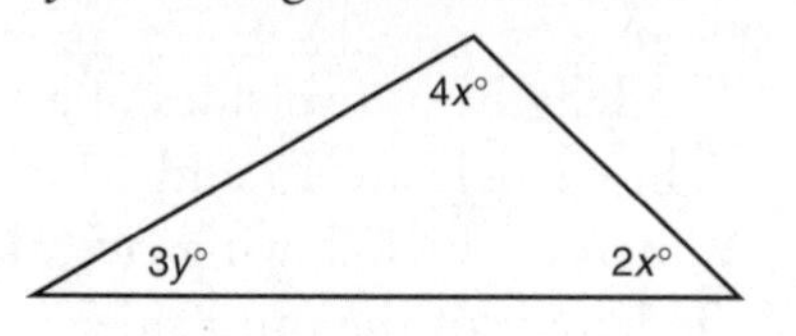

(A) $y = 60 - 2x$
(B) $y = 2x$
(C) $x + y = 90$
(D) $y = 180 - 3x$
(E) $x = 90 - 3y$

8. What is the value of AD in the triangle below?

(A) $5\sqrt{2}$
(B) 10
(C) 11
(D) 13
(E) $12\sqrt{2}$

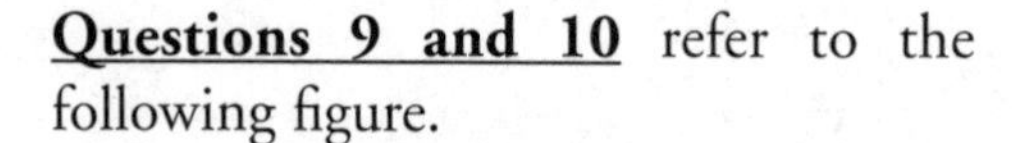

Questions 9 and 10 refer to the following figure.

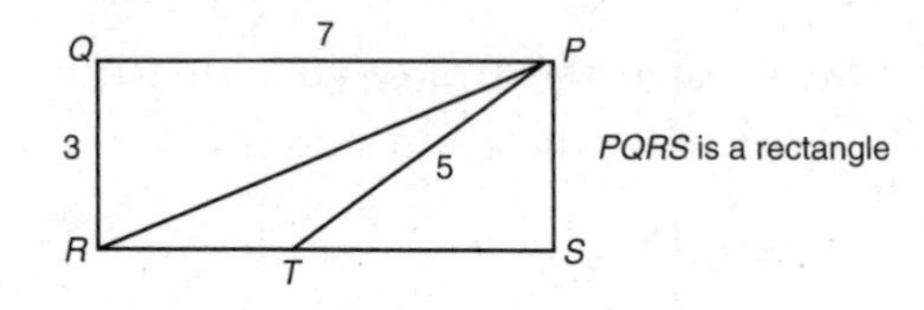

9. What is the area of ΔPRT?

(A) 3
(B) 4.5
(C) 6
(D) 7.5
(E) 10

10. What is the perimeter of ΔPRT?

(A) $8 + \sqrt{41}$
(B) $8 + \sqrt{58}$
(C) 16
(D) 17
(E) 18

11. If the measures of the angles of a triangle are in the ratio of $1 : 2 : 3$, what is the ratio of the length of the smallest side of the triangle to the length of the longest side of the triangle?

(A) $1 : 2$
(B) $1 : 3$
(C) $1 : 5$
(D) $2 : 3$
(E) $2 : 5$

12. If the difference between the measures of the two acute angles of a right triangle is 20°, what is the measure, in degrees, of the smaller one?

(A) 20°
(B) 35°
(C) 45°
(D) 55°
(E) 80°

13. If one side of ΔABC is 5, what is the smallest integer that could be the perimeter of the triangle?

(A) 8
(B) 10
(C) 11
(D) 12
(E) 15

14. In the figure below, what is the length of $\overline{BD}$?

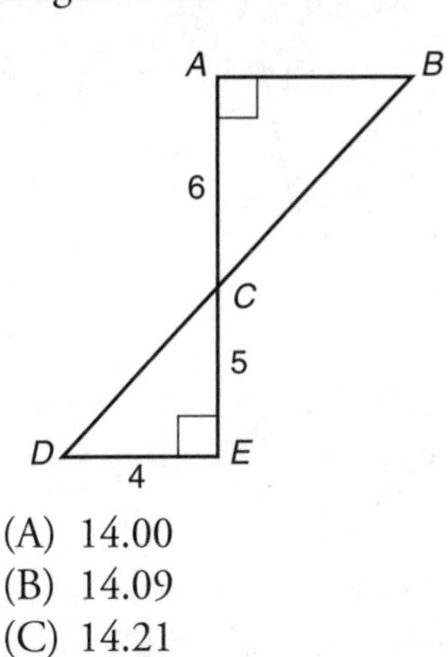

(A) 14.00
(B) 14.09
(C) 14.21
(D) 15.62
(E) 16.00

15. The lengths of the sides of a triangle are 3, 5, and x. How many possible values of x are there, if x must be an integer?

(A) 1
(B) 5
(C) 7
(D) 8
(E) Infinitely many

ANSWERS EXPLAINED

Answer Key

1. **(E)**	4. **(D)**	7. **(A)**	10. **(B)**	13. **(C)**
2. **(B)**	5. **(C)**	8. **(D)**	11. **(A)**	14. **(B)**
3. **(B)**	6. **(B)**	9. **(B)**	12. **(B)**	15. **(B)**

Solutions

Each of the problems in this set of exercises is typical of a question you could see on a Math 1 test. When you take the model tests in this book and, in particular, when you take the actual Math 1 test, if you get stuck on questions such as these, you do not have to leave them out—you can almost always answer them by using one or more of the strategies discussed in the "Tactics" chapter. The solutions given here do *not* depend on those strategies; they are the correct mathematical ones.

1. **(E)** First, draw ΔABC.

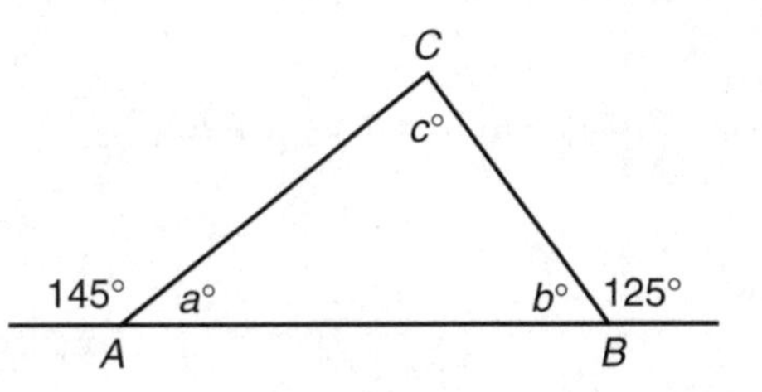

By KEY FACT G2, $a + 145 = 180$ and $b + 125 = 180 \Rightarrow a = 35$ and $b = 55$. Then by KEY FACT H1, $35 + 55 + c = 180 \Rightarrow c = 90$ (I is true). Since the measures of all three angles are different, by KEY FACT H3, the lengths of all three sides are different (II is false and III is true). Only I and III are true.

2. **(B)** $2x + 3x + 30 = 180 \Rightarrow 5x + 30 = 180 \Rightarrow 5x = 150 \Rightarrow x = 30$

3. **(B)** In the figure below, $50 + 90 + a = 180 \Rightarrow a = 40$. Since vertical angles are congruent, $b = 40$. So $40 + 30 + c = 180 \Rightarrow c = 110$.

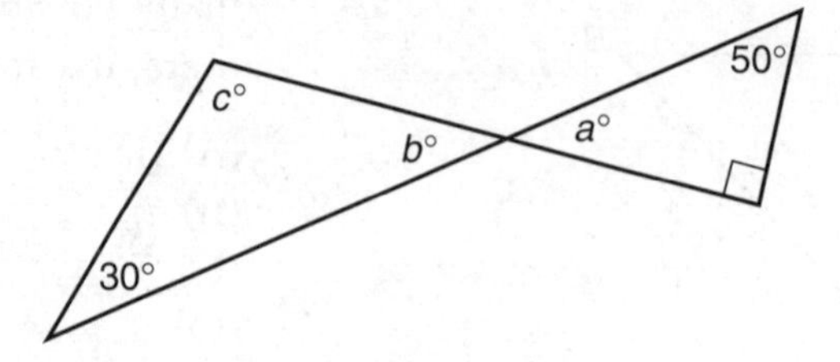

4. **(D)** If the triangle were not required to be right, by KEY FACT H9 *any* number greater than 2 and less than 32 could be the length of the third side. For a right triangle, though, there are only *two* possibilities:

   - 17 is the length of the hypotenuse, and the legs have lengths 15 and 8. I is true. (If you did not recognize the 8-15-17 triangle, use the Pythagorean theorem: solve the equation $15^2 + x^2 = 17^2$.)

   - 15 and 17 are the lengths of the two legs. Use the Pythagorean theorem to find the hypotenuse:

   $15^2 + 17^2 = c^2 \Rightarrow c^2 = 225 + 289 = 514 \Rightarrow c = \sqrt{514}$. III is true.

   - Since $15^2 + 17^2 \neq 19^2$, a 15-17-19 triangle is *not* a right triangle. II is false. Only I and III are true.

5. **(C)** Since the triangle is isosceles, in the figure below, a and b are equal. Since $a + b + 30 = 180$, a and b are each 75. Finally, $w = 180 - 75 = 105$.

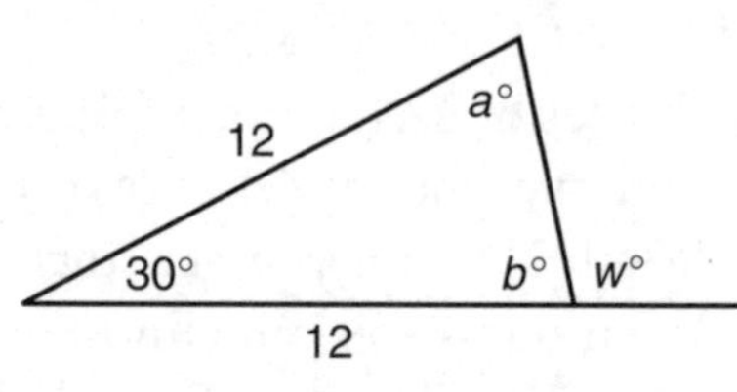

6. **(B)** Draw altitude $\overline{PS}$ in equilateral ΔPQR. By KEY FACT H8,

$SR = \frac{6}{\sqrt{3}} = \frac{6}{\sqrt{3}} \cdot \frac{\sqrt{3}}{\sqrt{3}} = \frac{6\sqrt{3}}{3} = 2\sqrt{3}$. Since, in ΔPQR, SR is one-half the base,

the area of the triangle is $\left(\frac{1}{2}b\right)h = 2\sqrt{3} \times 6 = 12\sqrt{3}$.

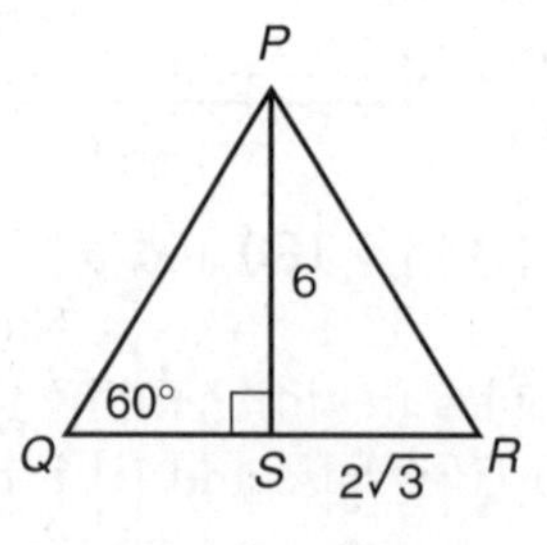

7. **(A)** $2x + 4x + 3y = 180 \Rightarrow 6x + 3y = 180 \Rightarrow 3y = 180 - 6x \Rightarrow y = 60 - 2x$

8. **(D)** Use the Pythagorean theorem twice, unless you recognize the common right triangles in this figure (*which you should*). Since $AB = 20$ and $BC = 16$, ΔABC is a $3x$-$4x$-$5x$ right triangle with $x = 4$. So $AC = 12$, and ΔACD is a right triangle whose legs are 5 and 12. Its hypotenuse, AD, is therefore 13.

9. **(B)** Since ΔPST is a right triangle whose hypotenuse is 5 and one of whose legs is 3, the other leg, $\overline{ST}$, is 4. Since $SR = PQ = 7$, $TR = 3$. Now, ΔPRT has a base of 3 ($\overline{TR}$) and a height of 3 ($\overline{PS}$), and so its area is $\frac{1}{2}(3)(3) = 4.5$.

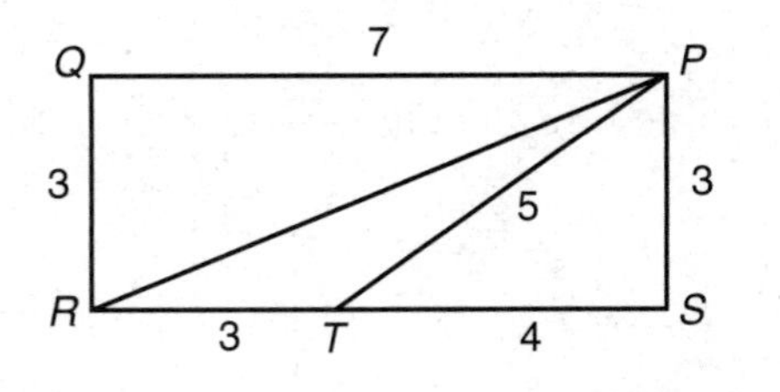

10. **(B)** In ΔPRT, we already have that $PT = 5$ and $TR = 3$; we need only find PR, which is the hypotenuse of ΔPQR. By the Pythagorean theorem:

$$3^2 + 7^2 = (PR)^2 \Rightarrow (PR)^2 = 9 + 49 = 58 \Rightarrow PR = \sqrt{58}.$$

So the perimeter is $3 + 5 + \sqrt{58} = 8 + \sqrt{58}$.

11. **(A)** If the measures of the three angles are in the ratio of 1 : 2 : 3, then for some number x, their values are x, $2x$, and $3x$.

$$x + 2x + 3x = 180° \Rightarrow 6x = 180° \Rightarrow x = 30°$$

So the triangle is a 30-60-90 right triangle, and for some number a, the lengths of the sides are a, $a\sqrt{3}$, and $2a$. The ratio of the length of the smallest side to the length of the largest side is $a : 2a = 1 : 2$.

12. **(B)** Draw a picture, label it, and then write the equations:

$$a + b = 90 \text{ and } a - b = 20$$

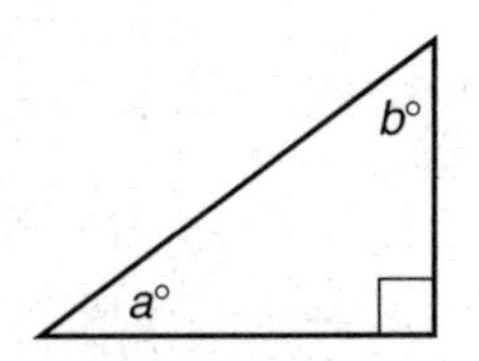

By adding the equations, we get $2a = 110$. So $a = 55$ and $b = 90 - 55 = 35$. The measure of the smallest angle is 35°.

13. **(C)** If x and y are the lengths of the other two sides, then by KEY FACT H9, $x + y > 5 \Rightarrow x + y + 5 > 10$. So the smallest integer that the perimeter could be is 11. (In fact, there is an isosceles triangle whose sides are 3, 3, and 5.)

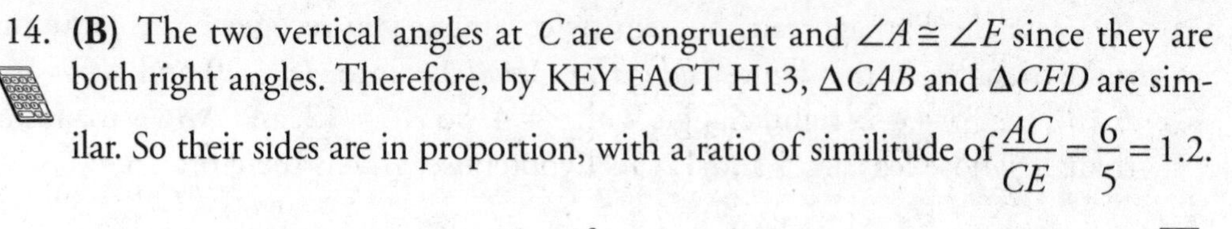

14. **(B)** The two vertical angles at C are congruent and $\angle A \cong \angle E$ since they are both right angles. Therefore, by KEY FACT H13, ΔCAB and ΔCED are similar. So their sides are in proportion, with a ratio of similitude of $\frac{AC}{CE} = \frac{6}{5} = 1.2$.

    By the Pythagorean theorem, $(CD)^2 = 4^2 + 5^2 = 16 + 25 = 41$. So $CD = \sqrt{41}$.

    $\frac{BC}{CD} = 1.2 \Rightarrow BC = 1.2\sqrt{41}$. Finally, $BD = \sqrt{41} + 1.2\sqrt{41} = 14.09$.

15. **(B)** By the triangle inequality (KEY FACT H9),

$$(5 - 3) < x < (5 + 3) \Rightarrow 2 < x < 8.$$

    So there are 5 possible integer values of x: 3, 4, 5, 6, 7.

Quadrilaterals and Other Polygons

A ***polygon*** is a closed geometric figure made up of line segments. The line segments are called ***sides***, and the endpoints of the line segments are called ***vertices*** (each one is called a ***vertex***). Line segments inside the polygon drawn from one vertex to another are called ***diagonals***.

Three-sided polygons, called triangles, were discussed in Chapter 9. Although in this section our main focus will be on four-sided polygons, which are called ***quadrilaterals***, we will discuss other polygons as well. There are special names for many polygons with more than four sides. The ones you need to know for the Math 1 test are given in the following chart.

Number of Sides	Name	Number of Sides	Name
5	Pentagon	8	Octagon
6	Hexagon	10	Decagon

A ***regular polygon*** is a polygon in which all the sides have the same length and all the angles have the same measure. A regular three-sided polygon is an equilateral triangle, and, as we shall see, a regular quadrilateral is a square. Pictured on the next page are a regular pentagon, regular hexagon, and regular octagon.

Remember

In a *regular* polygon, all the angles are congruent and all the sides are congruent.

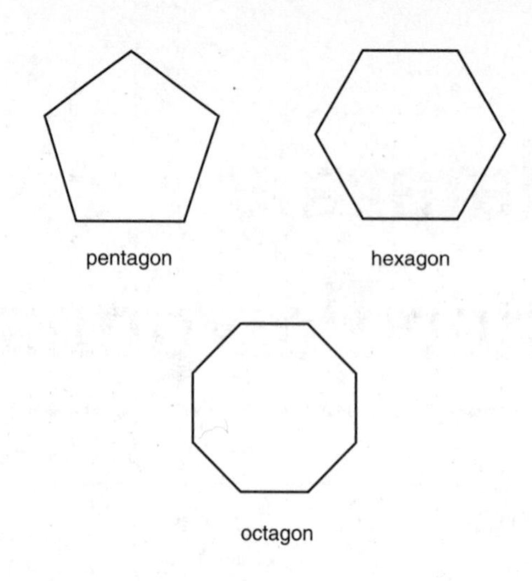

THE ANGLES OF A POLYGON

A diagonal of a quadrilateral divides it into two triangles. Since the sum of the measures of the three angles in each of the triangles is 180°, the sum of the measures of the angles in the quadrilateral is 360°.

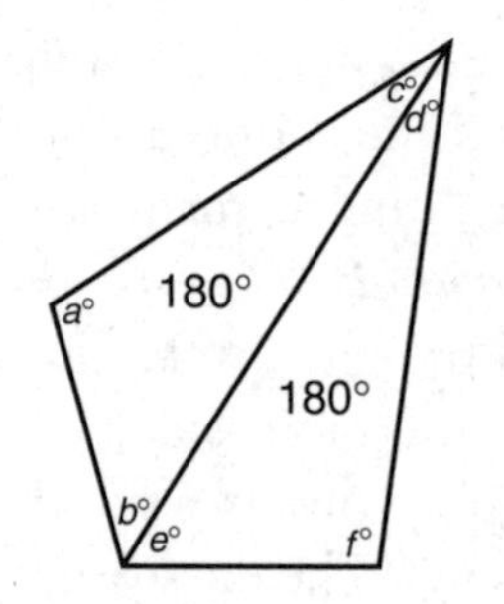

$$a + b + c = 180 \text{ and } d + e + f = 180$$
$$a + (b + e) + (c + d) + f = 360$$

Key Fact I1

In any quadrilateral, the sum of the measures of the four angles is 360°.

Similarly, any polygon can be divided into triangles by drawing in all of the diagonals emanating from one vertex.

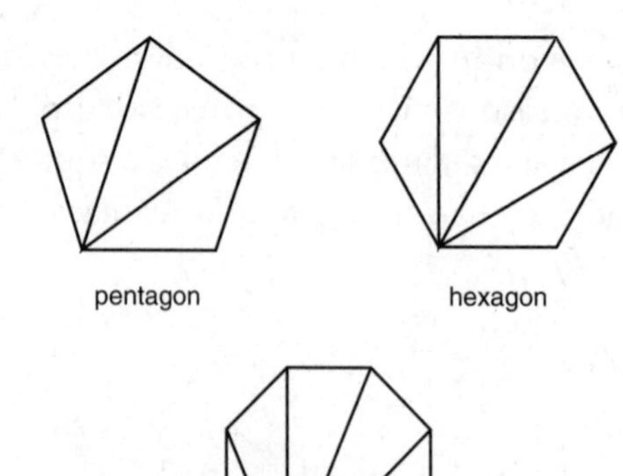

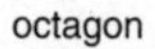

Notice that a five-sided polygon can be divided into three triangles, and a six-sided polygon can be divided into four triangles. In general, an n-sided polygon can be divided into $(n - 2)$ triangles, which leads to KEY FACT I2.

Key Fact I2

The sum of the measures of the n angles in a polygon with n sides is $(n - 2) \times 180°$.

EXAMPLE 1: To find the measure of each angle of a regular octagon, first use KEY FACT I2 to get that the sum of all eight angles is $(8 - 2) \times 180° = 6 \times 180° = 1{,}080°$. Then since in a regular octagon all eight angles have the same measure, the measure of each one is $1{,}080° \div 8 = 135°$.

An ***exterior angle*** of a polygon is formed by extending a side. Surprisingly, in all polygons, the sum of the measures of the exterior angles is the same.

Key Fact I3

In any polygon, the sum of the measures of the exterior angles, taking one at each vertex, is 360°.

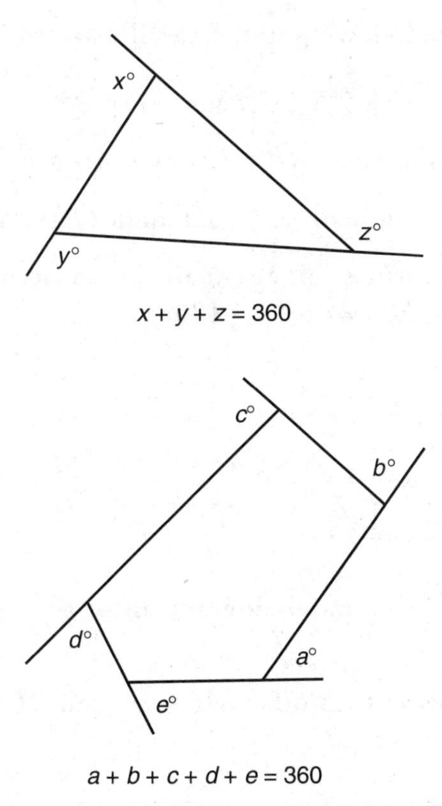

EXAMPLE 2: KEY FACT I3 gives us an alternative method of calculating the measure of each angle in a regular polygon. In Example 1 we used KEY FACT I2 to find the measure of each angle in a regular octagon. By KEY FACT I3, the

sum of the measures of the eight exterior angles of any octagon is 360°. As a result, in a regular octagon, the measure of each exterior angle is $360° \div 8 = 45°$. Therefore, the measure of each interior angle is $180° - 45° = 135°$.

SPECIAL QUADRILATERALS

We will now define five special quadrilaterals and review the important properties of each one.

A ***parallelogram*** is a quadrilateral in which both pairs of opposite sides are parallel. Any side of a parallelogram can be its ***base***, and a line segment drawn from a vertex perpendicular to the opposite base is called the ***height***.

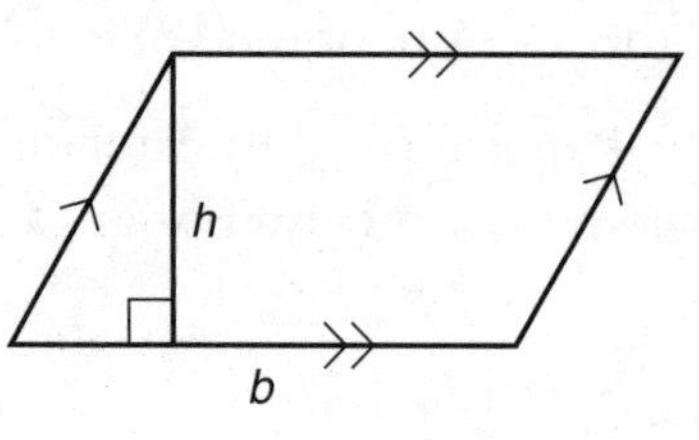

Key Fact I4

Parallelograms have the following properties illustrated in the figures below:

- **Opposite sides are parallel: $\overline{AB} \parallel \overline{CD}$ and $\overline{AD} \parallel \overline{BC}$.**
- **Opposite sides are congruent: $\overline{AB} \cong \overline{CD}$ and $\overline{AD} \cong \overline{BC}$.**
- **Opposite angles are congruent: $\angle A \cong \angle C$ and $\angle B \cong \angle D$.**
- **The sum of the measures of any pair of consecutive angles is 180°. For example, $a + b = 180$ and $b + c = 180$.**

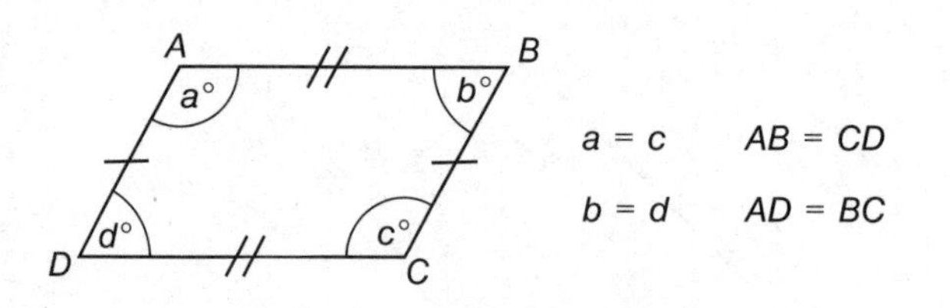

- **A diagonal divides the parallelogram into two congruent triangles: $\Delta ABC \cong \Delta ACD$.**
- **The two diagonals bisect each other: $AE = EC$ and $BE = ED$.**

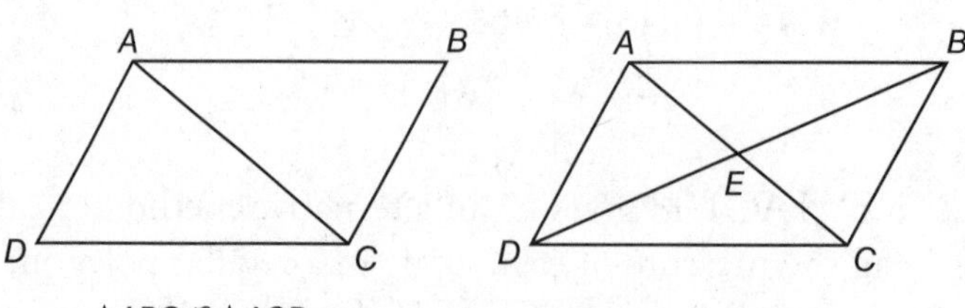

$\Delta ABC \cong \Delta ACD$

A ***rectangle*** is a parallelogram in which all four angles are right angles. Two adjacent sides of a rectangle are usually called the ***length*** (ℓ) and the ***width*** (w). Note that the length is not necessarily greater than the width.

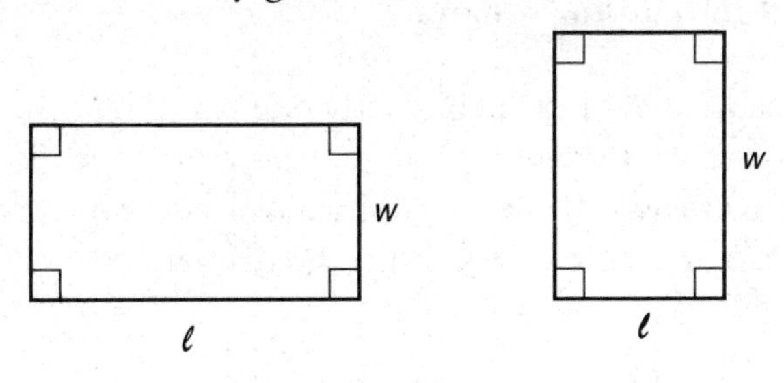

Key Fact I5

Since a rectangle is a parallelogram, all of the properties listed in KEY FACT I4 hold for rectangles. In addition:

- **The measure of each angle in a rectangle is 90°.**
- **The diagonals of a rectangle have the same length: $AC = BD$.**

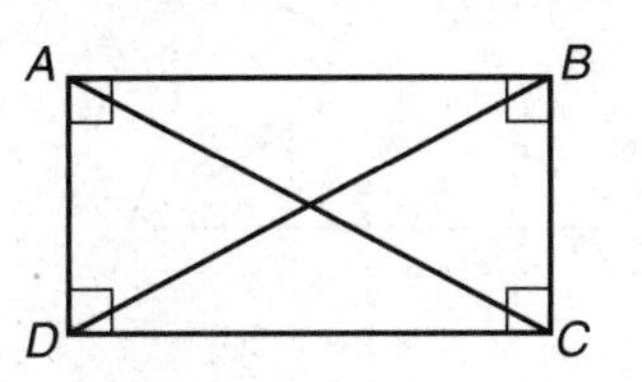

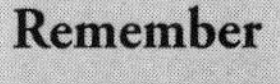

- A rectangle is a parallelogram.

A ***rhombus*** is a parallelogram in which all four sides have the same length.

Key Fact I6

Since a rhombus is a parallelogram, all of the properties listed in KEY FACT I4 hold for rhombuses. In addition:

- **The length of each side of a rhombus is the same.**
- **The two diagonals of a rhombus are perpendicular.**
- **The diagonals of a rhombus are angle bisectors.**

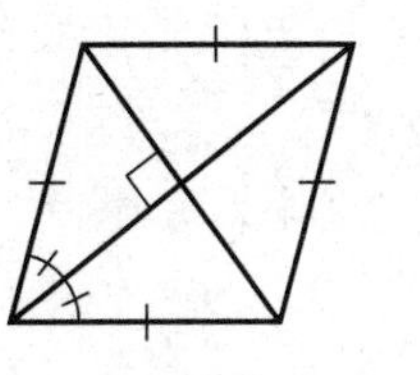

Remember

- A rhombus is a parallelogram.

A ***square*** is a rectangle in which all four sides have the same length. So a square is both a rectangle and a rhombus.

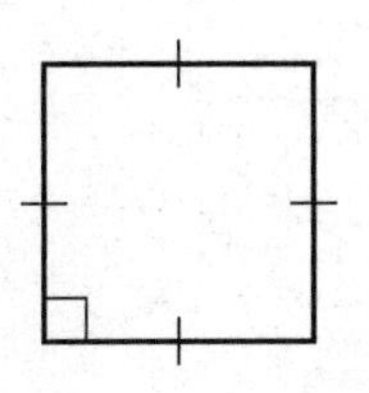

Key Fact I7

Since a square is a rectangle and a rhombus, all of the properties listed in KEY FACTS I4, I5, and I6 hold for squares.

Remember

A square is a special parallelogram, which is both a rectangle and a rhombus.

A ***trapezoid*** is a quadrilateral in which exactly one pair of opposite sides is parallel. The parallel sides are called the ***bases*** of the trapezoid, and the distance between the two bases is called the ***height***. If the two nonparallel sides are congruent, the trapezoid is called ***isosceles*** and, in that case only, the diagonals are congruent.

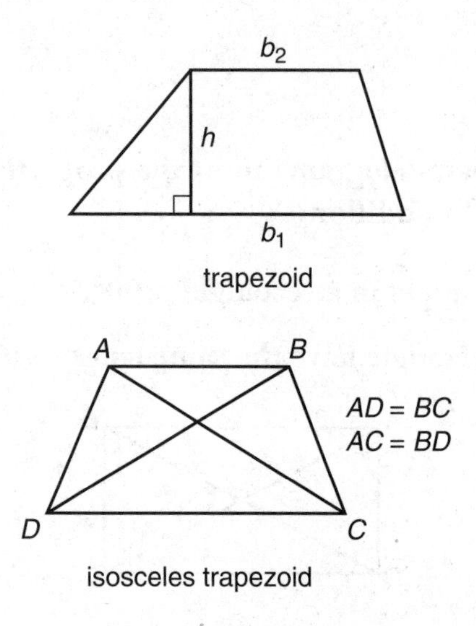

PERIMETER AND AREA OF QUADRILATERALS

The ***perimeter*** (P) of a polygon is the sum of the lengths of all of its sides. The ***area*** (A) of a polygon is the amount of space it encloses (measured in square units). The perimeter and area formulas you need to know for the Math 1 test are given in KEY FACTS I8 and I9.

Key Fact I8

PERIMETER

- **For a rectangle:** $P = 2(\ell + w)$
- **For a square:** $P = 4s$

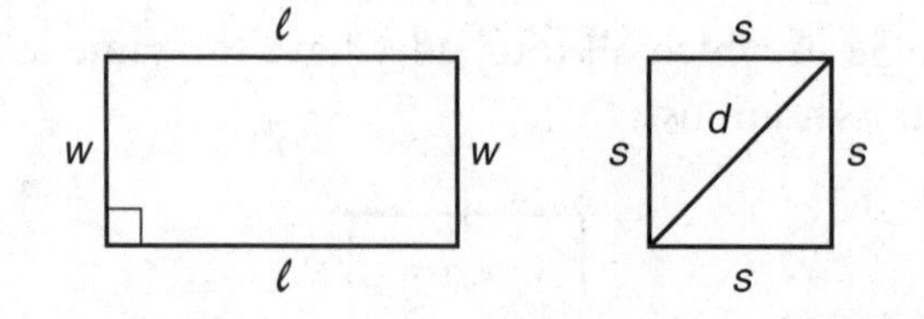

Key Fact I9

AREA

- **For a parallelogram:** $A = bh$
- **For a rectangle:** $A = \ell w$
- **For a square:** $A = s^2$ **or** $A = \frac{1}{2}d^2$
- **For a trapezoid:** $A = \frac{1}{2}(b_1 + b_2)h$

TIP

Here's a useful alternate formula for the area of a square:
If d is the diagonal of a square, then the area of the square is $\frac{d^2}{2}$.

EXAMPLE 3: What are the perimeter and area of a rhombus whose diagonals are 6 and 8? First draw and label a rhombus.

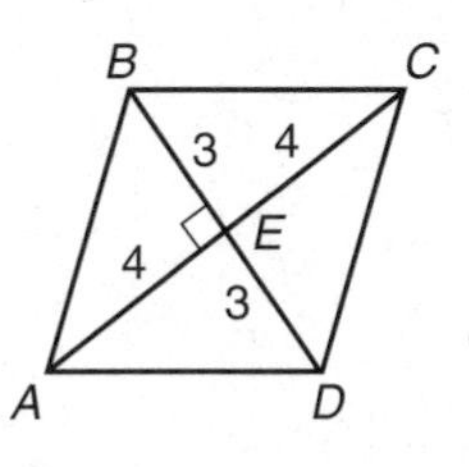

Since the diagonals bisect each other, $BE = ED = 3$ and $AE = EC = 4$. Also, since the diagonals of a rhombus are perpendicular, $\angle BEA$ is a right angle and ΔBEA is a 3-4-5 right triangle. So $AB = 5$ and the perimeter of the rhombus is $4 \times 5 = 20$. The easiest way to calculate the area of the rhombus is to recognize that it is the sum of the areas of four 3-4-5 right triangles. Since each triangle has an area of $\frac{1}{2}(3)(4) = 6$, the area of the rhombus is $4 \times 6 = 24$.

EXAMPLE 4: In the figure below, the area of parallelogram *ABCD* is 40. What are the areas of rectangle *AFCE*, trapezoid *AFCD*, and triangle *BCF*?

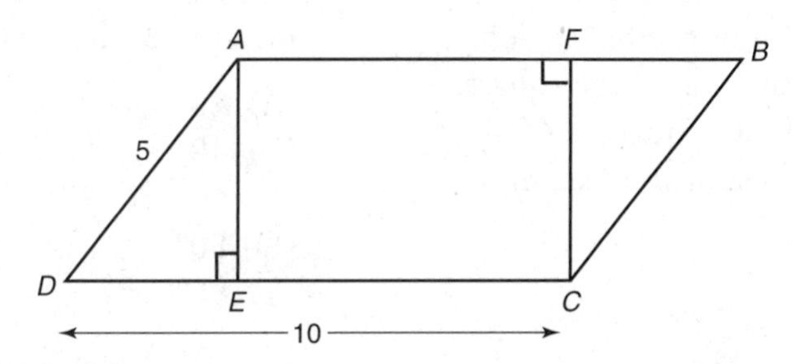

Since the base of parallelogram *ABCD* is 10 and its area is 40, its height, *AE*, must be 4. Then ΔAED must be a 3-4-5 right triangle with $DE = 3$, which implies that $EC = 7$. So the area of rectangle *AFCE* is $7 \times 4 = 28$; the area of trapezoid *AFCD* is $\frac{1}{2}(10 + 7)(4) = 34$; and the area of each small triangle is $\frac{1}{2}(3)(4) = 6$.

Exercises

Questions 1 and 2 refer to the following figure, in which M and N are midpoints of two of the sides of square $PQRS$.

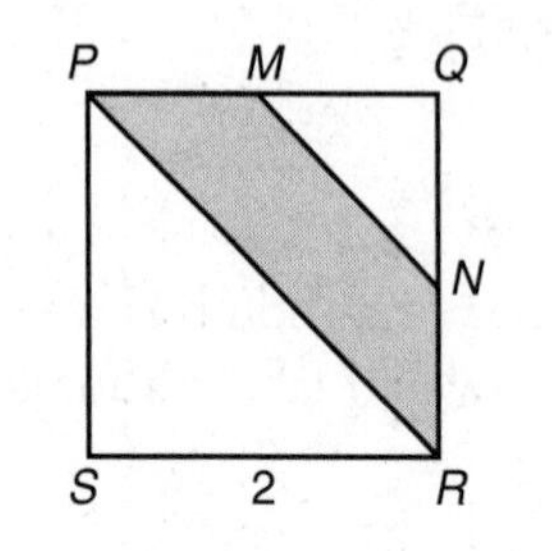

1. What is the perimeter of trapezoid $PMNR$?

   (A) 3
   (B) $2 + 3\sqrt{2}$
   (C) $3 + 2\sqrt{2}$
   (D) 5
   (E) 8

2. What is the area of trapezoid $PMNR$?

   (A) 1.5
   (B) 1.75
   (C) 3
   (D) $2\sqrt{2}$
   (E) $3\sqrt{2}$

3. The length of a rectangle is 5 more than the side of a square, and the width of the rectangle is 5 less than the side of the square. If the area of the square is 45, what is the area of the rectangle?

   (A) 20
   (B) 25
   (C) 45
   (D) 50
   (E) 70

4. In rhombus $PQRS$, the ratio of $m\angle P$ to $m\angle Q$ is 1 to 5 and $PQ = 6$. What is the area of the rhombus?

   (A) 6
   (B) 12
   (C) 18
   (D) 24
   (E) 30

5. If the length of a rectangle is 5 times its width and if its area is 180, what is its perimeter?

   (A) 6
   (B) 36
   (C) 60
   (D) 72
   (E) 144

6. What is the average (arithmetic mean) of the measures of all the interior angles in a decagon?

   (A) 18
   (B) 36
   (C) 72
   (D) 90
   (E) 144

7. If the interior angles of a pentagon are in the ratio of $2:3:3:5:5$, what is the measure of the smallest angle?

   (A) 20°
   (B) 40°
   (C) 60°
   (D) 80°
   (E) 90°

8. In the figure below, the two diagonals divide square *WXYZ* into four small triangles. What is the sum of the perimeters of those four triangles?

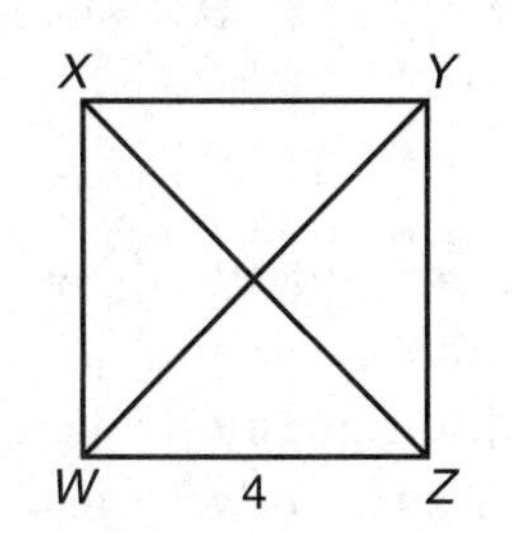

(A) $4 + 4\sqrt{2}$
(B) $16 + 8\sqrt{2}$
(C) $16 + 16\sqrt{2}$
(D) 32
(E) 48

9. How many sides does a polygon have if the measure of each interior angle is 9 times the measure of each exterior angle?

(A) 8
(B) 10
(C) 12
(D) 18
(E) 20

10. What is the area of trapezoid *ABCD* in the figure below?

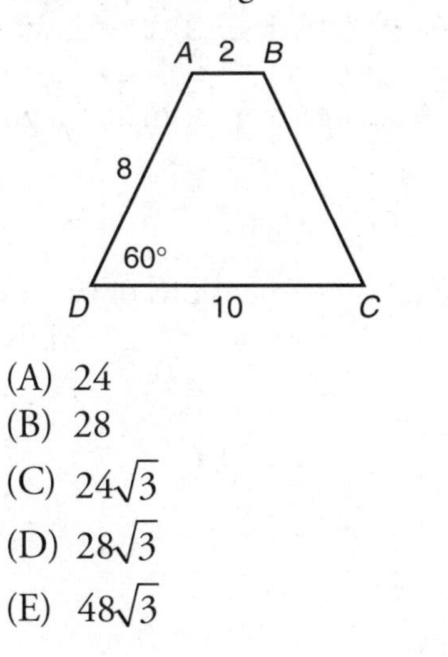

(A) 24
(B) 28
(C) $24\sqrt{3}$
(D) $28\sqrt{3}$
(E) $48\sqrt{3}$

11. What is the area of a regular hexagon whose sides are 4?

(A) 24
(B) 36
(C) $24\sqrt{3}$
(D) $36\sqrt{3}$
(E) $48\sqrt{3}$

12. The area of a regular octagon whose sides are 2 can be expressed as $a + b\sqrt{2}$. What is the value of $a + b$?

(A) 8
(B) 12
(C) 16
(D) 18
(E) 24

ANSWERS EXPLAINED

Answer Key

1. **(B)**	3. **(A)**	5. **(D)**	7. (C)	9. (E)	11. (C)
2. **(A)**	4. (C)	6. (E)	8. (C)	10. (C)	12. (C)

Solutions

Each of the problems in this set of exercises is typical of a question you could see on a Math 1 test. When you take the model tests in this book and, in particular, when you take the actual Math 1 test, if you get stuck on questions such as these, you do not have to leave them out—you can almost always answer them by using one or more of the strategies discussed in the "Tactics" chapter. The solutions given here do *not* depend on those strategies; they are the correct mathematical ones.

1. **(B)** Since M and N are midpoints of sides of length 2, $PM = MQ = QN = NR = 1$. Since $\overline{MN}$ is the hypotenuse of an isosceles right triangle whose legs are 1, $MN = \sqrt{2}$. Similarly, $PR = 2\sqrt{2}$, since it is the hypotenuse of an isosceles right triangle whose legs are 2. So the perimeter of trapezoid $PMNR$ is $1 + \sqrt{2} + 1 + 2\sqrt{2} = 2 + 3\sqrt{2}$.

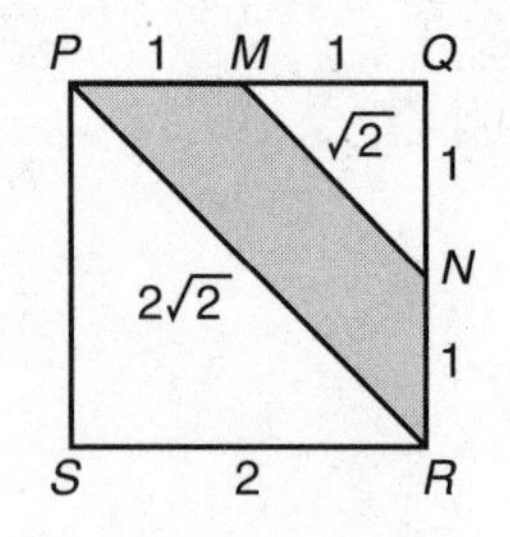

2. **(A)** Even if you know the formula for the area of a trapezoid (and you should), the best way to proceed is to subtract the areas of the two white triangles from 4, the area of the square. The area of $\Delta PSR = \frac{1}{2}(2)(2) = 2$, and the area of ΔMQN is $\frac{1}{2}(1)(1) = 0.5$. Therefore, the area of the shaded region is $4 - 2 - 0.5 = 1.5$.

3. **(A)** Let x represent the side of the square. Then the dimensions of the rectangle are $(x + 5)$ and $(x - 5)$, and its area is $(x + 5)(x - 5) = x^2 - 25$. Since the area of the square is 45, $x^2 = 45$, and so $x^2 - 25 = 20$.

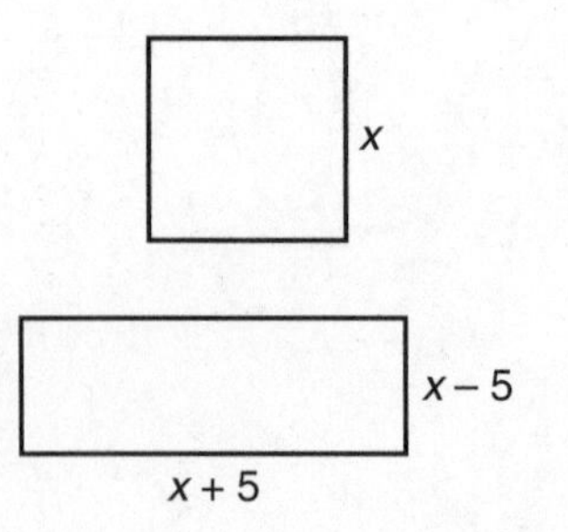

4. **(C)** Draw and label the rhombus.

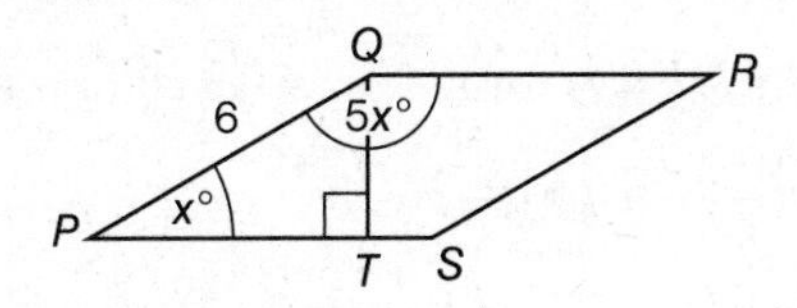

Since angles P and Q are supplementary:

$$m\angle P + m\angle Q = 180 \Rightarrow 5x + x = 180 \Rightarrow 6x = 180 \Rightarrow x = 30$$

Then since ΔPQT is a 30-60-90 right triangle, $QT = \frac{6}{2} = 3$. Finally, the area of rhombus $PQRS = bh = (PS)(QT) = (6)(3) = 18$.

5. **(D)** Draw a diagram and label it.

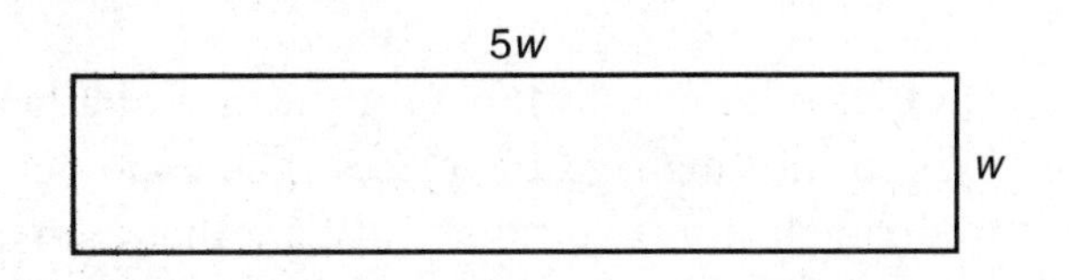

Since the area is 180, we have $180 = 5w^2 \Rightarrow w^2 = 36 \Rightarrow w = 6$. So the width is 6, the length is $5 \times 6 = 30$, and the perimeter is $2(6 + 30) = 2(36) = 72$.

6. **(E)** By KEY FACT I2, the sum of the measures of the 10 interior angles in a decagon is $(10 - 2) \times 180° = 8(180°) = 1{,}440°$. So the average of their measures is $1{,}440° \div 10 = 144°$.

7. **(C)** By KEY FACT I2, the sum of the angles of a pentagon is

$$(5 - 2) \times 180° = 3 \times 180° = 540°$$

Let the degree measures of the five angles be $2x$, $3x$, $3x$, $5x$, and $5x$. Then

$$540 = 2x + 3x + 3x + 5x + 5x = 18x \Rightarrow$$
$$x = (540 \div 18) = 30$$

The degree measure of the smallest angle is $2x$, and $2 \times 30° = 60°$.

8. **(C)** Since the diagonals of a square are perpendicular, congruent, and bisect each other, each small triangle is a 45-45-90 right triangle whose hypotenuse is 4. Therefore, the legs are each $\frac{4}{\sqrt{2}} = 2\sqrt{2}$. So the perimeter of each small triangle is $4 + 2\sqrt{2} + 2\sqrt{2} = 4 + 4\sqrt{2}$, and the sum of the perimeters is $4(4 + 4\sqrt{2}) = 16 + 16\sqrt{2}$.

9. **(E)** The sum of an interior and exterior angle is 180°. So $180 = 9x + x = 10x \Rightarrow x = 18$. Since the sum of all the exterior angles is 360, there are $360 \div 18 = 20$ exterior angles, 20 interior angles, and 20 sides.

9x°　x°

10. **(C)** Draw in height $\overline{AE}$. Then ΔAED is a 30-60-90 right triangle whose hypotenuse is 8. So $DE = 4$ and $AE = 4\sqrt{3}$. Then the area of trapezoid $ABCD$ is $\frac{1}{2}(10+2)(4\sqrt{3}) = 6(4\sqrt{3}) = 24\sqrt{3}$.

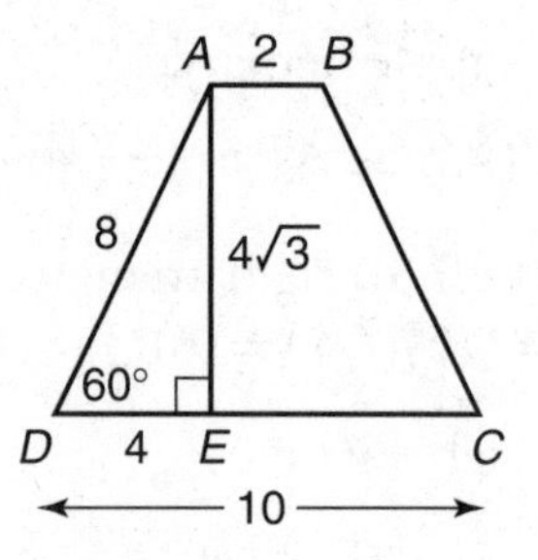

11. **(C)** Since you do not know a formula for the area of a hexagon, you have to divide the hexagon into manageable pieces. There are many ways to do this. One way is to divide it into two trapezoids; another is to divide it into a rectangle and two triangles. The simplest way, though, is to divide it into six equilateral triangles.

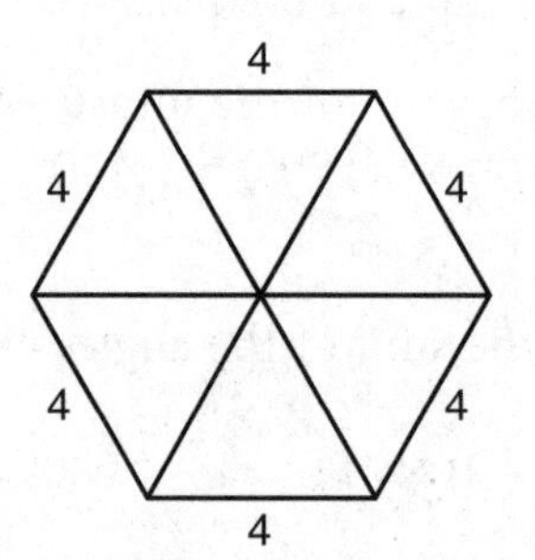

By KEY FACT H11, the area of an equilateral triangle whose sides are 4 is $\frac{4^2\sqrt{3}}{4} = \frac{16\sqrt{3}}{4} = 4\sqrt{3}$. So the area of the hexagon is $6(4\sqrt{3}) = 24\sqrt{3}$.

12. **(C)** Since you do not know a formula for the area of an octagon, you have to divide the octagon into manageable pieces. There are many ways to do that. One way is to draw four diagonals that divide the octagon into four isosceles right triangles, four rectangles, and a square.

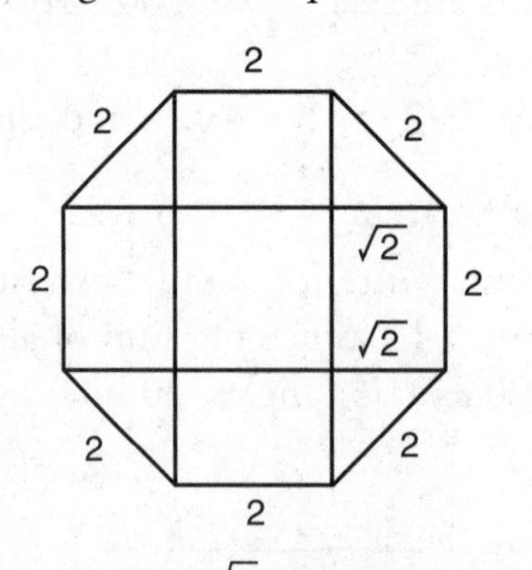

The legs of the triangles are each $\sqrt{2}$. Therefore, each triangle has area 1, each rectangle has area $2\sqrt{2}$, and the square has area 4. So the total area is $4(1) + 4(2\sqrt{2}) + 4 = 8 + 8\sqrt{2}$. So $a = 8$, $b = 8$, and $a + b = 16$.

Circles

CHAPTER 11

- Circumference and Area
- Tangents to a Circle
- Exercises
- Answers Explained

A ***circle*** consists of all the points that are the same distance from one fixed point called the ***center***. That distance is called the ***radius*** of the circle. The figure below is a circle of radius 1 unit whose center is at the point O. A, B, C, D, and E, which are each 1 unit from O, are all points on circle O. The word ***radius*** is also used to represent any of the line segments joining the center and a point on the circle. The plural of *radius* is *radii*. In circle O, below, $\overline{OA}$, $\overline{OB}$, $\overline{OC}$, $\overline{OD}$, and $\overline{OE}$ are all radii. If a circle has radius r, each of the radii is r units long. A point is ***inside*** a circle if the distance from the center to that point is less than the radius. A point is ***outside*** a circle if the distance from the center to that point is greater than the radius.

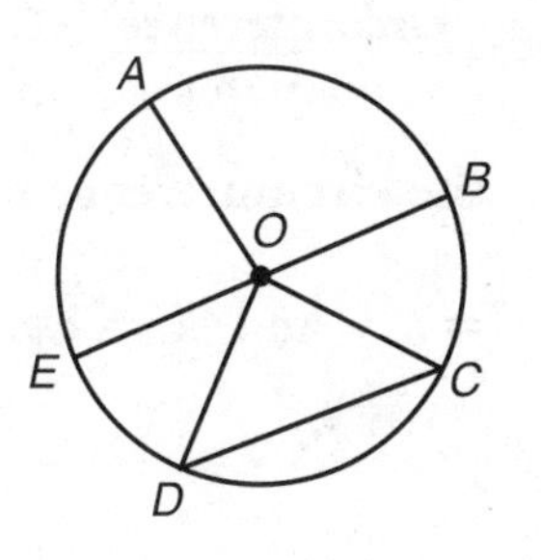

EXAMPLE 1: In the figure below, O is the center of the circle. To find $m\angle B$ and $m\angle O$, observe that since $\overline{OA}$ and $\overline{OB}$ are radii, $OA = OB$ and ΔAOB is isosceles. So $m\angle B = 25°$ and $m\angle O = 180° - (25° + 25°) = 130°$.

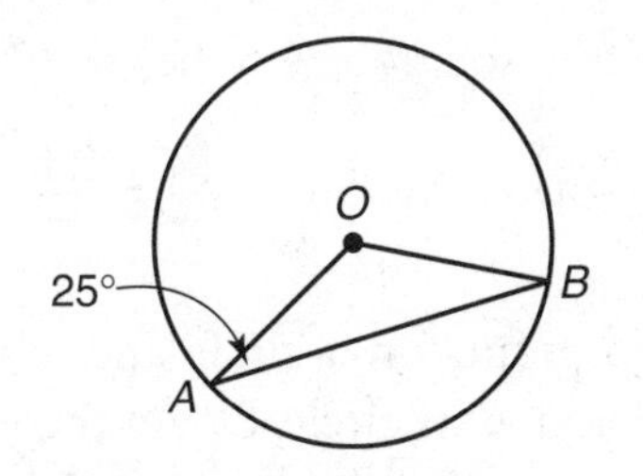

Key Fact J1

Any triangle formed by connecting the endpoints of two radii is isosceles.

A ***chord*** of a circle is a line segment that has both endpoints on the circle. In the figure at the beginning of this chapter, $\overline{CD}$ and $\overline{BE}$ are chords. A chord such as $\overline{BE}$ that passes through the center is called a ***diameter***. Since $BE = EO + OB$, a diameter is twice as long as a radius.

Key Fact J2

If d is the diameter and r is the radius of a circle: $d = 2r$.

Key Fact J3

Diameters are the longest line segments that can be drawn that have both endpoints on or inside a circle.

CIRCUMFERENCE AND AREA

The total length around a circle is called the ***circumference***. In every circle, the ratio of the circumference to the diameter is exactly the same and is denoted by the symbol π (the Greek letter "pi").

Key Fact J4

$$\pi = \frac{\text{circumference}}{\text{diameter}} = \frac{C}{d}.$$

So there are two formulas for the circumference of a circle:

$$C = \pi d \text{ and } C = 2\pi r$$

Key Fact J5

The value of π is *approximately* 3.14.

Smart Strategy

On almost all questions on the Math 1 test that involve circles, you are expected to leave your answer in terms of π. If you need an approximation to test the values of the choices, for example, then use your calculator. To avoid rounding errors, use the π key rather than punching in 3.14.

EXAMPLE 2: If the circumference of a circle is equal to the perimeter of a square whose sides are 12, what is the radius of the circle?

Since the perimeter of the square is $4 \times 12 = 48$:

$$2\pi r = 48 \Rightarrow r = \frac{48}{2\pi} = \frac{24}{\pi} \approx 7.64$$

An ***arc*** consists of two points on a circle and all the points between them. If two points, such as A and B in circle O, are the endpoints of a diameter, they divide the circle into two arcs called ***semicircles***. On the Math 1 test, arc $\widehat{XY}$ always refers to the smaller arc joining X and Y. In the figure at the top of page 157, if we wanted to refer to the larger arc going from X to Y, the one through A and B, we would refer to it as arc $\widehat{XAY}$ or arc $\widehat{XBY}$.

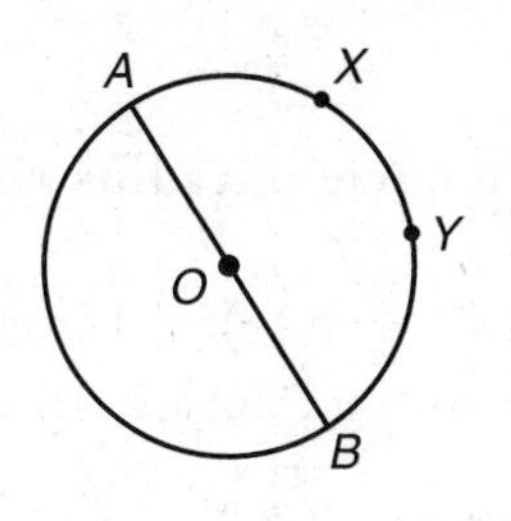

Key Fact J6

The degree measure of a circle is 360°.

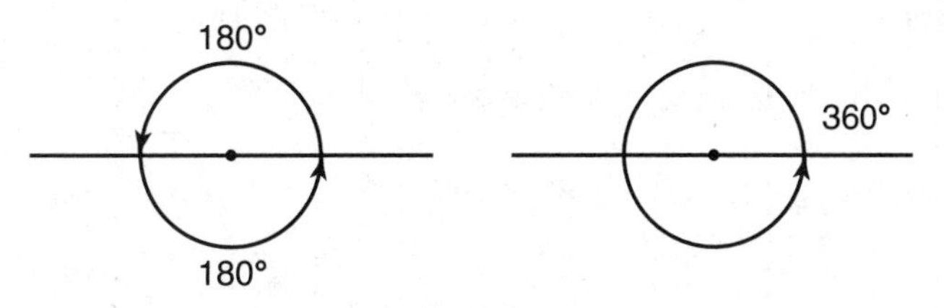

An angle such as $\angle AOB$ in the figure below, whose vertex is at the center of a circle, is called a ***central angle***.

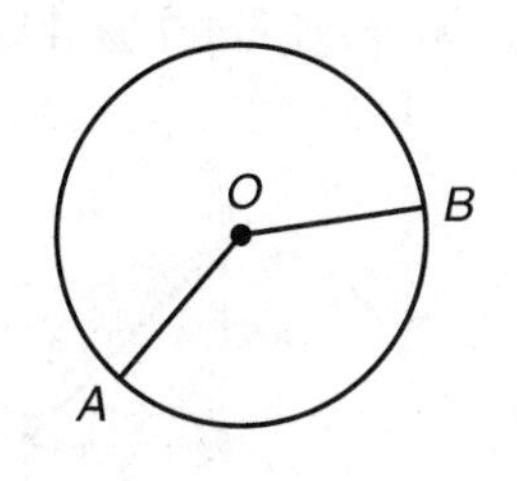

Key Fact J7

The degree measure of an arc equals the degree measure of the central angle that intercepts it.

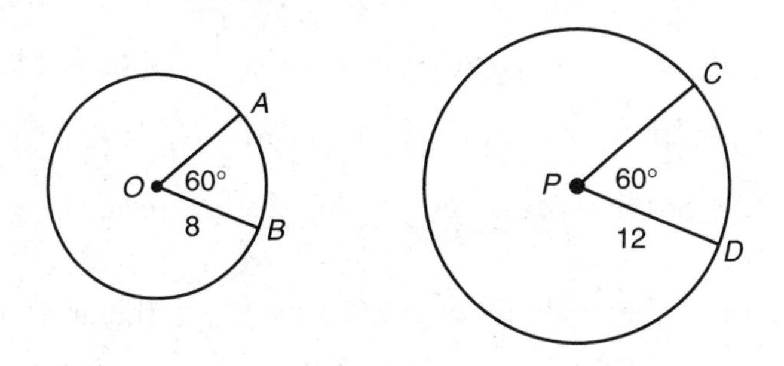

Remember

Degree measure is not a measure of length. In the circles at the left, arc $\widehat{AB}$ and arc $\widehat{CD}$ each measure 60° even though arc $\widehat{CD}$ is much longer.

In the figure above, how long is arc $\widehat{CD}$? Since the radius of circle P is 12, its diameter is 24 and its circumference is 24π. Since there are 360° in a circle, arc $\widehat{CD}$ is $\frac{60}{360}$, or $\frac{1}{6}$, of the circumference: $\frac{1}{6}(24\pi) = 4\pi$.

Key Fact J8

The formula for the area of a circle of radius r is $A = \pi r^2$.

The area of circle *P*, on page 157, is $\pi(12)^2 = 144\pi$ square units. A ***sector*** is a region of a circle bounded by two radii and an arc. In that circle, since the measure of $\angle CPD$ is 60° ($\frac{1}{6}$ of the circle), the area of sector *CPD* is $\frac{1}{6}$ the area of the circle:

$\frac{1}{6}(144\pi) = 24\pi$.

Key Fact J9

In a circle of radius r, if an arc measures $x°$:

- **The length of the arc is $\frac{x}{360}(2\pi r)$**
- **The area of the sector formed by the arc and two radii is $\frac{x}{360}(\pi r^2)$.**

EXAMPLE 3: What are the perimeter and area of the shaded region in the figure below?

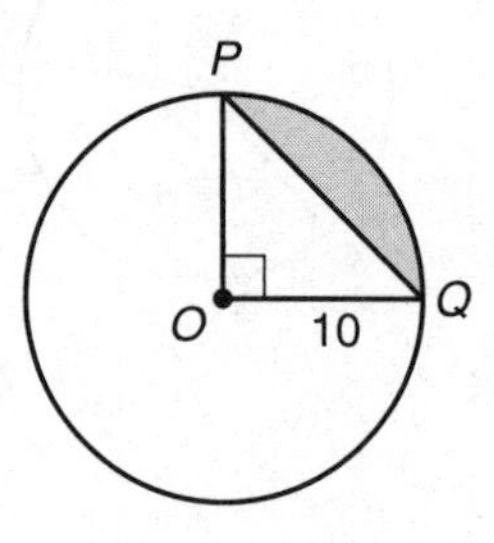

Smart Strategy

On the Math 1 test, the answer choices to questions such as these are almost always given in terms of π and square roots, so you do not need to use your calculator to evaluate them.

The circumference of the circle is $2\pi r = 2(10)\pi = 20\pi$. Since arc $\overset{\frown}{PQ}$ is $\frac{90}{360} = \frac{1}{4}$ of the circle, the length of arc $\overset{\frown}{PQ} = \frac{1}{4}(20\pi) = 5\pi$. Since $\overline{PQ}$ is the hypotenuse of isosceles right triangle *POQ*, by KEY FACT H7, $PQ = 10\sqrt{2}$. So the perimeter of the shaded region is $10\sqrt{2} + 5\pi$.

Since the area of the circle is $\pi r^2 = \pi(10^2) = 100\pi$, the area of sector *POQ* is $\frac{1}{4}(100\pi) = 25\pi$. The area of $\Delta POQ = \frac{1}{2}(10)(10) = 50$. So the area of the shaded region is $25\pi - 50$.

An angle formed by two chords with a common endpoint is called an ***inscribed angle***. In the figure below, $\angle ABC$, $\angle ADC$, $\angle BAD$, and $\angle BCD$ are all inscribed angles.

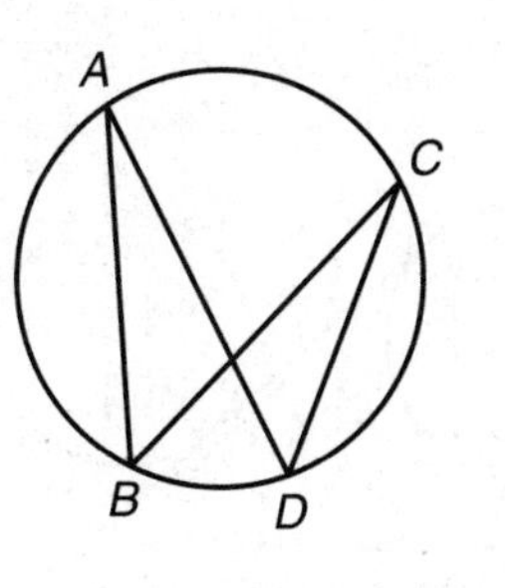

Key Fact J10

The measure of an inscribed angle is one-half the measure of its intercepted arc.

EXAMPLE 4: To find $m\angle ABC$ in circle O in the figure below, observe that since the measure of an arc is equal to the measure of the central angle that intercepts it, the measure of arc $\overparen{AC}$ is 110°. Since $\angle ABC$ is an inscribed angle, its measure is one-half the measure of arc $\overparen{AC}$: $\frac{1}{2}(110°) = 55°$.

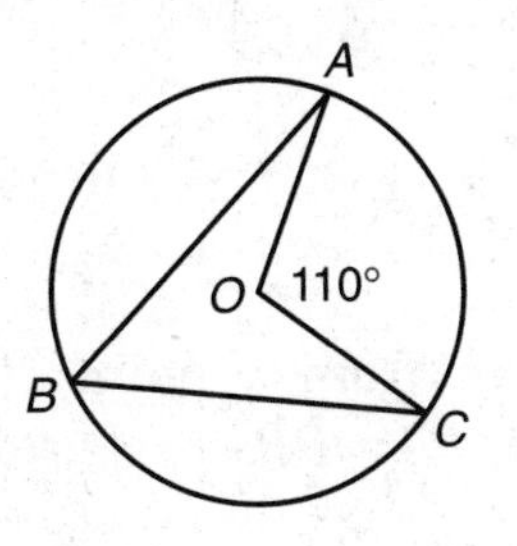

TANGENTS TO A CIRCLE

A line or line segment is ***tangent*** to a circle if it intersects the circle exactly once.

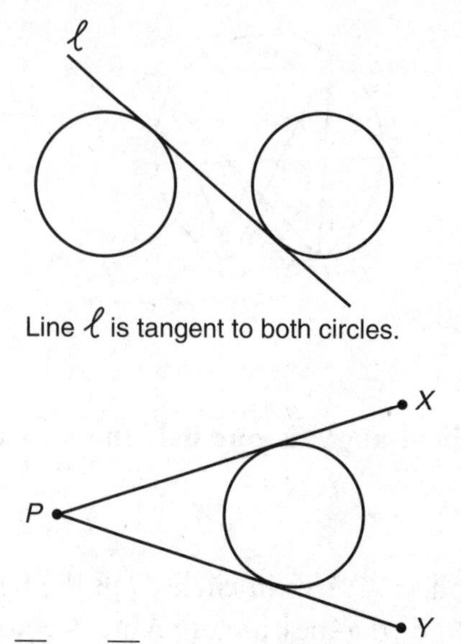

Line ℓ is tangent to both circles.

$\overline{PX}$ and $\overline{PY}$ are each tangent to the circle.

The following KEY FACT lists four important theorems about tangents.

Key Fact J11

- **From any point outside a circle, exactly two tangents can be drawn to the circle.**
- **If two tangents are drawn from a point P outside a circle, intersecting the circle at A and B, then $PA = PB$.**
- **The measure of the angle formed by two tangents drawn from the same point is one-half the difference of the two intercepted arcs.**
- **A line tangent to a circle is perpendicular to the radius (or diameter) drawn to the point of contact.**

EXAMPLE 5: In the diagram at the top of page 161, $\overline{PA}$ and $\overline{PB}$ are tangent to circle O. What is $m\angle P$?

By the fourth theorem in KEY FACT J11, $m\angle PAO = m\angle PBO = 90°$. Since the sum of the measures of the four angles in quadrilateral $PAOB$ is 360°:

$$90° + 140° + 90° + m\angle P = 360° \Rightarrow m\angle P = 40°$$

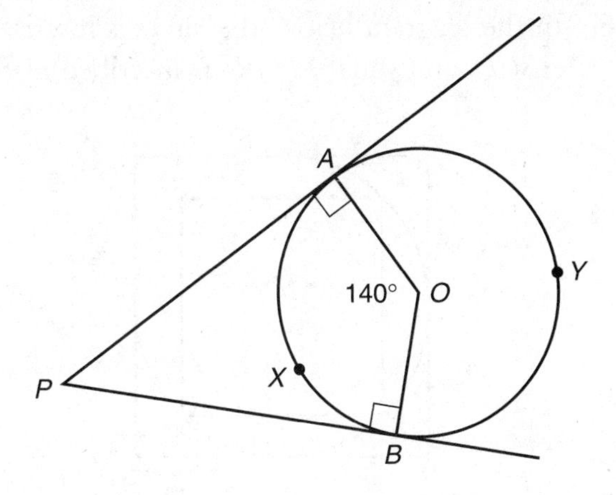

Another way to evaluate $m\angle P$ is to observe that since $\angle AOB$ is a central angle, the measure of arc $\overset{\frown}{AXB}$ is 140° and the measure of arc $\overset{\frown}{AYB}$ is 360° – 140° = 220°. Then, by the third theorem in KEY FACT J11:

$$m\angle P = \frac{1}{2}(220° - 140°) = \frac{1}{2}(80°) = 40°$$

Key Fact J12

- **When a square is inscribed in a circle, the diagonals of the square are diameters of the circle. ($\overline{AB}$ is a diagonal and a diameter.)**

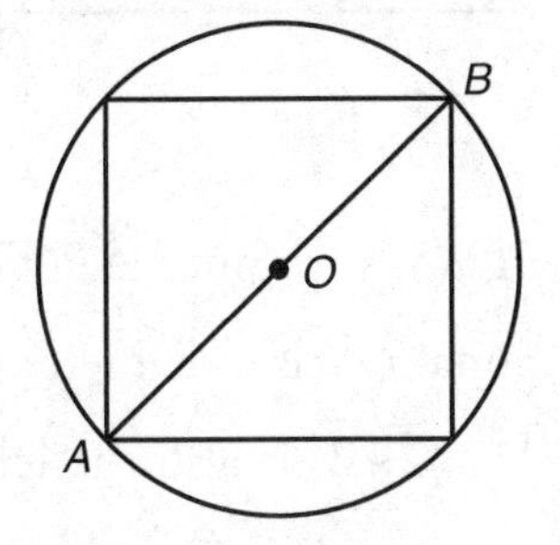

- **When a circle is inscribed in a square, the length of a diameter is equal to the length of a side of the square. ($AB = WX$)**

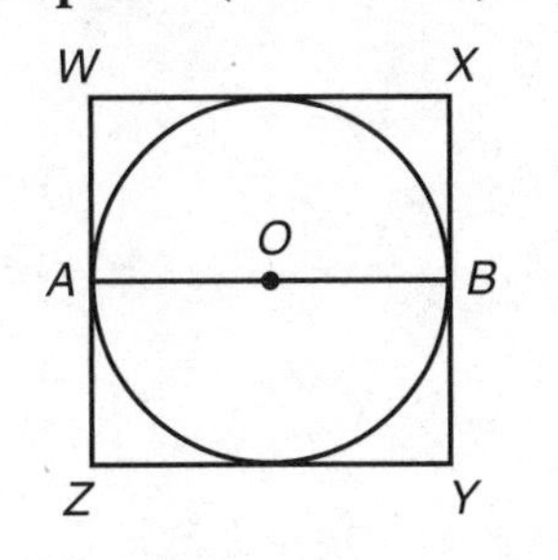

EXAMPLE 6: In the diagram below, the circle is inscribed in square *ABCD*, and square *PQRS* is inscribed in the circle.

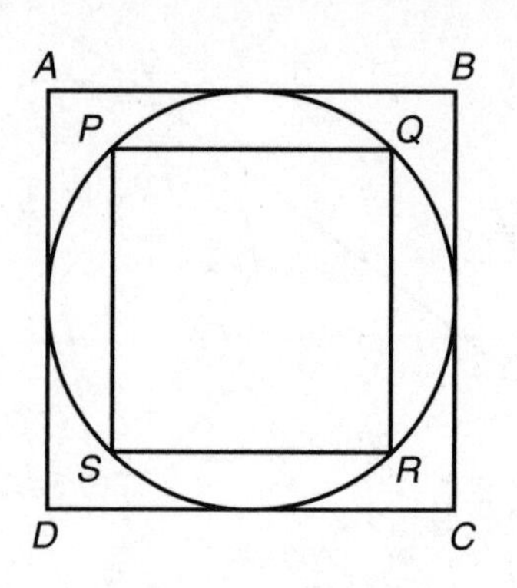

What is the ratio of the area of the large square to the area of the small square?

Choose any number for the sides of square *PQRS*, say $PQ = 2$.

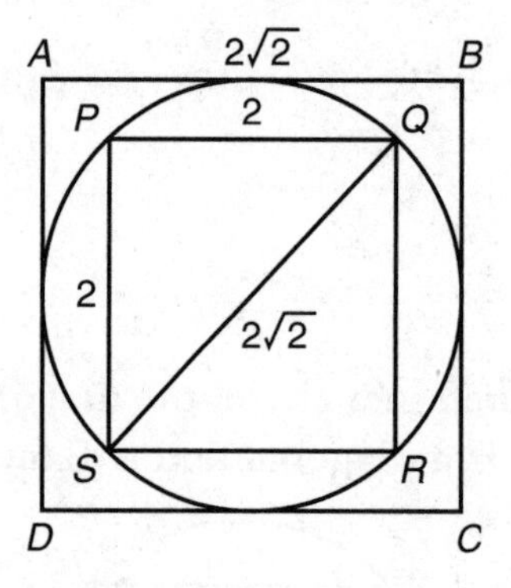

Then the area of square *PQRS* is 4. Since ΔPQS is an isosceles right triangle, $QS = 2\sqrt{2}$. Since diagonal $\overline{QS}$ is a diameter of the circle, by the second statement in KEY FACT J12, $2\sqrt{2}$ is also the length of side $\overline{AB}$ of square *ABCD*. So the area of square *ABCD* is $(2\sqrt{2})^2 = 8$. Finally, the ratio of the area of square *ABCD* to the area of square *PQRS* is 8 : 4 or 2 : 1.

Exercises

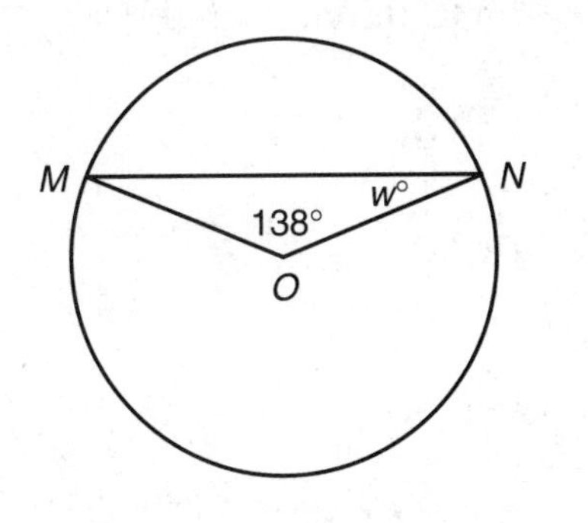

1. In the figure above, if O is the center of the circle, what is the value of w?

   (A) 21
   (B) 38
   (C) 42
   (D) 69
   (E) 111

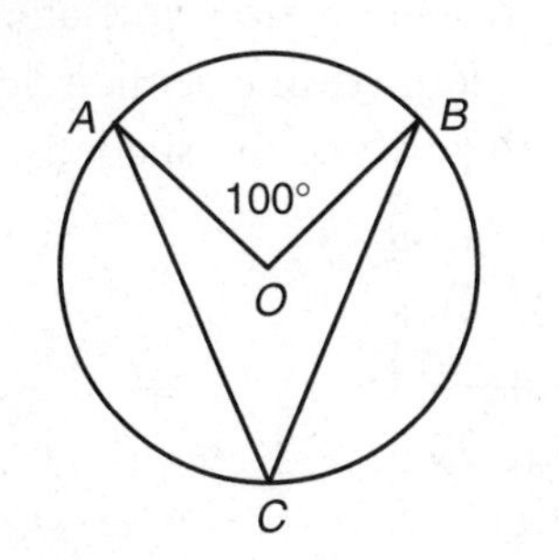

2. In the figure above, if O is the center of the circle, what is $m\angle C$?

   (A) 40°
   (B) 50°
   (C) 80°
   (D) 100°
   (E) It cannot be determined from the given information.

Questions 3 and 4 refer to the following figure:

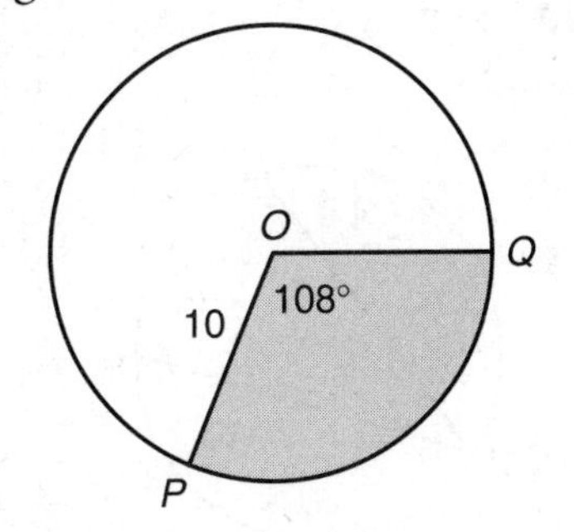

3. What is the length of arc $\overset{\frown}{PQ}$?

   (A) 6
   (B) 12
   (C) 6π
   (D) 12π
   (E) 18π

4. What is the area of the shaded sector?

   (A) 12
   (B) 18
   (C) 12π
   (D) 18π
   (E) 30π

5. What is the area of a circle whose circumference is $\frac{\pi}{2}$?

   (A) $\frac{\pi}{16}$
   (B) $\frac{\pi}{4}$
   (C) $\frac{\pi}{2}$
   (D) π
   (E) 4π

6. What is the circumference of a circle whose area is 100π?

(A) 10
(B) 20
(C) 10π
(D) 20π
(E) 25π

Note: Figure not drawn to scale.

7. In the figure above, the ratio of the measure of arc $\widehat{AB}$ to the measure of arc $\widehat{BC}$ to the measure of arc $\widehat{CA}$ is 4 to 3 to 5. What is $m\angle A$?

(A) 30°
(B) 45°
(C) 60°
(D) 75°
(E) 90°

8. A square of area 2 is inscribed in a circle. What is the area of the circle?

(A) $\frac{\pi}{4}$
(B) $\frac{\pi}{2}$
(C) π
(D) $\pi\sqrt{2}$
(E) 2π

9. What is the area of a circle that is inscribed in a square of area 2?

(A) $\frac{\pi}{4}$
(B) $\frac{\pi}{2}$
(C) π
(D) $\pi\sqrt{2}$
(E) 2π

10. If A is the area and C is the circumference of a circle, which of the following is an expression for C in terms of A?

(A) $2\sqrt{\frac{\pi}{A}}$
(B) $\frac{2\pi}{\sqrt{A}}$
(C) $2\sqrt{\frac{A}{\pi}}$
(D) $2\pi\sqrt{A}$
(E) $2\sqrt{\pi A}$

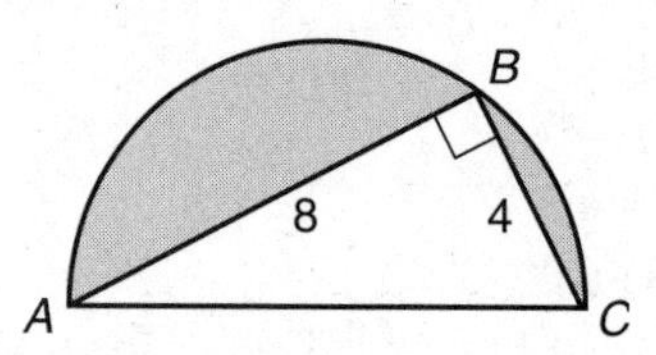

11. In the figure above, triangle ABC is inscribed in a semicircle. What is the area of the shaded region?

(A) 15.4
(B) 48.6
(C) 71.4
(D) 93.7
(E) 109.7

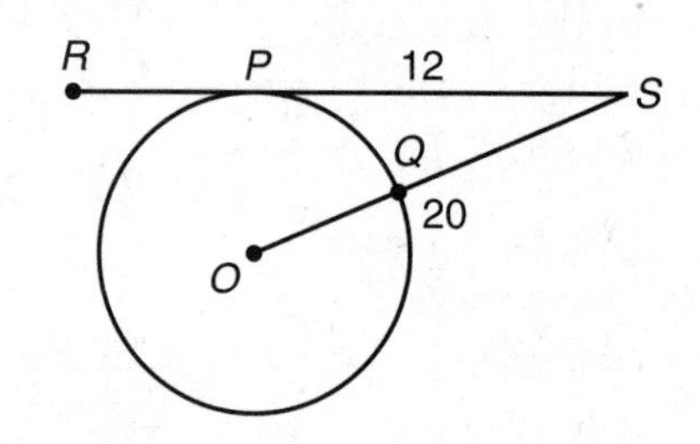

Note: Figure not drawn to scale.

12. In the figure above, $\overline{RS}$ is tangent to circle O at point P. If $OS = 20$ and $PS = 12$, what is QS?

(A) 4
(B) 8
(C) 9
(D) 10
(E) 11

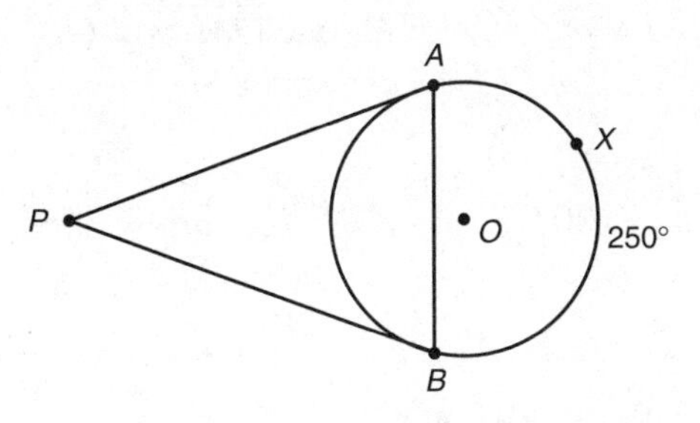

<u>Note:</u> Figure not drawn to scale

13. In the figure above, $\overline{PA}$ and $\overline{PB}$ are tangents to circle O and the measure of arc $\widehat{AXB}$ is 250°. What is $m\angle PAB$?
(A) 35°
(B) 55°
(C) 70°
(D) 90°
(E) 110°

14. The circumference of a circle is $x\pi$ units, and the area of the circle is $y\pi$ square units. If $y = 2x$, what is the radius of the circle?

(A) 1
(B) 2
(C) 4
(D) 6
(E) 8

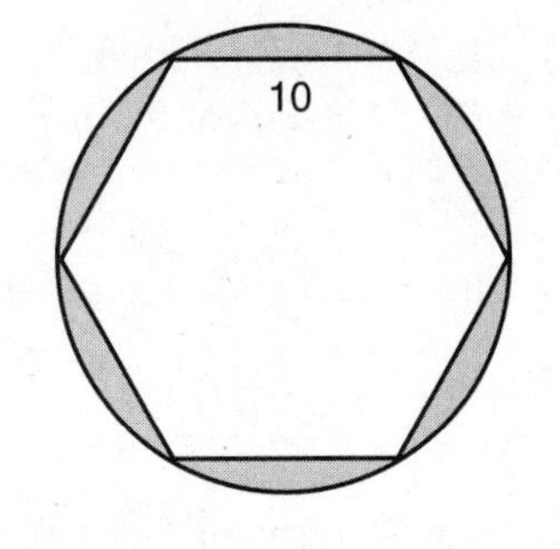

15. The figure above shows a regular hexagon whose sides are 10 inscribed in a circle. What is the area of the shaded region, rounded to the nearest whole number?

(A) 24
(B) 34
(C) 54
(D) 64
(E) 164

ANSWERS EXPLAINED

Answer Key

1. **(A)**	4. **(E)**	7. **(B)**	10. **(E)**	13. **(B)**
2. **(B)**	5. **(A)**	8. **(C)**	11. **(A)**	14. **(C)**
3. **(C)**	6. **(D)**	9. **(B)**	12. **(A)**	15. **(C)**

Solutions

Each of the problems in this set of exercises is typical of a question you could see on a Math 1 test. When you take the model tests in this book and, in particular, when you take the actual Math 1 test, if you get stuck on questions such as these, you do not have to leave them out—you can almost always answer them by using one or more of the strategies discussed in the "Tactics" chapter. The solutions given here do *not* depend on those strategies; they are the correct mathematical ones.

1. **(A)** $m\angle M + m\angle N + 138 = 180 \Rightarrow m\angle M + m\angle N = 42$. Since $\overline{OM}$ and $\overline{ON}$ are radii, $OM = ON$, ΔOMN is isosceles, and $m\angle M = m\angle N = w$. So $2w = 42 \Rightarrow w = 21$.

2. **(B)** Since $\angle AOB$ is a central angle, $m\widehat{AB} = 100°$. Since $\angle ACB$ is an inscribed angle, $m\angle C = \frac{1}{2}(100°) = 50°$.

3. **(C)** The length of arc $\widehat{PQ}$ is $\frac{108}{360} = \frac{3}{10}$ of the circumference, which is $2\pi(10) = 20\pi$. So the length of arc $\widehat{PQ} = \frac{3}{10}(20\pi) = 6\pi$.

4. **(E)** The area of the shaded sector is $\frac{108}{360} = \frac{3}{10}$ of the area of the circle, which is $\pi(10^2) = 100\pi$. So the area of the shaded sector is $\frac{3}{10}(100\pi) = 30\pi$.

5. **(A)** $C = 2\pi r = \frac{\pi}{2} \Rightarrow 2r = \frac{1}{2} \Rightarrow r = \frac{1}{4} \Rightarrow A = \pi r^2 = \pi\left(\frac{1}{4}\right)^2 = \frac{1}{16}\pi = \frac{\pi}{16}$

6. **(D)** $A = \pi r^2 = 100\pi \Rightarrow r^2 = 100 \Rightarrow r = 10 \Rightarrow C = 2\pi r = 2\pi(10) = 20\pi$

7. **(B)** The sum of the measures of the three arcs is 360°. So:

$$4x + 3x + 5x = 360° \Rightarrow 12x = 360° \Rightarrow x = 30°$$

So the measure of arc $\widehat{BC}$ is $3 \times 30° = 90°$, and $m\angle A = \frac{1}{2}(90°) = 45°$.

8. **(C)** Draw a picture. Since the area of square $ABCD$ is 2, $AD = \sqrt{2}$. So diagonal $BD = \sqrt{2} \times \sqrt{2} = 2$. $\overline{BD}$ is also a diameter of the circle. So the diameter is 2 and the radius is 1. Therefore, the area is $\pi(1)^2 = \pi$.

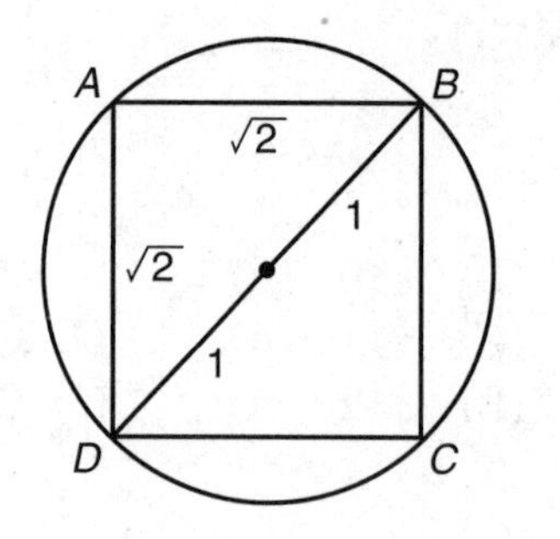

9. **(B)** Draw a picture. Since the area of square *ABCD* is 2, $AD = \sqrt{2}$. Then diameter $EF = \sqrt{2} \Rightarrow$ radius $OE = \frac{\sqrt{2}}{2} \Rightarrow$ the area of circle *O* is $\pi\left(\frac{\sqrt{2}}{2}\right)^2 =$ $\frac{2}{4}\pi = \frac{\pi}{2}$.

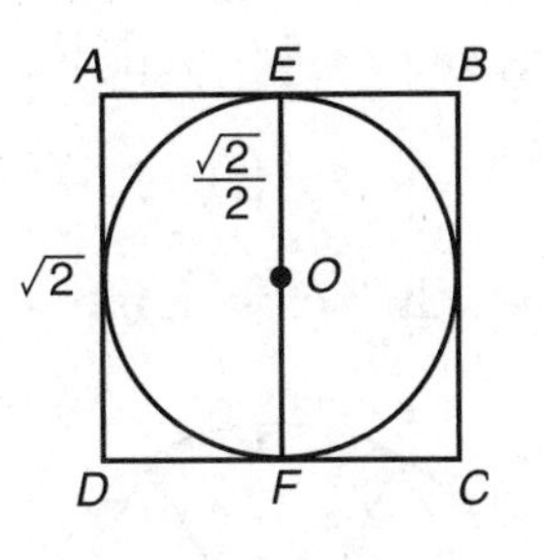

10. **(E)** $A = \pi r^2 \Rightarrow r^2 = \frac{A}{\pi} \Rightarrow r = \sqrt{\frac{A}{\pi}} = \frac{\sqrt{A}}{\sqrt{\pi}}$

$$C = 2\pi r = 2\pi\frac{\sqrt{A}}{\sqrt{\pi}} = \frac{2\pi\sqrt{A}}{\sqrt{\pi}} \cdot \frac{\sqrt{\pi}}{\sqrt{\pi}} = \frac{2\not{\pi}\sqrt{A\pi}}{\not{\pi}} = 2\sqrt{\pi}\sqrt{A} = 2\sqrt{\pi A}$$

11. **(A)** Since an angle inscribed in a semicircle is a right angle, ΔABC is a right triangle. By the Pythagorean theorem:

$$(AC)^2 = 4^2 + 8^2 = 16 + 64 = 80 \Rightarrow AC = \sqrt{80}$$

Since $\overline{AC}$ is a diameter, the radius is $\frac{\sqrt{80}}{2}$ and the area of the semicircle is $\frac{1}{2}\pi\left(\frac{\sqrt{80}}{2}\right)^2 = \frac{1}{2}\pi\frac{80}{4} = 10\pi$. The area of right triangle *ABC* is $\frac{1}{2}(8)(4) = 16$.

So the area of the shaded region is $10\pi - 16 \approx 31.4 - 16 = 15.4$.

12. **(A)** Since $\overline{OP} \perp \overline{RS}$, ΔOPS is a right triangle. By the Pythagorean theorem:

$$OP^2 + PS^2 = OS^2 \Rightarrow OP^2 + 144 = 400 \text{ and so } OP^2 = 256 \Rightarrow OP = 16$$

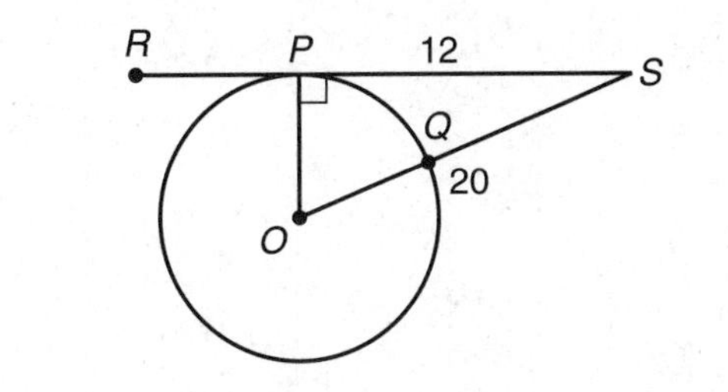

Since all radii are equal, $OQ = 16 \Rightarrow QS = 20 - 16 = 4$.

13. **(B)** The measure of the smaller arc from A to B is $360° - 250° = 110°$. So $m\angle P = \frac{1}{2}(250° - 110°) = \frac{1}{2}(140°) = 70°$. Since the two tangents are congruent, ΔPAB is isosceles and the measure of each base angle is $\frac{1}{2}(180° - 70°) = \frac{1}{2}(110°) = 55°$.

14. **(C)** Since $C = x\pi$, we have $x\pi = 2\pi r \Rightarrow x = 2r$. Since $A = y\pi$, we have $y\pi = \pi r^2 \Rightarrow y = r^2$. Finally, since it is given that $y = 2x$,

$$y = 2x \Rightarrow r^2 = 2(2r) = 4r \Rightarrow r = 4.$$

15. **(C)** Three diameters divide the hexagon into 6 equilateral triangles.

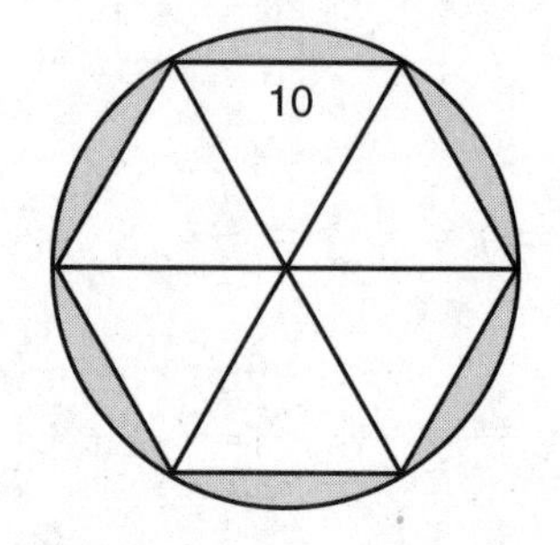

By using the formula for the area of an equilateral triangle, we find that the area of the hexagon is $6\left(\frac{10^2\sqrt{3}}{4}\right) = 6\left(\frac{100\sqrt{3}}{4}\right) = 150\sqrt{3}$. Since the radius of the circle is 10, its area is $\pi(10^2) = 100\pi$. So the area of the shaded region is $100\pi - 150\sqrt{3} \approx 314 - 260 = 54$.

SOLID AND COORDINATE GEOMETRY

Solid Geometry

CHAPTER 

- Rectangular Solids
- Cylinders
- Prisms
- Cones
- Pyramids
- Spheres
- Exercises
- Answers Explained

On a typical Math 1 test, there are about three questions on solid geometry. Most of them can be answered if you know the formulas for the volumes and surface areas of rectangular solids and cylinders. Although these formulas are given to you on the PSAT and SAT, they are *not* provided on the Math 1 test, so you must learn them.

Occasionally, a Math 1 test has one question about spheres, cones, or pyramids. The formulas you would need to answer such a question *are* given to you in a box labeled "Reference Information" at the beginning of the test. However, many Math 1 tests have no questions at all on these topics, so if your study time is limited, you may want to skip the discussion of these solids.

RECTANGULAR SOLIDS

A ***rectangular solid*** or ***box*** is a solid formed by six rectangles, called ***faces***. The sides of the rectangles are called ***edges***. As in the diagram below, the edges are called the ***length***, ***width***, and ***height***. A ***cube*** is a rectangular solid in which the length, width, and height are equal, so all the edges are the same length.

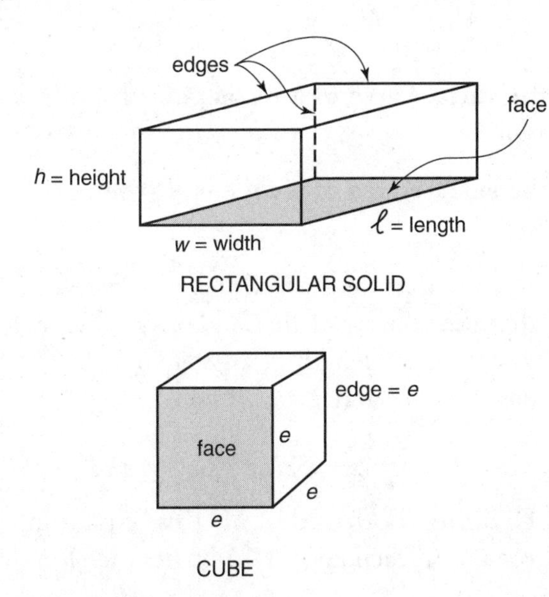

RECTANGULAR SOLID

CUBE

The ***volume*** of a solid is measured in ***cubic units***. One cubic unit is the amount of space taken up by a cube all of whose edges are one unit long. In the figure on page 171, if each edge of the cube is 1 inch, the area of each face is 1 square inch, and the volume of the cube is 1 cubic inch.

Key Fact K1

- **The formula for the volume of a rectangular solid is $V = \ell wh$, where ℓ, w, and h are the length, width, and height, respectively.**
- **Since all the edges of a cube are equal, if e is an edge, the formula for the volume of a cube is $V = e^3$.**

The ***surface area*** of a rectangular solid is the sum of the areas of the six rectangular faces. The areas of the top and bottom faces are equal, the areas of the front and back faces are equal, and the areas of the left and right faces are equal. Therefore, to get the total surface area, we can calculate the area of one face from each pair, add them up, and then double the sum. In a cube, each of the six faces has the same area, so the surface area is six times the area of any face.

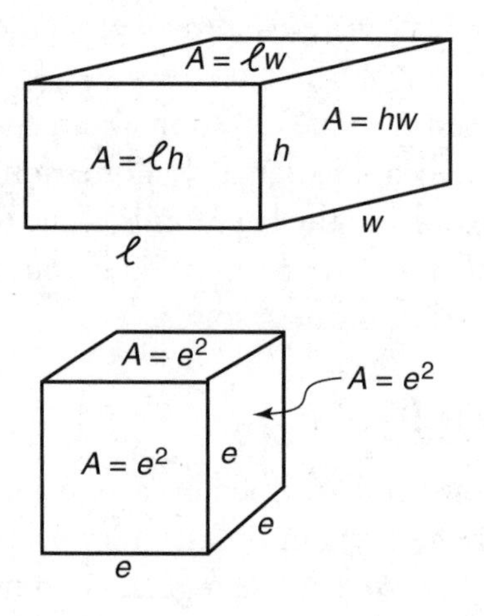

Key Fact K2

- **The formula for the surface area of a rectangular solid is $A = 2(\ell w + \ell h + wh)$.**
- **The formula for the surface area of a cube is $A = 6e^2$.**

EXAMPLE 1: Assume that the surface area of a cube is x square inches, that the volume of that cube is y cubic inches, and that $y = 2x$. If the length of an edge of the cube is e inches, we have $x = 6e^2$ and $y = e^3$. Then

$$y = 2x \Rightarrow e^3 = 2(6e^2) = 12e^2$$

Dividing both sides of this equation by e^2, we get $e = 12$. So each edge is 12 inches or 1 foot long.

A ***diagonal*** of a rectangular solid is a line segment joining a vertex on one face of the box to the vertex on the opposite face that is furthest away. A rectangular solid has four diagonals, all the same length. In the following box, diagonals $\overline{AG}$ and $\overline{BH}$ are drawn in. The other two diagonals are $\overline{CE}$ and $\overline{DF}$.

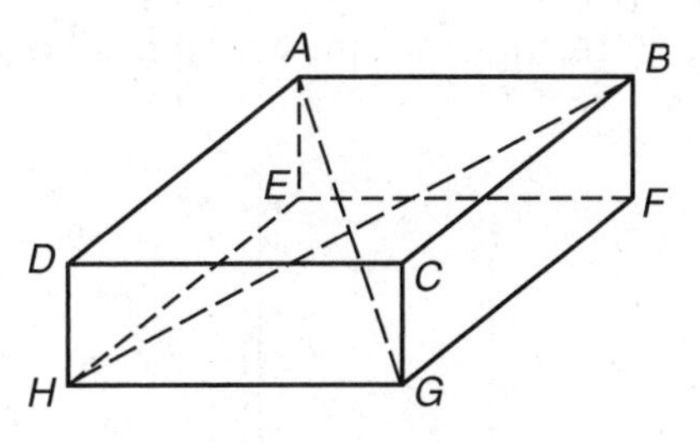

Key Fact K3

If the dimensions of a rectangular solid are ℓ, w, and h and if d is the length of a diagonal, then $d = \sqrt{\ell^2 + w^2 + h^2}$.

The formula given in KEY FACT K3 is obtained by using the Pythagorean theorem twice. In the figure below, $\overline{BD}$ is the diagonal of rectangular face $BCDE$. By the Pythagorean theorem, $(BD)^2 = \ell^2 + w^2$. Now $\triangle ADB$ is a right triangle, and by the Pythagorean theorem:

$$d^2 = (AB)^2 = (BD)^2 + (AD)^2 = (\ell^2 + w^2) + h^2 \Rightarrow$$
$$d = \sqrt{\ell^2 + w^2 + h^2}$$

The distance between *any* two points can *always* be calculated using the Pythagorean theorem.

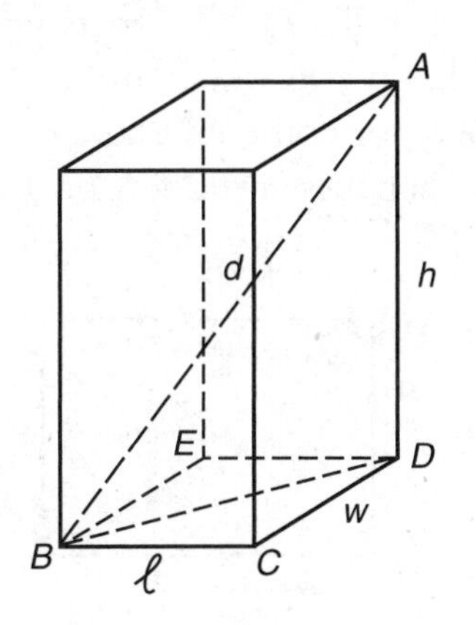

EXAMPLE 2: To determine the length of a diagonal of a cube whose sides are e, use the formula from KEY FACT K3, noting that $\ell = w = h = e$. Then, $d = \sqrt{e^2 + e^2 + e^2} = \sqrt{3e^2} = e\sqrt{3}$.

CYLINDERS

A ***cylinder*** is similar to a rectangular solid except that the base is a circle instead of a rectangle. To find the volume of a rectangular solid, we multiply the area of its rectangular base, ℓw, by its height, h. For a cylinder, we do exactly the same thing. The volume of a cylinder is the area of its circular base, πr^2, times its height, h. The surface area of a cylinder depends on whether you are envisioning a tube, such as a straw without a top and bottom, or a can, which has both a top and bottom.

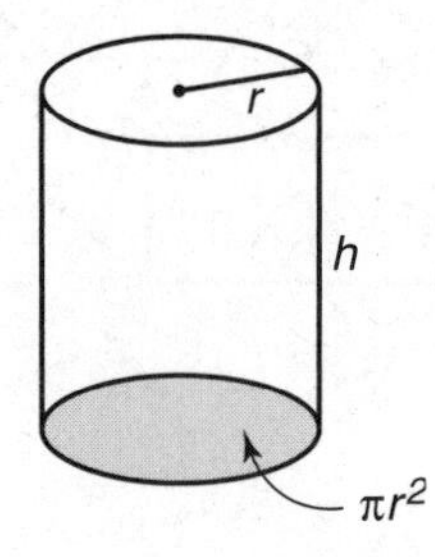

Key Fact K4

- **The formula for the volume, V, of a cylinder whose circular base has radius r and whose height is h is $V = \pi r^2 h$.**
- **The formula for the surface area, A, of the side of a cylinder is the product of the circumference of its circular base and its height: $A = 2\pi rh$.**
- **The areas of the top and bottom of a cylinder are each πr^2, so the total surface area of a cylindrical can is $2\pi rh + 2\pi r^2$.**

EXAMPLE 3: You can roll an 8×12 rectangular piece of paper into a cylinder in two ways. You could tape the 8-inch sides together, or you could tape the 12-inch sides together. Note that these cylinders do *not* have the same volume.

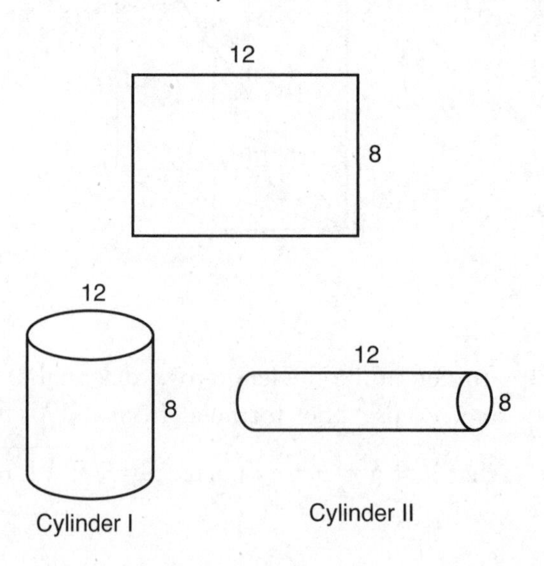

In cylinder I, $C = 12 \Rightarrow 2\pi r = 12 \Rightarrow r = \frac{6}{\pi}$, and so

$$V = \pi\left(\frac{6}{\pi}\right)^2(8) = \pi\left(\frac{36}{\pi^2}\right)8 = \frac{288}{\pi}.$$

In cylinder II, $C = 8 \Rightarrow 2\pi r = 8 \Rightarrow r = \frac{4}{\pi}$, and so

$$V = \pi\left(\frac{4}{\pi}\right)^2(12) = \pi\left(\frac{16}{\pi^2}\right)12 = \frac{192}{\pi}.$$

PRISMS

Rectangular solids and cylinders are special cases of geometric solids called prisms. A ***prism*** is a three-dimensional figure that has two congruent parallel ***bases.*** The (perpendicular) distance between the two bases is called the ***height.*** Four prisms are depicted in the figure below.

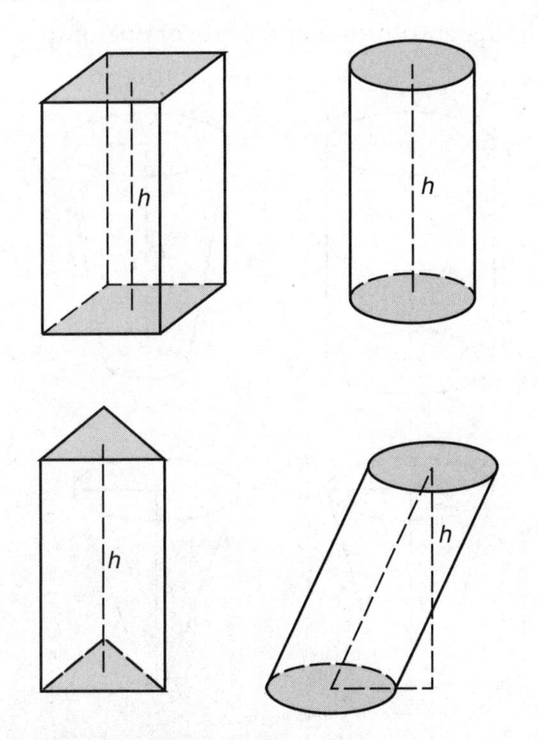

On the Math 1 test, all of the prisms are ***right prisms***, which means that any line segment joining corresponding points on the bases is perpendicular to the bases. In the figure above, the first three prisms are right prisms; the fourth one is not. The volume formulas given in KEY FACTS K1 and K4 are special cases of the following formula.

Key Fact K5

The formula for the volume of any right prism is $V = Bh$, where B is the area of one of the bases and h is the height.

EXAMPLE 4: What is the volume of the triangular prism below?

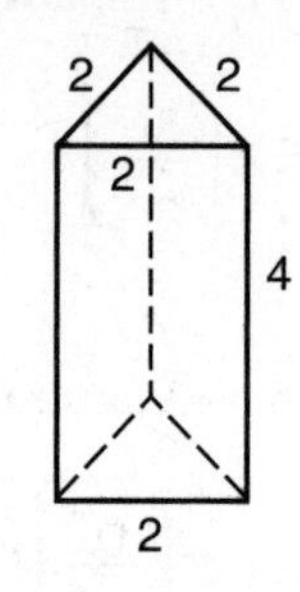

First calculate the area of a base and multiply it by the height. By KEY FACT H11, the area of an equilateral triangle whose sides are 2 is $\frac{2^2\sqrt{3}}{4} = \sqrt{3}$. So the volume is $(\sqrt{3})(4) = 4\sqrt{3}$.

CONES

Imagine taking a cylinder and shrinking the size of one of the circular bases.

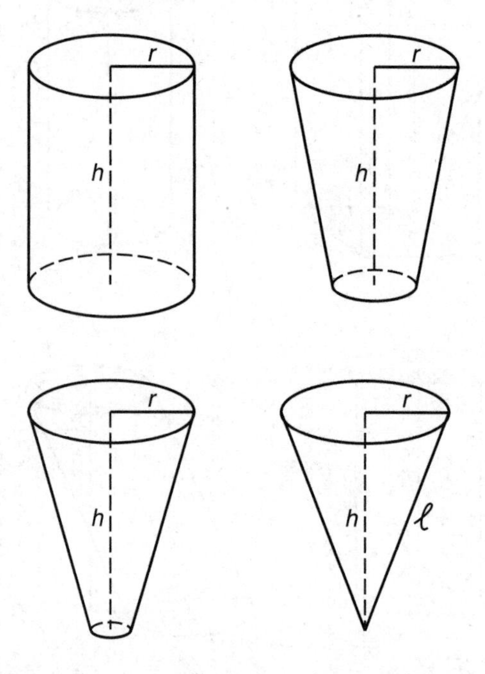

When it shrinks to a point, we call the resulting solid a cone. If you picture an ice cream cone (without the ice cream) or a dunce cap, you have the right idea. Notice that the radius of the circular base and the height of the cone are the same as in the original cylinder. In the figure above, ℓ is called the slant height of the cone.

In the Math 1 "Reference Information," cones are referred to as "right circular cones." This is done to emphasize that the base is a circle and that the height (the line segment joining the vertex and the center of the circular base) is perpendicular to the base. KEY FACT K6 lists the two formulas concerning cones that are given to you in the "Reference Information" box on the first page of the test and gives an alternative for one of them. Note that in the second formula, the circumference of the circular base is denoted by a lowercase "*c*," whereas we usually use an uppercase "*C*" to represent circumference. Also, ℓ, the ***slant height***, is the distance from any point on the circumference to the vertex of the cone.

Key Fact K6

- **The formula for the volume of a right circular cone is $V = \frac{1}{3}\pi r^2 h$.**
- **The formula for the lateral surface area of a right circular cone is $A = \frac{1}{2}c\ell$, where c is the circumference and ℓ is the slant height.**
- **Since $c = 2\pi r$, $\frac{1}{2}c = \pi r$, so an alternative formula for the lateral surface area is $A = \pi r\ell$.**
- **The total surface area of a cone is the sum of its lateral area and the area of its circular base: $SA = \pi r\ell + \pi r^2$.**

Remember

The first two formulas in KEY FACT K6 are given to you on the Math 1 test.

EXAMPLE 5: Assume that the volumes of a right circular cone and a right circular cylinder are equal and that the radius of the cone is twice the radius of the cylinder. How do their heights compare?

Let r be the radius of the cylinder, $2r$ the radius of the cone, and h and H the heights of the cone and the cylinder, respectively. Then

$$\frac{1}{3}\pi(2r)^2 h = \pi r^2 H \Rightarrow \frac{1}{3}\cancel{\pi}4\cancel{r^2}h = \cancel{\pi}\cancel{r^2}H \Rightarrow \frac{4}{3}h = H$$

PYRAMIDS

A ***pyramid*** is very similar to a cone. The difference is that the base of a pyramid is a polygon, not a circle. If there is a question concerning pyramids on the Math 1 test you take, the polygonal base will almost surely be a square or a triangle.

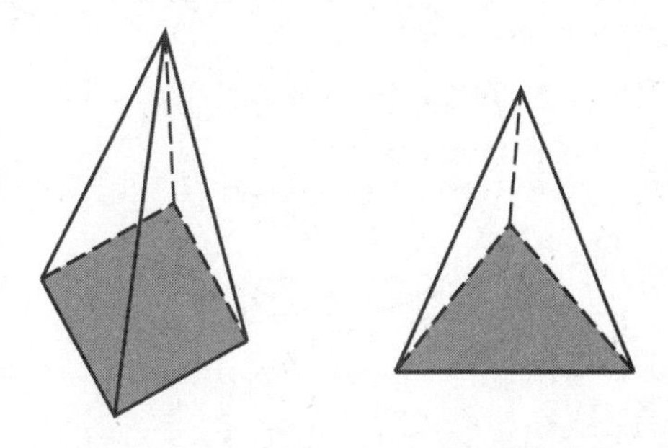

The formula for the volume of a pyramid, which is one of the five formulas in the "Reference Information" box on the Math 1 test, is given in KEY FACT K7.

Key Fact K7

The formula for the volume of a pyramid is $V = \frac{1}{3}Bh$, where B is the area of the base and h is the height.

Remember

The formula in KEY FACT K7 is given to you on the Math 1 test.

Note that since the area of the circular base of a cone is πr^2, this formula applies to cones, as well.

EXAMPLE 6: What is the volume of a pyramid whose base is a square of side 3 feet and whose height is 6 feet?

Sketch the pyramid and use the formula given in KEY FACT K7.

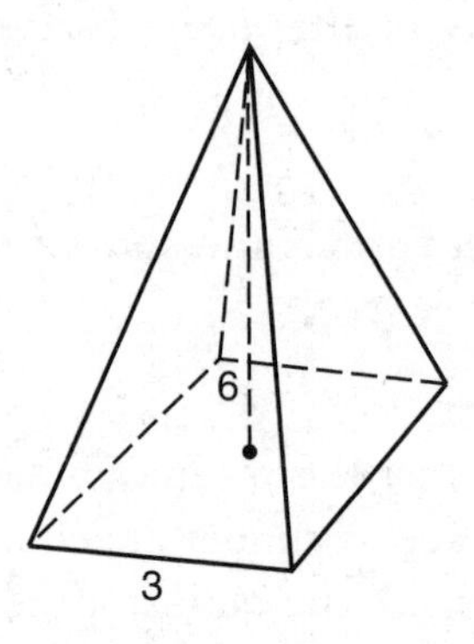

The area of the square base is 9 square feet, and so $V = \frac{1}{3}(9)(6) = 18$ cubic feet.

SPHERES

A sphere is the set of all points in space that are a fixed distance, r, from a given point, O. O is called the center of the sphere, and r is the radius.

The two formulas about spheres that you might need to use on the Math 1 test are those for volume and surface area. These two formulas, which are given to you in the "Reference Information" box, constitute the next KEY FACT.

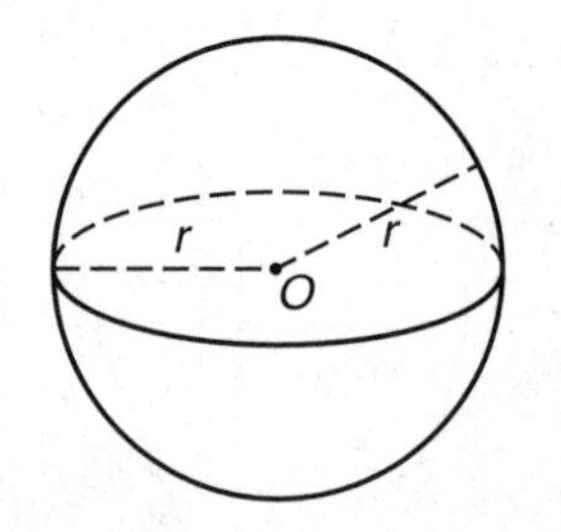

Key Fact K8

- **The formula for the volume of a sphere of radius r is $V = \frac{4}{3}\pi r^3$.**
- **The formula for the surface area of a sphere of radius r is $A = 4\pi r^2$.**

Remember

The formulas in KEY FACT K8 are given to you on the Math 1 test.

EXAMPLE 7: What is the radius of a sphere whose surface area is equal to the surface area of a cube whose edges are 1?

By KEY FACT K2, the surface area of the cube is $6e^2 = 6(1)^2 = 6$. Then by KEY FACT K8,

$$4\pi r^2 = 6 \Rightarrow r^2 = \frac{6}{4\pi} \Rightarrow r = \sqrt{\frac{6}{4\pi}} \approx 0.69$$

Exercises

1. What is the volume of a cube whose surface area is 54?

   (A) 27
   (B) 36
   (C) 54
   (D) 64
   (E) 216

2. What is the surface area of a cube whose volume is 216?

   (A) 6
   (B) 36
   (C) 72
   (D) 144
   (E) 216

3. What is the volume of a sphere whose surface area is 36π?

   (A) $\frac{4}{3}\pi$
   (B) 9π
   (C) 36π
   (D) 72π
   (E) 108π

4. What is the surface area of a sphere whose volume is 288π?

   (A) 6π
   (B) 36π
   (C) 72π
   (D) 144π
   (E) 288π

5. What is the lateral surface area of a right circular cone whose radius is 3 and whose height is 4?

   (A) 12π
   (B) 15π
   (C) 24π
   (D) 30π
   (E) 36π

6. The height, h, of a right circular cylinder is equal to the edge of a cube. If the cylinder and cube have the same volume, what is the radius of the cylinder?

   (A) $\frac{h}{\sqrt{\pi}}$
   (B) $h\sqrt{\pi}$
   (C) $\frac{\sqrt{\pi}}{h}$
   (D) $\frac{h^2}{\pi}$
   (E) πh^2

7. What is the volume of a pyramid whose base is a square of area 36 and whose four faces are equilateral triangles?

   (A) $24\sqrt{2}$
   (B) 36
   (C) $36\sqrt{2}$
   (D) 108
   (E) $108\sqrt{2}$

8. An isosceles right triangle whose legs are 6 is rotated about one of its legs to generate a right circular cone. What is the volume of that cone?

   (A) $48\sqrt{2}\pi$
   (B) 72π
   (C) $72\sqrt{2}\pi$
   (D) 144π
   (E) 216π

9. If the height of a right circular cylinder is 4 times its circumference, what is the volume of the cylinder in terms of its circumference, C?

(A) $\frac{C^3}{\pi}$

(B) $\frac{2C^3}{\pi}$

(C) $\frac{2C^2}{\pi^2}$

(D) $\frac{\pi C^2}{4}$

(E) $4\pi C^3$

10. Three identical balls fit snugly into a cylindrical can: the radius of the spheres equals the radius of the can, and the balls just touch the bottom and top of the can. What fraction of the volume of the can is taken up by the balls?

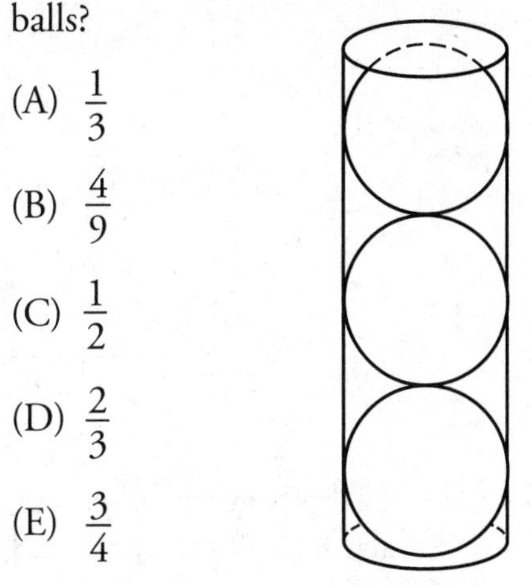

(A) $\frac{1}{3}$

(B) $\frac{4}{9}$

(C) $\frac{1}{2}$

(D) $\frac{2}{3}$

(E) $\frac{3}{4}$

ANSWERS EXPLAINED

Answer Key

1. **(A)**	3. **(C)**	5. **(B)**	7. **(C)**	9. **(A)**
2. **(E)**	4. **(D)**	6. **(A)**	8. **(B)**	10. **(D)**

Solutions

Each of the problems in this set of exercises is typical of a question you could see on a Math 1 test. When you take the model tests in this book and, in particular, when you take the actual Math 1 test, if you get stuck on questions such as these, you do not have to leave them out—you can almost always answer them by using one or more of the strategies discussed in the "Tactics" chapter. The solutions given here do *not* depend on those strategies; they are the correct mathematical ones.

1. **(A)** Since the surface area is 54, each of the 6 faces is a square whose area is $54 \div 6 = 9$. So the edges are all 3, and by KEY FACT K1, the volume is $3^3 = 27$.

2. **(E)** Since the volume of the cube is 216, we have $e^3 = 216 \Rightarrow e = 6$. By KEY FACT K2, the surface area is $6e^2 = 6 \times 6^2 = 6 \times 36 = 216$.

3. **(C)** Since the surface area of the sphere is 36π, by KEY FACT K8:

$$4\pi r^2 = 36\pi \Rightarrow r^2 = 9 \Rightarrow r = 3.$$

$$\text{So } V = \frac{4}{3}\pi r^3 = \frac{4}{3}\pi(27) = 36\pi$$

4. **(D)** Since the volume of the sphere is 288π, by KEY FACT K8, we have:

$$\frac{4}{3}\pi r^3 = 288\pi \Rightarrow r^3 = \frac{3}{4}(288) = 216 \Rightarrow r = \sqrt[3]{216} = 6.$$

$$\text{So } A = 4\pi r^2 = 4\pi(36) = 144\pi$$

5. **(B)** First sketch the cone.

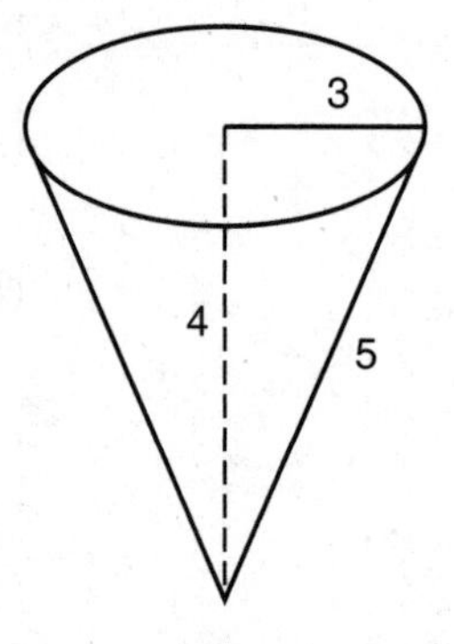

By the Pythagorean theorem, the slant height is 5. The circumference of the circular base is $2\pi(3) = 6\pi$. So by KEY FACT K6, the lateral surface area is: $\frac{1}{2}c\ell = \frac{1}{2}(6\pi)5 = 15\pi$.

6. **(A)** Since the volumes are equal, $\pi r^2 h = e^3 = h^3$. So:

$$\pi r^2 = h^2 \Rightarrow r^2 = \frac{h^2}{\pi} \Rightarrow r = \frac{h}{\sqrt{\pi}}$$

7. **(C)** Sketch the pyramid.

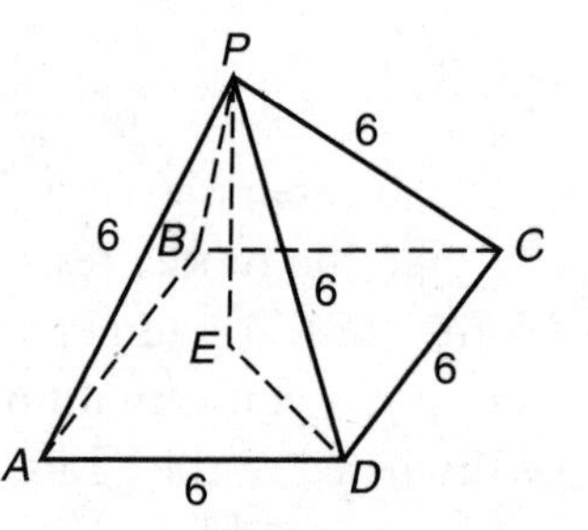

Since the area of the square is 36, each side is 6; and since each triangular face is equilateral, each edge of the pyramid is 6. To find the height of the pyramid, use the Pythagorean theorem on $\triangle PED$. Since $\overline{ED}$ is half of diagonal $\overline{BD}$,

$ED = \frac{1}{2}(6\sqrt{2}) = 3\sqrt{2}$. So:

$$(PD)^2 = (PE)^2 + (ED)^2 \Rightarrow (6)^2 = h^2 + (3\sqrt{2})^2 \Rightarrow$$
$$36 = h^2 + 18 \Rightarrow h^2 = 18 \Rightarrow h = \sqrt{18} = 3\sqrt{2}$$

Finally, by KEY FACT K7, the volume of the pyramid is:

$$\frac{1}{3}Bh = \frac{1}{\cancel{3}}(36)(\cancel{3}\sqrt{2}) = 36\sqrt{2}$$

8. **(B)** Sketch the triangle and cone.

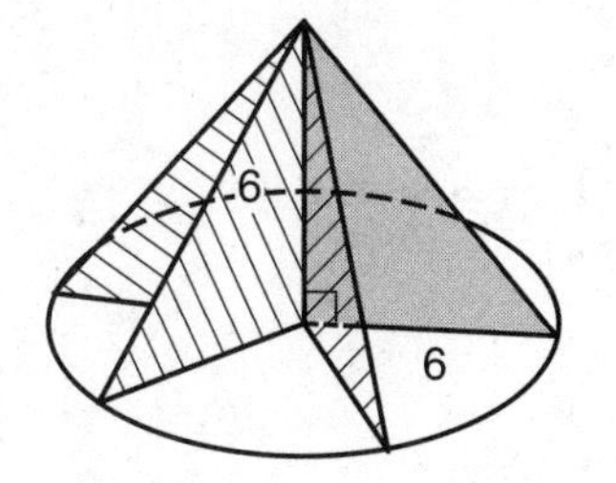

Since the radius and height of the cone are 6, by KEY FACT K6:

$$V = \frac{1}{3}\pi r^2 h = \frac{1}{3}\pi (6)^2(6) = 72\pi$$

9. **(A)** Since $V = \pi r^2 h$, we need to express r and h in terms of C. It is given that $h = 4C$, and since $C = 2\pi r$, $r = \frac{C}{2\pi}$. So:

$$V = \pi\left(\frac{C}{2\pi}\right)^2(4C) = \pi\left(\frac{C^2}{\not{4}\pi^2}\right)(\not{4}C) = \frac{C^3}{\pi}$$

10. **(D)** The volume of each ball is $\frac{4}{3}\pi r^3$, and the total volume of the 3 balls is $\not{3}\left(\frac{4}{\not{3}}\pi r^3\right) = 4\pi r^3$. The height of the can is equal to 3 times the diameter, and hence 6 times the radius, of a ball. So the volume of the can is $\pi r^2(6r) = 6\pi r^3$. Therefore, the balls take up $\frac{4\pi r^3}{6\pi r^3} = \frac{2}{3}$ of the can.

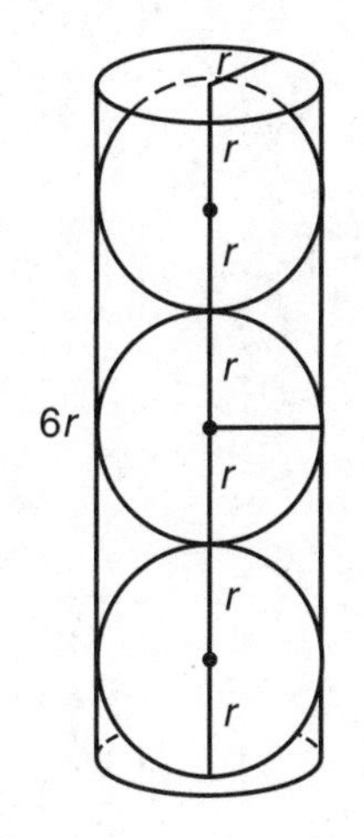

Coordinate Geometry

Of the 50 questions on the Math 1 test, approximately six are on the coordinate geometry topics specifically discussed in this chapter. In addition, three or four more questions that the College Board would classify under other topics, such as functions and their graphs or even probability or trigonometry, could be considered as coordinate geometry questions. So this is an important topic that you should review thoroughly.

The coordinate plane is formed by two perpendicular number lines called the ***x-axis*** and ***y-axis***, which intersect at the ***origin***. The axes divide the plane into four ***quadrants***, labeled I, II, III, and IV as shown in the figure below. Each point in the plane is assigned two numbers, an ***x-coordinate*** and a ***y-coordinate***, which are written as an ordered pair, (x, y). The point (2, 3) has an x-coordinate of 2 and a y-coordinate of 3 and is located at the intersection of the vertical line that crosses the x-axis at 2 and the horizontal line that crosses the y-axis at 3.

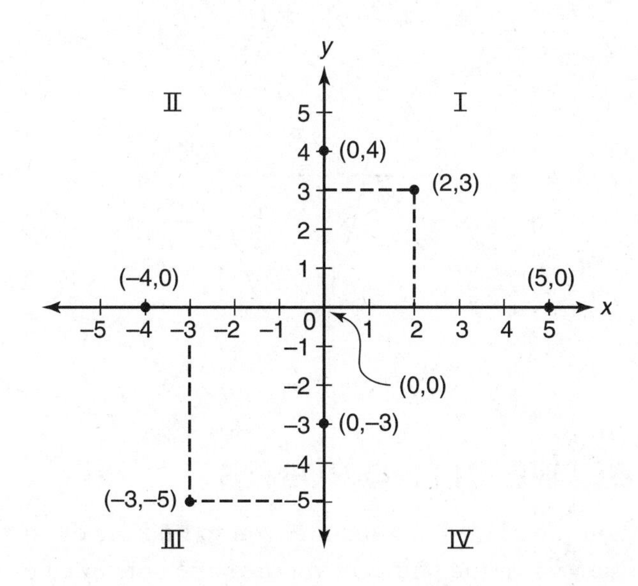

Key Fact L1

- **The y-coordinate of every point on the x-axis is 0.**
- **The x-coordinate of every point on the y-axis is 0.**
- **To find the x-intercepts of a graph, replace y by 0 in the equation of the graph.**
- **To find the y-intercepts of a graph, replace x by 0 in the equation of the graph.**

EXAMPLE 1: Where does the graph of $y = x^2 - 2x - 3$ cross the x-axis and where does it cross the y-axis?

By KEY FACT L1, the x-intercepts are the points whose y-coordinates are 0. So replace y with 0:

$$0 = x^2 - 2x - 3 \Rightarrow (x - 3)(x + 1) = 0 \Rightarrow x = -1 \text{ or } x = 3$$

The x-intercepts of this graph are the points (–1, 0) and (3, 0).

By KEY FACT L1, the y-intercept is the point whose x-coordinate is 0. So replace x with 0:

$$y = 0^2 - 2(0) - 3 = -3 \Rightarrow y = -3$$

The y-intercept is the point (0, –3).

The graph of $y = x^2 - 2x - 3$ is the parabola in the figure below, on which the x- and y-intercepts are indicated.

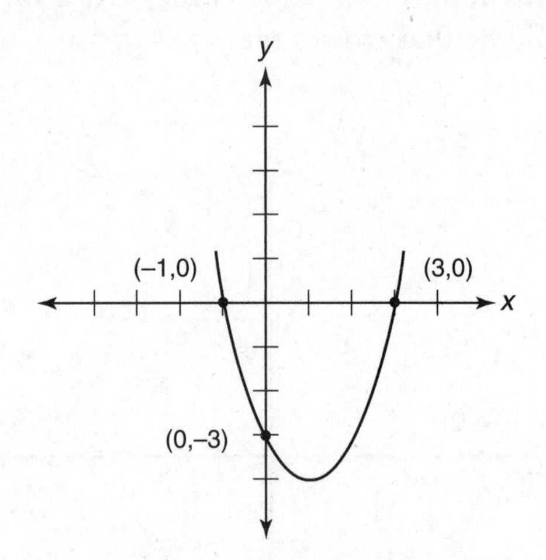

DISTANCE BETWEEN TWO POINTS

Often a question on the Math 1 test requires you to find the distance between two points. This is easiest when the points lie on the same horizontal or vertical line.

Key Fact L2

- **All the points on a horizontal line have the same *y*-coordinate. To find the distance between them, subtract their *x*-coordinates.**
- **All the points on a vertical line have the same *x*-coordinate. To find the distance between them, subtract their *y*-coordinates.**

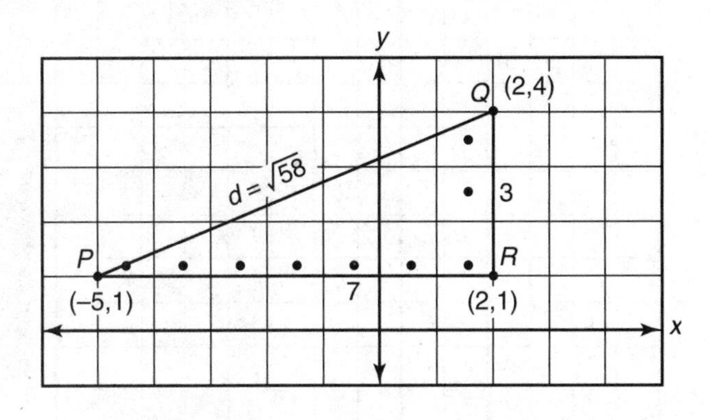

In the diagram above, the distance from P to R is $2 - (-5) = 7$ and the distance from R to Q is $4 - 1 = 3$. If you want to find the distance between two points, such as P and Q, that do not lie on the same horizontal or vertical line, sketch the line segment joining those points and then create a right triangle with that segment as the hypotenuse by drawing a horizontal line through one of the points and a vertical line through the other. Use KEY FACT L2 to find the length of the legs. To calculate the hypotenuse, use the Pythagorean theorem. For example, in the figure above, if d represents the distance from P to Q,

$$d^2 = 7^2 + 3^2 = 49 + 9 = 58 \Rightarrow d = \sqrt{58}$$

An alternative method is to use the distance formula given in KEY FACT L3.

Key Fact L3

The distance, d, between two points, $P_1(x_1, y_1)$ and $P_2(x_2, y_2)$, can be calculated using the distance formula: $d = \sqrt{(x_2 - x_1)^2 + (y_2 - y_1)^2}$.

From the diagram below, you should see that the distance formula is really just the Pythagorean theorem.

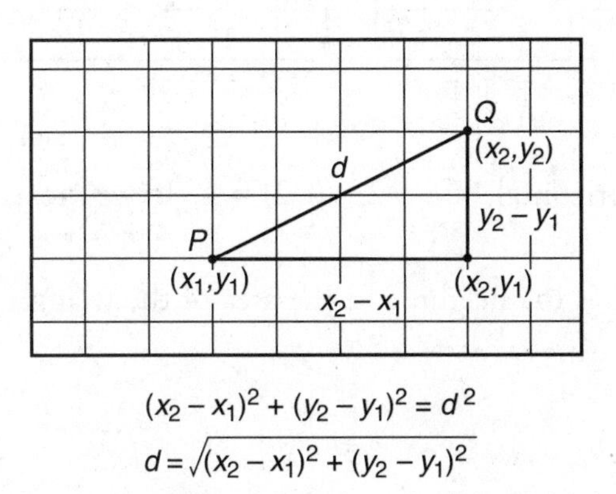

$$(x_2 - x_1)^2 + (y_2 - y_1)^2 = d^2$$
$$d = \sqrt{(x_2 - x_1)^2 + (y_2 - y_1)^2}$$

EXAMPLE 2: If $A(-2, 5)$, $B(-2, -3)$, and $C(2, 2)$ are the vertices of $\triangle ABC$, what are the area and perimeter of the triangle?

Start by making a sketch, and then calculate the lengths of the three line segments.

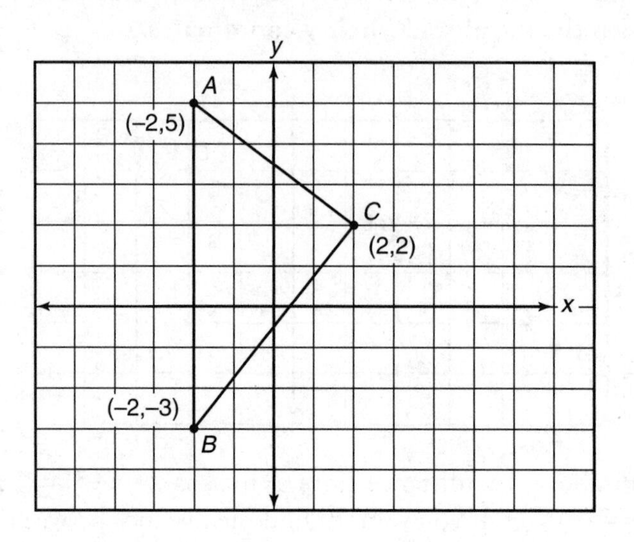

Since A and B lie on the same vertical line, we can find AB by subtracting their y-coordinates: $AB = 5 - (-3) = 8$.

Now draw in altitude $\overline{CE}$.

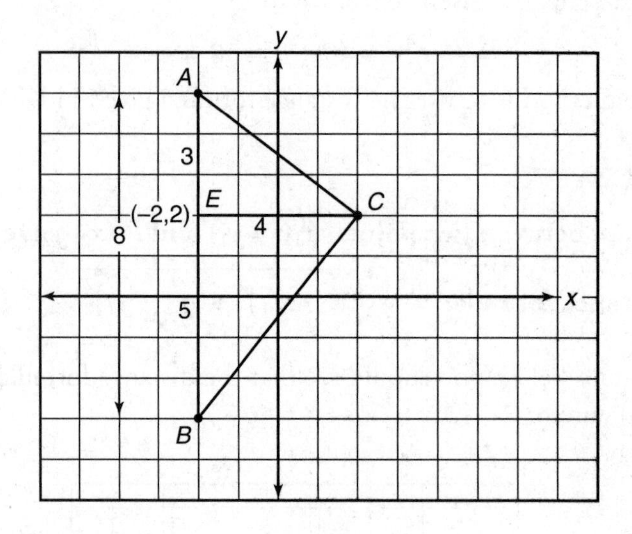

Since $\overline{CE}$ is horizontal, $CE = 2 - (-2) = 4$. If we treat $\overline{AB}$ as the base of $\triangle ABC$, then $\overline{CE}$ is the height, and the area of the triangle is $\frac{1}{2}(8)(4) = 16$.

On page 188, we found that $AB = 8$. To find the perimeter of $\triangle ABC$, we need the lengths of the other two sides. Each of these can be calculated using the distance formula or the Pythagorean theorem:

$$(AC)^2 = 3^2 + 4^2 = 9 + 16 = 25 \Rightarrow AC = 5$$
$$(BC)^2 = 5^2 + 4^2 = 25 + 16 = 41 \Rightarrow BC = \sqrt{41}$$

So the perimeter of $\triangle ABC$ is $8 + 5 + \sqrt{41} = 13 + \sqrt{41}$.

THE MIDPOINT OF A SEGMENT

Recall that the midpoint, M, of line segment $\overline{PQ}$ is the point on $\overline{PQ}$ such that $PM = MQ$. In coordinate geometry, the x-coordinate of the midpoint is the average of the x-coordinates of the two endpoints, and the y-coordinate of the midpoint is the average of the y-coordinates of the two endpoints.

Key Fact L4

If $P_1(x_1, y_1)$ and $P_2(x_2, y_2)$ are any two points, then the midpoint, M, of segment $\overline{P_1P_2}$ is the point whose coordinates are $\left(\frac{x_1 + x_2}{2}, \frac{y_1 + y_2}{2}\right)$.

EXAMPLE 3: If $C(3, -4)$ and $D(7,2)$ are the endpoints of diameter CD of circle O, what are the coordinates of O?

First, sketch the circle (see diagram on top of page 190). Since the center of a circle is the midpoint of any diameter, we can use the midpoint formula to find the coordinates of O.

$$\left(\frac{3+7}{2}, \frac{-4+2}{2}\right) = \left(\frac{10}{2}, \frac{-2}{2}\right) = (5, -1)$$

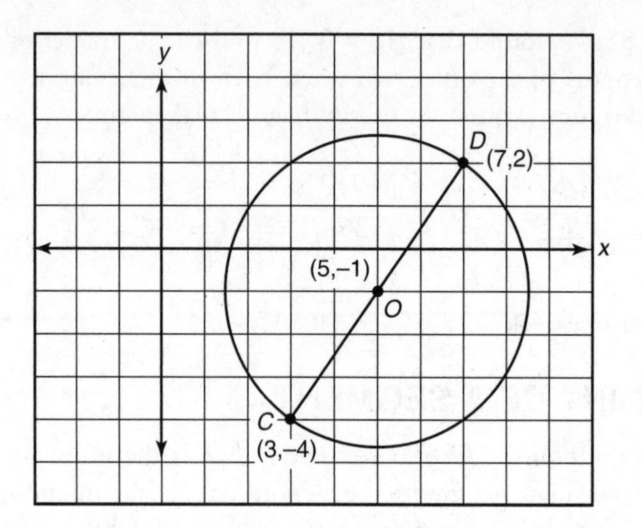

EXAMPLE 4: To find the area of circle O in Example 3, we have to use the formula $A = \pi r^2$, which means we have to determine r. To do this, we can either find the length of diameter $\overline{CD}$ and divide it by 2 or find the length of radius $\overline{OD}$.

$$CD = \sqrt{(7-3)^2 + (2-(-4))^2} = \sqrt{4^2 + 6^2} = \sqrt{16+36}$$
$$= \sqrt{52} \Rightarrow r = \frac{\sqrt{52}}{2}, \text{ or}$$

$$OD = \sqrt{(7-5)^2 + (2-(-1))^2} = \sqrt{2^2 + 3^2} = \sqrt{4+9} \Rightarrow \sqrt{13}$$

Then, depending on which expression you found,

$$A = \pi\left(\frac{\sqrt{52}}{2}\right)^2 = \pi\left(\frac{52}{4}\right) = 13\pi \quad \text{or } A = \pi\left(\sqrt{13}\right)^2 = 13\pi.$$

SLOPE

The ***slope*** of a line is a number that represents how steep it is: the larger the absolute value of the slope, the steeper the line. This intuitive definition is made more precise in KEY FACT L5.

Key Fact L5

- **Vertical lines *do not have slopes.***
- **To find the slope of any other line, proceed as follows:**
  (1) Choose any two points $P_1(x_1, y_1)$ and $P_2(x_2, y_2)$ on the line.
  (2) Determine the differences of their y-coordinates, $y_2 - y_1$, and their x-coordinates, $x_2 - x_1$.
  (3) Divide: slope $= \dfrac{y_2 - y_1}{x_2 - x_1}$.

Since $y_2 - y_1$ is the difference between the y-coordinates of the two points and $x_2 - x_1$ is the difference between the x-coordinates of the two points, we often say that the slope is "the change in y" over "the change in x":

$$\text{slope} = \frac{\text{change in } y}{\text{change in } x}.$$

We will illustrate the next KEY FACT by using the slope formula to calculate the slopes of $\overline{AB}$, $\overline{BC}$, and $\overline{AC}$ in the next figure.

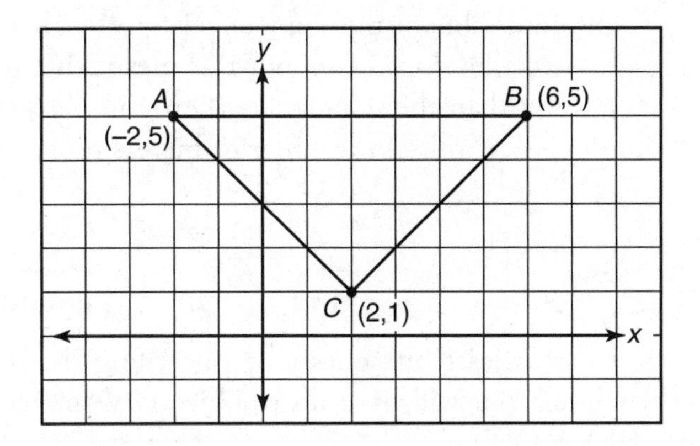

Key Fact L6

- **The slope of any horizontal line is 0:**

$$\textbf{slope of } \overline{AB} = \frac{5-5}{6-(-2)} = \frac{0}{8} = 0$$

- **The slope of any line that goes up as you move from left to right is positive:**

$$\textbf{slope of } \overline{CB} = \frac{5-1}{6-2} = \frac{4}{4} = 1$$

- **The slope of any line that goes down as you move from left to right is negative:**

$$\textbf{slope of } \overline{AC} = \frac{1-5}{2-(-2)} = \frac{-4}{4} = -1$$

On the Math 1 test, you may be asked to compare the slopes of lines without being given any coordinates.

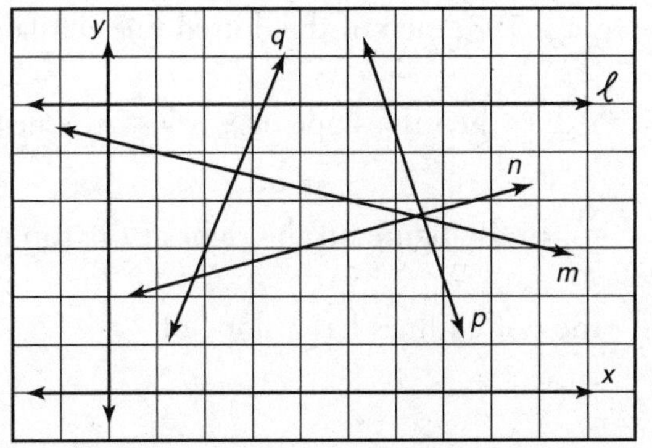

EXAMPLE 5: To list the lines shown in the previous figure in order of increasing slope, first use KEY FACT L6. Since line ℓ is horizontal, its slope is 0. Since as you move from left to right lines m and p go down and lines n and q go up, the slopes of m and p are negative, whereas the slopes of n and q are positive. Since q is steeper than n, the slope of q is greater than the slope of n. Be careful as you compare the slopes of m and p. Line p is steeper than line m, so the absolute value of the slope of p is greater than the absolute value of the slope of m, but this means that the slope of p is less than the slope of m. (For example, $|-5| > |-1|$, but $-5 < -1$.) So listed in order of increasing slope we have:

$$\underbrace{p < m}_{\text{negative}} < \underbrace{\ell}_{0} < \underbrace{n < q}_{\text{positive}}$$

Often, a Math 1 test has at least one question concerning the slopes of parallel and/or perpendicular lines. You will have no trouble answering such questions if you know the next KEY FACT.

Key Fact L7

- **If two nonvertical lines are parallel, their slopes are equal.**
- **If two nonvertical lines are perpendicular, the product of their slopes is −1.**

If the product of two numbers, a and b, is -1, then $ab = -1 \Rightarrow a = -\frac{1}{b}$. So another way to express the second part of KEY FACT L7 is to say that **if two nonvertical lines are perpendicular, then the slope of one is the negative reciprocal of the slope of the other**.

EXAMPLE 6: Let ℓ be the line that passes through points (1, 3) and (3, 6). Then the slope of ℓ is $\frac{6-3}{3-1} = \frac{3}{2}$. If $m \parallel \ell$, then the slope of m is also $\frac{3}{2}$. In the figure on the left at the top of page 193, each of the dotted lines has a slope of $\frac{3}{2}$. If $n \perp \ell$, then the slope of n is $-\frac{2}{3}$, the negative reciprocal of $\frac{3}{2}$. In the figure on the right at the top of page 193, each of the dotted lines has a slope of $-\frac{2}{3}$.

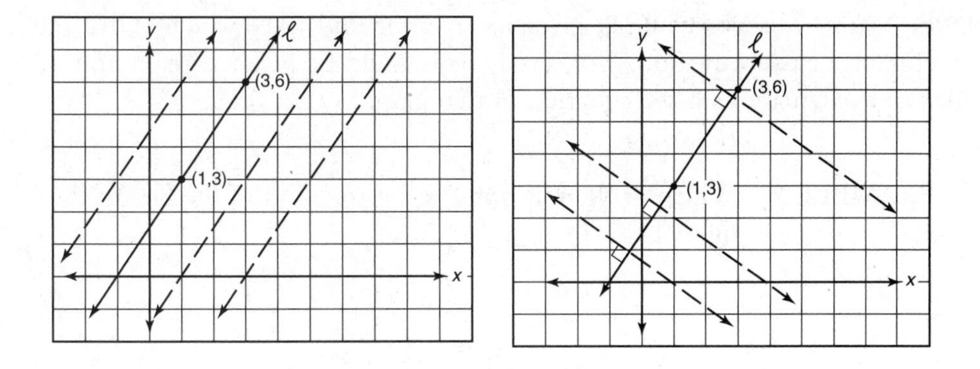

EQUATIONS OF LINES

Every line that is drawn in a coordinate plane has an equation. A point (x, y) is on a line if and only if the values of x and y satisfy that equation. All the points on a horizontal line have the same y-coordinate. For example, in the figure below, horizontal line ℓ passes through $(-3, 4)$, $(0, 4)$, $(2, 4)$, $(5, 4)$, and $(10, 4)$. The equation of line ℓ is $y = 4$. Similarly, every point on a vertical line has the same x-coordinate. Each point on line m has an x-coordinate of 3. The equation of m is $x = 3$.

Every other line in the coordinate plane has an equation that can be written in the form $y = mx + b$, where m is the slope of the line and b is the ***y-intercept***—the y-coordinate of the point where the line crosses the y-axis. These facts are summarized in KEY FACT L8.

Key Fact L8

- **For any real number a, $x = a$ is the equation of the vertical line that crosses the x-axis at $(a, 0)$.**
- **For any real number b, $y = b$ is the equation of the horizontal line that crosses the y-axis at $(0, b)$.**
- **For any real numbers b and m, $y = mx + b$ is the equation of the line that crosses the y-axis at $(0, b)$ and whose slope is m.**

On the Math 1 test, you could be given an equation and asked which of five graphs is the graph of that equation. Conversely, you could be given a graph and asked which of five equations is the equation of that graph.

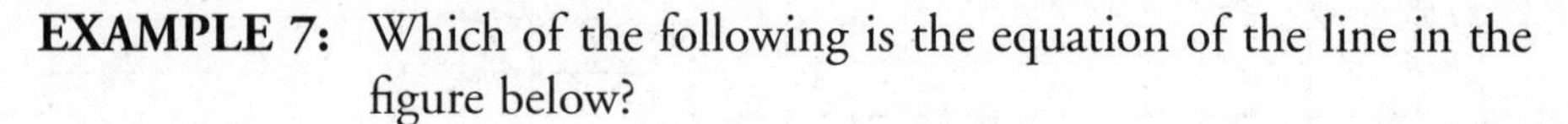

EXAMPLE 7: Which of the following is the equation of the line in the figure below?

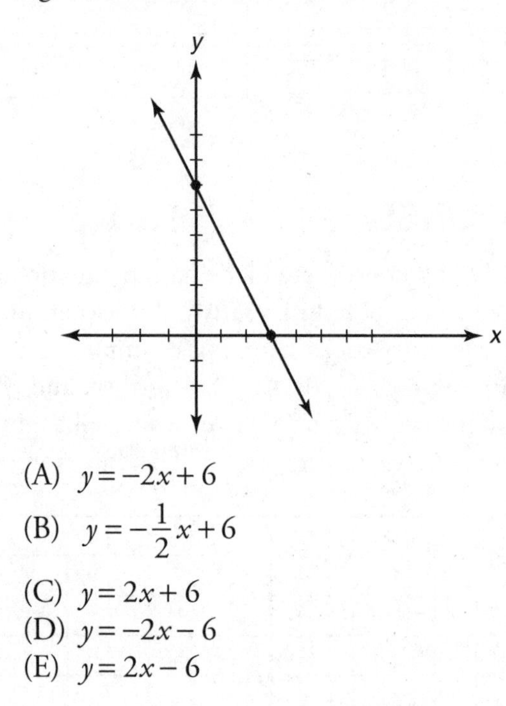

(A) $y = -2x + 6$
(B) $y = -\frac{1}{2}x + 6$
(C) $y = 2x + 6$
(D) $y = -2x - 6$
(E) $y = 2x - 6$

On the Math 1 test, there are two ways to answer the question:

Solution 1. Since the line is neither horizontal nor vertical, its equation has the form $y = mx + b$. Since the line crosses the y-axis at 6, $b = 6$. Since the line passes through (0, 6) and (3, 0), its slope, m, is $\frac{0-6}{3-0} = \frac{-6}{3} = -2$. The equation is $y = -2x + 6$, choice A.

Solution 2. The second way to answer this question takes advantage of the fact that the Math 1 test is a multiple-choice test. Since the line passes through (0, 6), $y = 6$ when $x = 0$. Plug in 0 for x in the five choices, and eliminate choices D and E, the two choices for which $y \neq 6$. Then note that the line also passes through (3, 0), so when $x = 3$, $y = 0$. Replace x by 3 in the remaining choices (A, B, and C) and see which one equals 0. Only choice A works.

EXAMPLE 8: What is the equation of the line that passes through (1, −3) and (5, 5)?

First find the slope of the line:

$$m = \frac{5-(-3)}{5-1} = \frac{8}{4} = 2$$

So the equation is $y = 2x + b$.

To find b, replace x and y by the x- and y-coordinates of either of the two given points, say (5, 5):

$$5 = 2(5) + b \Rightarrow 5 = 10 + b \Rightarrow b = -5$$

The equation is $y = 2x - 5$.

Any equation in which the only variables are x and y, and neither x nor y is written with an exponent, is the equation of a line.

EXAMPLE 9: $3y - 4x + 6 = 0$ is the equation of a line. To determine the slope and y-intercept of this line, first solve for y:

$$3y - 4x + 6 = 0 \Rightarrow 3y = 4x - 6 \Rightarrow y = \frac{4}{3}x - 2$$

The equation is now written in the form $y = mx + b$. We can see that the slope is $\frac{4}{3}$ and the y-intercept is −2.

In order to write the equation of a line, you need to know two things: the slope of the line and a point on the line. If the point you know happens to be the y-intercept, then, of course, you can use the equation $y = mx + b$. If the point is not the y-intercept, you have two choices. The first is to use the equation $y = mx + b$ and proceed as in Example 8. The second is to use the equation given in KEY FACT L9.

Key Fact L9

The equation of the line that passes through (x_1, y_1) and has slope m is $y - y_1 = m(x - x_1)$.

EXAMPLE 10: Line n passes through (1, 1) and is parallel to line ℓ, whose equation is $y = 2x - 3$. What is the equation of n?

Note that the slope of ℓ is 2, and since parallel lines have equal slopes, the slope of n is also 2. So the equation of n is:

$$y - 1 = 2(x - 1) \Rightarrow y - 1 = 2x - 2 \Rightarrow y = 2x - 1$$

Occasionally on the Math 1 test, there is a question concerning a linear inequality, such as $y > 2x + 1$. The graph of this inequality consists of all the points that are above the line $y = 2x + 1$. Note that the point (2, 5) is on the line, whereas (2, 6), (2, 7), and (2, 8) are all above the line. We indicate the set of all points satisfying the inequality by shading or striping the region above the line. To indicate that the points on the line itself do not satisfy the inequality, we draw a dotted line. To indicate that the points on a line are included in a graph, we draw a solid line. For example, (2, 5) is not on the graph of $y > 2x + 1$, but it is on the graph of $y \geq 2x + 1$. Similarly, the graph of $y < 2x + 1$ and $y \leq 2x + 1$ are shaded or striped regions below the line $y = 2x + 1$. These inequalities are shown in the following graphs.

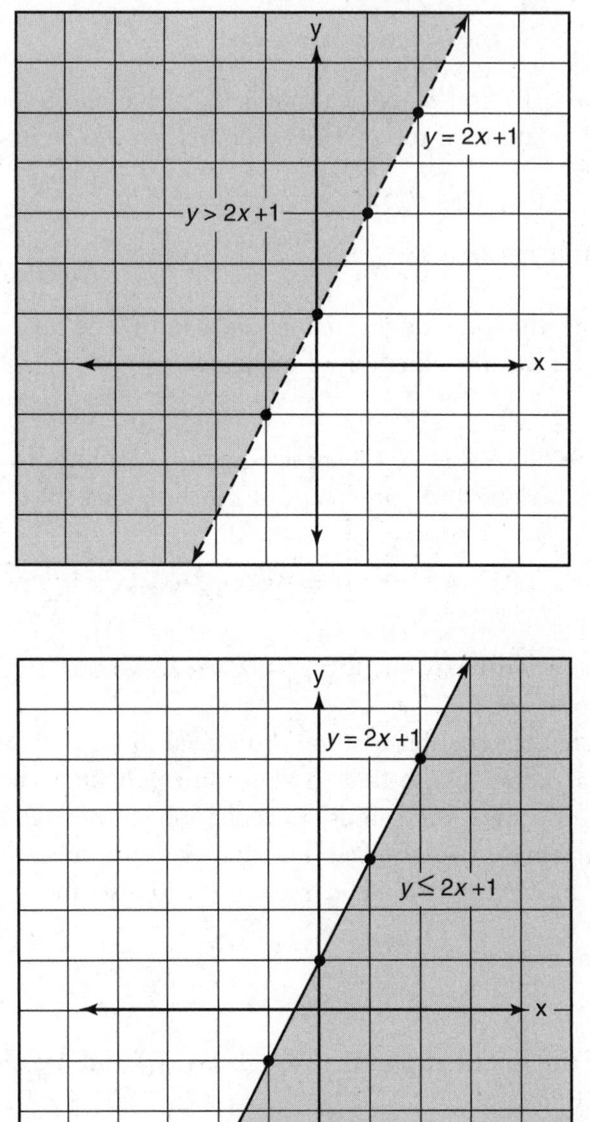

CIRCLES AND PARABOLAS

Besides the line, you need to know the equations for two other geometric shapes for the Math 1 test: the circle and the parabola.

Chapter 11 discussed all the facts you need to know about circles. In this chapter, we review the standard equation of a circle, which is given in KEY FACT L10.

Key Fact L10

- **The equation of the circle whose center is the point (h, k) and whose radius is r is $(x - h)^2 + (y - k)^2 = r^2$.**

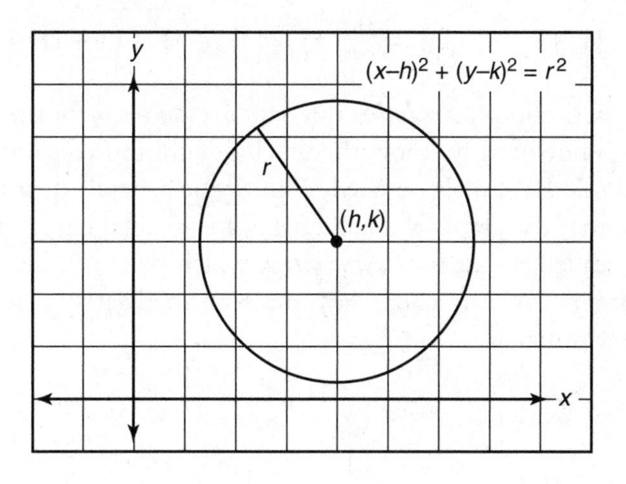

- **If the center is at the origin, (0, 0), then $h = k = 0$ and the equation reduces to $x^2 + y^2 = r^2$.**

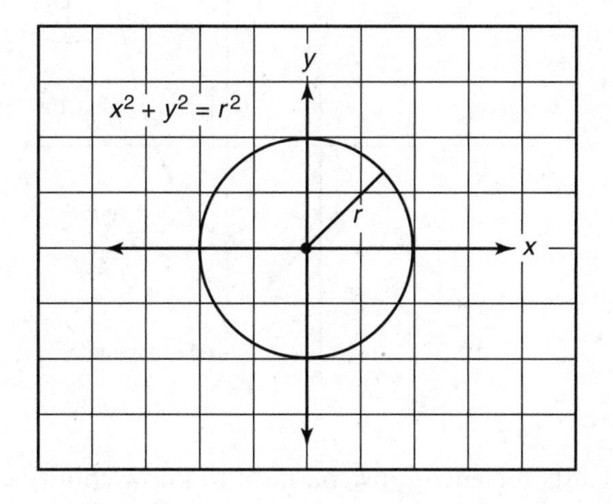

To write the equation of a circle, you need to know its center and its radius. Either they will be specifically given to you or you will be given some other information that will enable you to determine them.

EXAMPLE 11: What is the equation of the circle whose center is at (–3, 2) and whose radius is 5?

Plug $h = -3$, $k = 2$, and $r = 5$ into the standard equation:

$$(x - (-3))^2 + (y - 2)^2 = 5^2 \Rightarrow (x + 3)^2 + (y - 2)^2 = 25$$

EXAMPLE 12: In Examples 3 and 4, we considered the circle in which $C(3, -4)$ and $D(7, 2)$ are the endpoints of a diameter. What is the equation of this circle?

In Example 3, we used the midpoint formula to find that the center is at (5, –1). In Example 4, we used the distance formula to get that the radius is $\sqrt{13}$. Therefore, the equation for this circle is:

$$(x-5)^2+(y-(-1))^2=\left(\sqrt{13}\right)^2 \Rightarrow (x-5)^2+(y+1)^2=13$$

There are many facts about ***parabolas*** that you do *not* need for the Math 1 test. In particular, you do not need to know the precise definition of parabola in terms of its *focus* and *directrix*. Basically, you need to know the general equation of a parabola and to recognize that the graph of a parabola is a U-shaped curve that is symmetrical about a line, called the ***axis of symmetry***, which passes through the parabola's ***vertex***, or ***turning point***. Any parabola you see on the Math 1 test will likely have a vertical axis of symmetry.

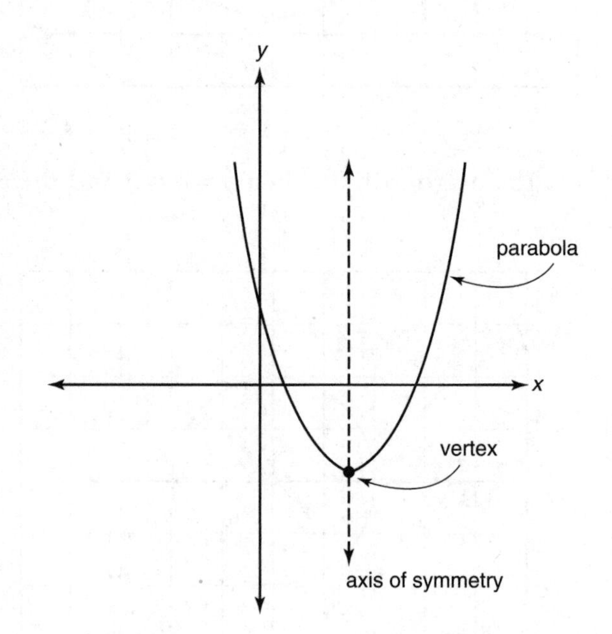

KEY FACT L11 lists the equations you need to know about parabolas.

Key Fact L11

For any real numbers a, b, c with $a \neq 0$:

- **$y = ax^2 + bx + c$ is the equation of a parabola whose axis of symmetry is a vertical line.**
- **Conversely, the equation of any parabola with a vertical axis of symmetry has an equation of the form $y = ax^2 + bx + c$.**
- **The equation of the parabola's axis of symmetry is $x = \frac{-b}{2a}$.**
- **The vertex of the parabola is the point on the parabola whose x-coordinate is $\frac{-b}{2a}$.**

- **If $a > 0$, the parabola opens upward and the vertex is the lowest point on the parabola.**
- **If $a < 0$, the parabola opens downward and the vertex is the highest point on the parabola.**

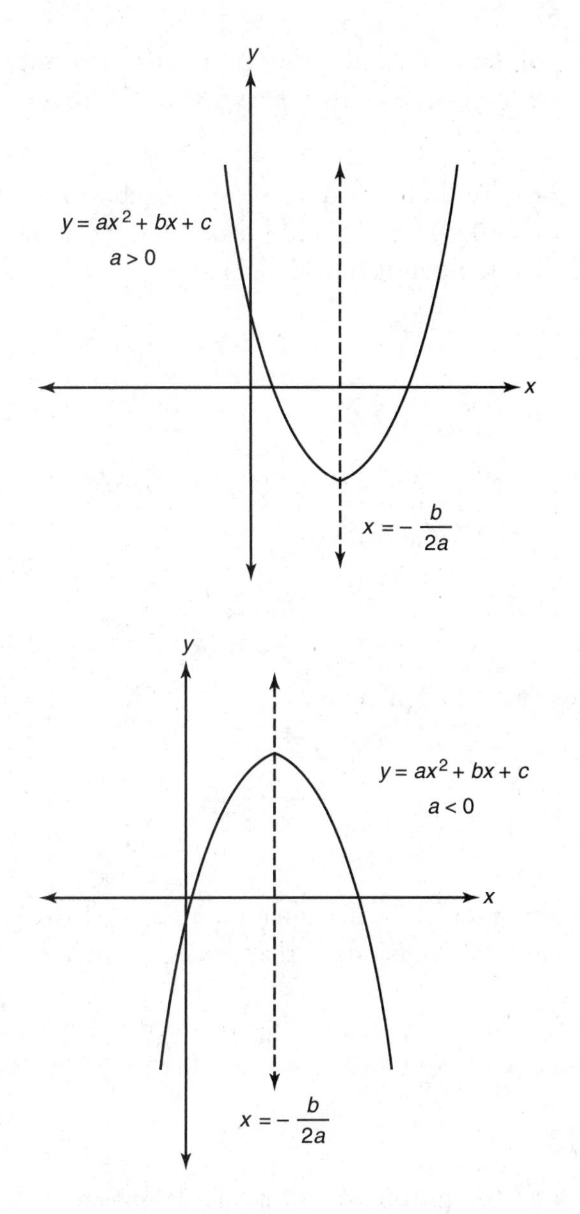

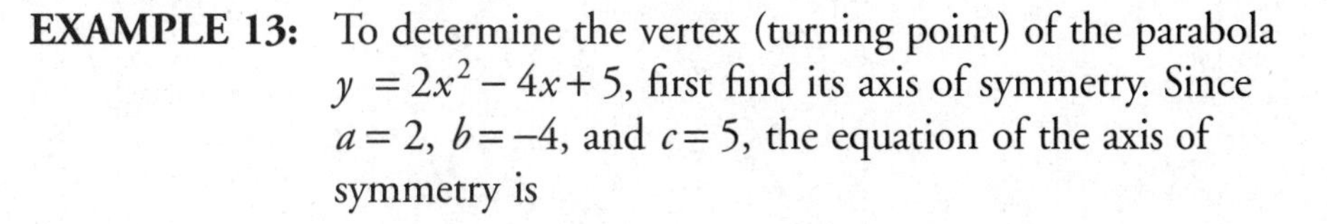

EXAMPLE 13: To determine the vertex (turning point) of the parabola $y = 2x^2 - 4x + 5$, first find its axis of symmetry. Since $a = 2$, $b = -4$, and $c = 5$, the equation of the axis of symmetry is

$x = \frac{-b}{2a} = \frac{4}{4} = 1$. So the x-coordinate of the turning point is 1 and the y-coordinate is

$$2(1)^2 - 4(1) + 5 = 2 - 4 + 5 = 3$$

The turning point is (1, 3).

Of course, an alternative solution would be to graph $y = 2x^2 - 4x + 5$ on a graphing calculator and to trace along the parabola until the cursor is at the turning point.

Key Fact L12

Given any three points, *P*, *Q*, and *R* that do not lie on a line, there is a parabola with a vertical axis of symmetry that passes through them.

EXAMPLE 14: To find the equation of a parabola that passes through (0, 4), (1, 3), and (2, 6), plug the x and y coordinates of the points into the equation $y = ax^2 + bx + c$.

$x = 0$ and $y = 4$:

$$4 = a(0)^2 + b(0) + c \Rightarrow 4 = c$$

$x = 1$ and $y = 3$:

$$3 = a(1)^2 + b(1) + c = a + b + 4 \Rightarrow a + b = -1$$

$x = 2$ and $y = 6$:

$$6 = a(2)^2 + b(2) + c = 4a + 2b + 4 \Rightarrow$$

$$4a + 2b = 2 \Rightarrow 2a + b = 1$$

Subtract these last two equations:

$$\begin{array}{r} 2a + b = 1 \\ \underline{a + b = -1} \\ a \quad\;\; = 2 \end{array}$$

Then by replacing a by 2 in the equation $a + b = -1$, we get that $b = -3$.

So the equation of the parabola that passes through the three given points is $y = 2x^2 - 3x + 4$.

There is an important relationship between the parabola $y = ax^2 + bx + c$ and the quadratic equation $ax^2 + bx + c = 0$.

Key Fact L13

The x-intercepts of the graph of the parabola $y = ax^2 + bx + c$ are the (real) solutions of the equation $ax^2 + bx + c = 0$.

Consider the graphs of the following six parabolas.

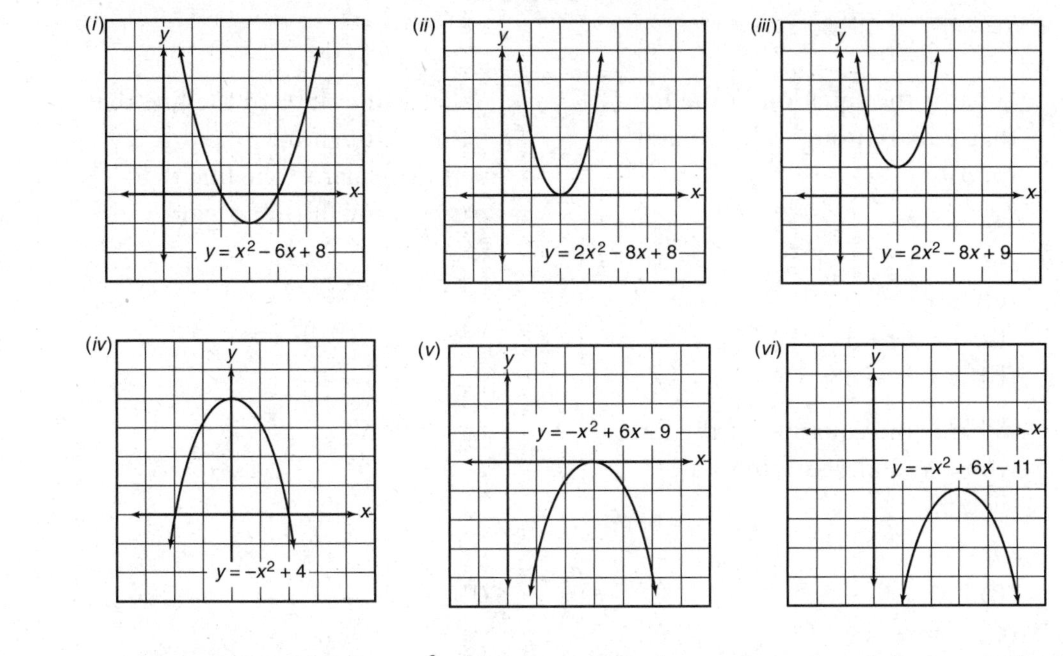

I. Since the graph of $y = x^2 - 6x + 8$ crosses the x-axis at 2 and 4, the quadratic equation $x^2 - 6x + 8 = 0$ has two real solutions: $x = 2$ and $x = 4$.

II. Since the graph of $y = 2x^2 - 8x + 8$ crosses the x-axis only at 2, the quadratic equation $2x^2 - 8x + 8 = 0$ has only one real solution: $x = 2$.

III. Since the graph of $y = 2x^2 - 8x + 9$ does not cross the x-axis, the quadratic equation $2x^2 - 8x + 9 = 0$ has no real solutions.

Similarly, from graphs iv, v, and vi, you can see that the equation $-x^2 + 4 = 0$ has two real solutions; the equation $-x^2 + 6x - 9 = 0$ has only one real solution; and the equation $-x^2 + 6x - 11 = 0$ has no real solutions.

Exercises

1. What is the equation of the line that passes through (3, –3) and (3, 3)?

   (A) $x = 3$
   (B) $y = 3$
   (C) $y = 3x$
   (D) $y = 3x + 3$
   (E) $y = 3x - 3$

2. What is the equation of the line whose *x*- and *y*-intercepts are each 3?

   (A) $x = 3$
   (B) $y = 3$
   (C) $y = x + 3$
   (D) $y = x - 3$
   (E) $y = -x + 3$

3. What is the slope of the line that passes through (4, 3) and is parallel to the line that passes through (–3, 4) and (3, –4)?

   (A) $-\frac{4}{3}$
   (B) $-\frac{3}{4}$
   (C) $\frac{3}{4}$
   (D) 1
   (E) $\frac{4}{3}$

4. What is the slope of the line that passes through (4, 3) and is perpendicular to the line that passes through (–3, 4) and (3, –4)?

   (A) $-\frac{4}{3}$
   (B) $-\frac{3}{4}$
   (C) $\frac{3}{4}$
   (D) 1
   (E) $\frac{4}{3}$

5. If *A* (–1, 0) and *B* (3, –2) are two adjacent vertices of square *ABCD*, what is the area of the square?

   (A) 12
   (B) 16
   (C) 20
   (D) 25
   (E) 36

Questions 6 and 7 refer to circle *O*, in which *A* (–1, 0) and *B* (3, –2) are the endpoints of a diameter.

6. What is the area of circle *O*?

   (A) 2.5π
   (B) 5π
   (C) 6.25π
   (D) 10π
   (E) 20π

7. Which of the following is the equation of circle *O*?

   (A) $(x + 1)^2 + (y - 1)^2 = \sqrt{5}$
   (B) $(x - 1)^2 + (y + 1)^2 = \sqrt{5}$
   (C) $(x + 1)^2 + (y - 1)^2 = 5$
   (D) $(x - 1)^2 + (y + 1)^2 = 5$
   (E) $(x - 1)^2 + (y + 1)^2 = 25$

8. If M (3, 1) and N (7, 1) are two adjacent vertices of a rectangle, which of the following could *not* be one of the rectangle's other vertices?

   (A) (3, 7)
   (B) (7, 3)
   (C) (3, −7)
   (D) (−3, 7)
   (E) (7, 7)

9. Which of the following is the equation of a parabola that does NOT intersect the x-axis?

   (A) $y = x^4 + 1$
   (B) $y = x^2 + 2x - 3$
   (C) $y = (x - 1)^2$
   (D) $y = x^2 - 1$
   (E) $y = x^2 + 1$

10. What is the slope of the line that passes through (a, b) and $\left(\frac{1}{a}, b\right)$?

    (A) 0
    (B) $\frac{1}{b}$
    (C) $\frac{1 - a^2}{a}$
    (D) $\frac{a^2 - 1}{a}$
    (E) The slope is undefined.

11. Line ℓ is tangent to the circle whose center is at (3, 2). If the point of tangency is (6, 6), what is the slope of line ℓ?

    (A) $-\frac{4}{3}$
    (B) $-\frac{3}{4}$
    (C) 0
    (D) $\frac{3}{4}$
    (E) $\frac{4}{3}$

Questions 12–14 concern the parabola whose equation is $y = x^2 - 20x - 69$.

12. Where does the graph of the parabola cross the y-axis?

    (A) −69
    (B) −3
    (C) 23
    (D) 69
    (E) The graph does not cross the y-axis.

13. What is the sum of the x-coordinates of the points where the graph of the parabola crosses the x-axis?

    (A) −69
    (B) 0
    (C) 20
    (D) 69
    (E) The graph does not cross the x-axis.

14. Which of the following is the parabola's turning point?

    (A) (−10, 231)
    (B) (0, −69)
    (C) (10, −169)
    (D) (−3, 0)
    (E) (23, 0)

15. What is the equation of the circle whose center is at the origin and that passes through the point (8, 15)?

    (A) $x^2 + y^2 = 15$
    (B) $x^2 + y^2 = 17$
    (C) $x^2 + y^2 = 289$
    (D) $(x - 8)^2 + (y - 15)^2 = 17$
    (E) $(x - 8)^2 + (y - 15)^2 = 289$

ANSWERS EXPLAINED

Answer Key

1. **(A)**	4. **(C)**	7. **(D)**	10. **(A)**	13. **(C)**
2. **(E)**	5. **(C)**	8. **(D)**	11. **(B)**	14. **(C)**
3. **(A)**	6. **(B)**	9. **(E)**	12. **(A)**	15. **(C)**

Solutions

Each of the problems in this set of exercises is typical of a question you could see on a Math 1 test. When you take the model tests in this book and, in particular, when you take the actual Math 1 test, if you get stuck on questions such as these, you do not have to leave them out—you can almost always answer them by using one or more of the strategies discussed in the "Tactics" chapter. The solutions given here do *not* depend on those strategies; they are the correct mathematical ones.

1. **(A)** A quick sketch shows that the line that passes through (3, –3) and (3, 3) is vertical. By KEY FACT L8, its equation is $x = 3$.

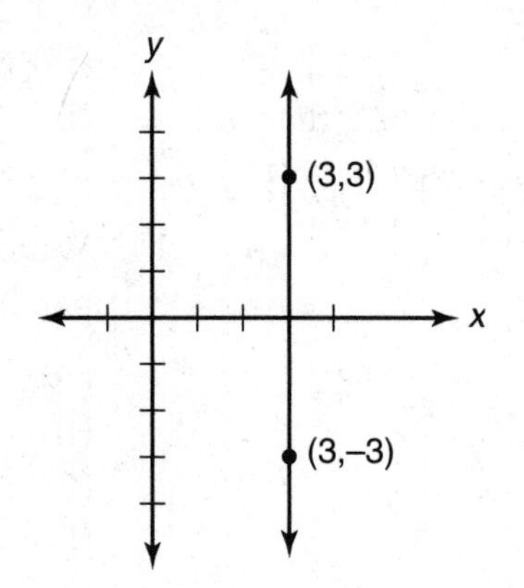

2. **(E)** Quickly sketch the line.

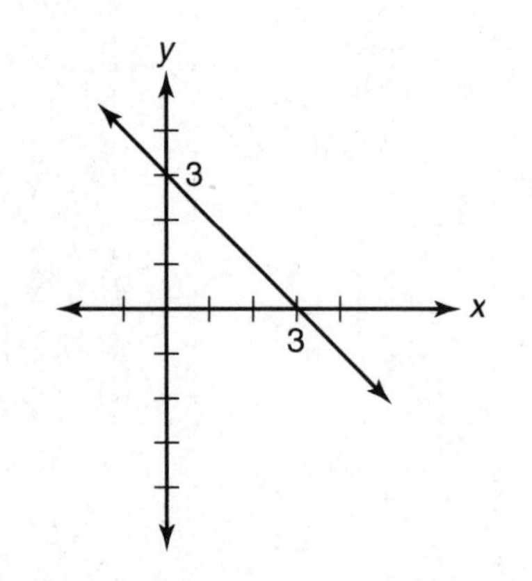

Since the line is neither horizontal nor vertical, its equation has the form $y = mx + b$. Since it crosses the y-axis at (0, 3), $b = 3$.

Since it passes through (0, 3) and (3, 0), its slope is $\frac{0-3}{3-0} = \frac{-3}{3} = -1$, and so $y = -1x + 3$ or $y = -x + 3$.

<u>Solutions 3 and 4.</u> Before answering either question, quickly sketch line ℓ that goes through (–3, 4) and (3, –4) and lines ℓ_1 and ℓ_2, that go through (4, 3) and are parallel to ℓ and perpendicular to ℓ, respectively.

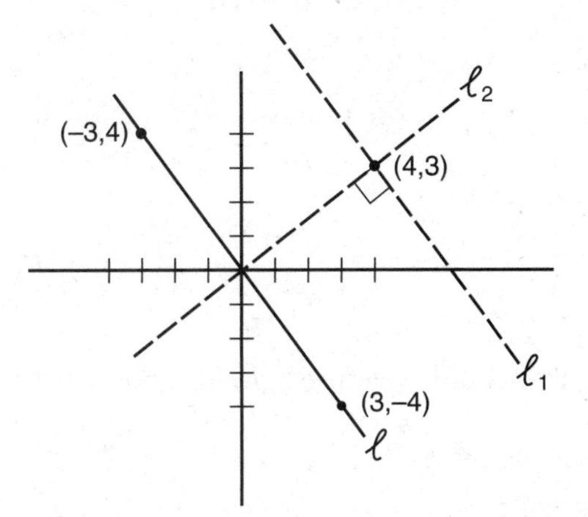

You should immediately see that the slope of ℓ_1 is negative and the slope of ℓ_2 is positive. So the answer to Question 3 must be A or B, and the answer to Question 4 must be C, D, or E.

3. **(A)** By KEY FACT L5, the slope of ℓ, the line through (–3, 4) and (3, –4) is

$$\frac{-4-4}{3-(-3)} = \frac{-8}{6} = -\frac{4}{3}$$

By KEY FACT L7, parallel lines have equal slopes, so the slope of ℓ_1 is $-\frac{4}{3}$. Note that it is irrelevant that ℓ_1 passes through (4, 3).

4. **(C)** From Solution 3, above, the slope of ℓ is $-\frac{4}{3}$. By KEY FACT L7, if two lines are perpendicular, the product of their slopes is –1. So if m is the slope of ℓ_2, then:

$$-\frac{4}{3}m = -1 \Rightarrow -4m = -3 \Rightarrow m = \frac{3}{4}$$

As in Question 3, it is irrelevant that ℓ_2 passes through (4, 3).

5. **(C)** From the distance formula (KEY FACT L3):

$$AB = \sqrt{3-(-1)^2 + (-2-0)^2} = \sqrt{4^2 + (-2)^2}$$
$$= \sqrt{16+4} = \sqrt{20}$$

The area of square $ABCD = (AB)^2 = \left(\sqrt{20}\right)^2 = 20$.

6. **(B)** By the distance formula (KEY FACT L3):

$$AB = \sqrt{(3-(-1))^2 + (-2-0)^2} = \sqrt{4^2 + (-2)^2}$$
$$= \sqrt{16+4} = \sqrt{20}$$

Since the diameter is $\sqrt{20}$, the radius is $\frac{\sqrt{20}}{2}$. Now use the area formula:

$$A = \pi r^2 = \pi\left(\frac{\sqrt{20}}{2}\right)^2 = \pi\left(\frac{20}{4}\right) = 5\pi$$

**Alternatively, you could have used the midpoint formula (KEY FACT L4) to determine that the center O is at

$$\left(\frac{-1+3}{2}, \frac{0+(-2)}{2}\right) = \left(\frac{2}{2}, \frac{-2}{2}\right) = (1, -1)$$

and then used the distance formula to find the length of radius $\overline{OA}$.

7. **(D)** From the solution to Question 6, above, the radius of circle O is $\frac{\sqrt{20}}{2} = \frac{2\sqrt{5}}{2} = \sqrt{5}$. From the alternative solution to Question 6, the center is at $(1, -1)$. Finally, by KEY FACT J11, the equation of circle O is:

$$(x-1)^2 + (y-(-1))^2 = (\sqrt{5})^2$$
$$(x-1)^2 + (y+1)^2 = 5$$

8. **(D)** Any point whose x-coordinate is 3 or 7 could be another vertex. Of the choices, only $(-3, 7)$ could not be one of the vertices.

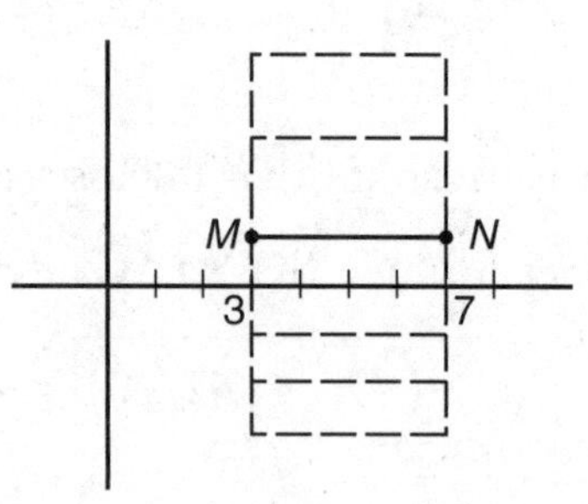

9. **(E)** A graph intersects the x-axis at points whose y-coordinates are 0. Choices B, C, and D all intersect the x-axis at $(1, 0)$. Choices A and E do not intersect the x-axis, but Choice A is not the equation of a parabola.

10. **(A)** By KEY FACT L5, the formula for slope is $\frac{y_2 - y_1}{x_2 - x_1}$. Before using it, though, look. Since the y-coordinates are equal, the numerator, and thus the fraction, equals 0.

11. **(B)** Draw a rough sketch

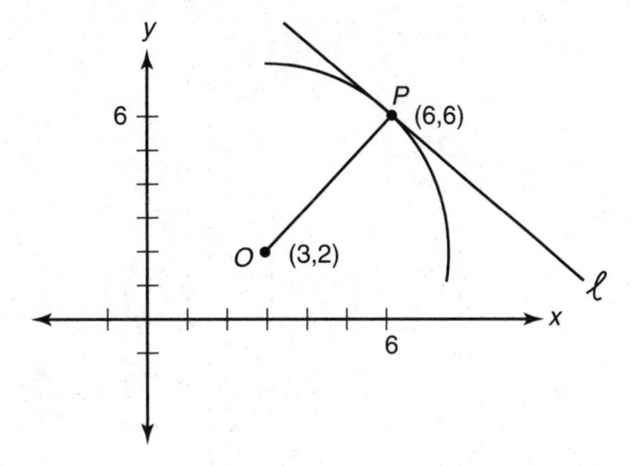

Line segment $\overline{OP}$, joining (3, 2) and (6, 6) is a radius and so by KEY FACT L10 is perpendicular to line ℓ. The slope of $\overline{OP}$ is $\frac{6-2}{6-3}=\frac{4}{3}$. Therefore, by KEY FACT L7, the slope of ℓ is $-\frac{3}{4}$.

12. **(A)** Since the y-intercept of a graph is a point whose x-coordinate is 0, just evaluate the equation when $x = 0$:

$$y = 0^2 - 20(0) - 69 = -69$$

13. **(C)** The x-intercepts of the graph are the solutions of the equation $x^2 - 20x - 69 = 0$.

$$x^2 - 20x - 69 = 0 \Rightarrow (x - 23)(x + 3) = 0 \Rightarrow$$
$$x = 23 \text{ or } x = -3$$

The sum of the roots is $23 + (-3) = 20$.

Since the question asks for the sum of the roots and not the individual roots, you can use KEY FACT E3—the formula for the sum of the roots:

$$\text{sum of the roots} = \frac{-b}{a} = \frac{-(-20)}{1} = 20$$

14. **(C)** The turning point of the parabola is on the parabola's axis of symmetry, whose equation is $x = \frac{-b}{2a} = \frac{20}{2} = 10$. So the x-coordinate of the turning point is 10; to get the y-coordinate, replace x by 10 in the equation of the parabola:

$$y = (10)^2 - 20(10) - 69 = 100 - 200 - 69 = -169.$$

The turning point is $(10, -169)$.

15. **(C)** The equation of a circle whose center is at the origin is $x^2 + y^2 = r^2$, where r is the radius. Since (8, 15) is on the circle, $r^2 = 8^2 + 15^2 = 64 + 225 = 289$. So the equation of this circle is $x^2 + y^2 = 289$.

TRIGONOMETRY

Basic Trigonometry

- Sine, Cosine, and Tangent
- What You *Don't* Need to Know
- Exercises
- Answers Explained

This chapter is very short, because it does not review all the trigonometry you learned in school. It reviews only the trigonometry you need for the Math 1 test—which is not very much. The ONLY trigonometry you need to know for the Math 1 test are the meanings of the trigonometric ratios—sine, cosine, and tangent—and one simple identity.

SINE, COSINE, AND TANGENT

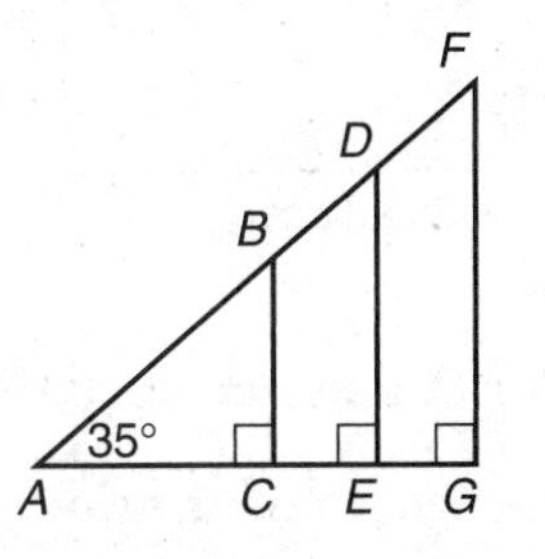

In the figure above, right triangles *ABC*, *ADE*, and *AFG* each have a 90° angle and a 35° angle, so they are all similar to one another. Therefore, their sides are in proportion:

$$\frac{BC}{AB} = \frac{DE}{AD} = \frac{FG}{AF}$$
$$= \frac{\text{length of the side opposite the } 35^\circ \text{ angle}}{\text{length of the hypotenuse}}$$

This ratio is called the sine of 35° and is written sin 35°. To evaluate sin 35°, you could very carefully measure the lengths of $\overline{FG}$ and $\overline{AF}$ and divide. In the given figure, $FG \approx 1$ inch, $AF \approx 1.75$ inches, so $\frac{FG}{AF} \approx \frac{1}{1.75} = 0.57$.

Fortunately, you do not have to do this. You can use your calculator. Depending on what calculator you use, you would either enter 35 and then press the [sin] button or press the [sin] button and then enter 35. Regardless, in a fraction of a second, you will see the answer correct to several decimal places: $\sin 35° = 0.573576436$, far greater accuracy than you need for the Math 1 test.

EXAMPLE 1: To find the length of hypotenuse $\overline{PQ}$ in the triangle below, use the sine ratio (and your calculator):

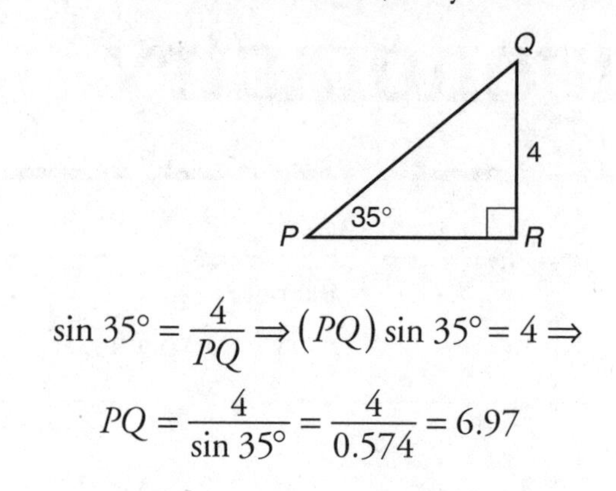

$$\sin 35° = \frac{4}{PQ} \Rightarrow (PQ)\sin 35° = 4 \Rightarrow$$

$$PQ = \frac{4}{\sin 35°} = \frac{4}{0.574} = 6.97$$

The formal definitions of the three trigonometric ratios you need are given in KEY FACT M1.

Key Fact M1

Let θ be one of the acute angles in a right triangle.

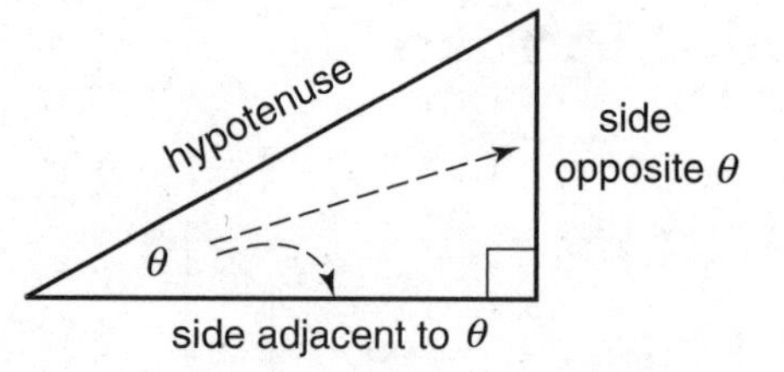

- **The formula for the sine of θ, denoted $\sin\theta$, is:**

$$\sin\theta = \frac{\text{the length of the side opposite } \theta}{\text{the length of the hypotenuse}} = \frac{\text{opposite}}{\text{hypotenuse}}$$

- **The formula for the cosine of θ, denoted $\cos\theta$, is:**

$$\cos\theta = \frac{\text{the length of the side adjacent to } \theta}{\text{the length of the hypotenuse}} = \frac{\text{adjacent}}{\text{hypotenuse}}$$

- **The formula for the tangent of θ, denoted $\tan\theta$, is:**

$$\tan\theta = \frac{\text{the length of the side opposite } \theta}{\text{the length of the side adjacent to } \theta} = \frac{\text{opposite}}{\text{adjacent}}$$

- **From the definitions of the three trigonometry ratios, it follows immediately that for any acute angle θ, $\tan\theta = \dfrac{\sin\theta}{\cos\theta}$.**

For decades, students have remembered these definitions by memorizing the "word" SOHCAHTOA. For example, the "S" in "Soh" stands for "sine" and the "OH" reminds you that sine is Opposite over Hypotenuse.

The only trigonometric equation or identity you need to know other than those in KEY FACT M1, for the Math 1 test, is given in KEY FACT M2.

Key Fact M2

For any angle θ, $\sin^2\theta + \cos^2\theta = 1$.

Note that $\sin^2\theta$ is an abbreviation for $(\sin\theta)^2$. So the identity $\sin^2\theta + \cos^2\theta = 1$ means that if you take the sine of any angle and square it and then take the cosine of that same angle and square it, the sum of those two squares is 1.

For example, you can use your calculator to verify that $(\sin 37°)^2 + (\cos 37°)^2 = 1$. Note that in this identity, θ can be replaced by any expression at all. So $\sin^2(4x + 1) + \cos^2(4x + 1) = 1$.

EXAMPLE 2: $(3\sin^2 3\theta + 3\cos^2 3\theta - 1)^2 = (3(\sin^2 3\theta + \cos^2 3\theta) - 1)^2 = (3(1) - 1)^2 = (3 - 1)^2 = 2^2 = 4$.

If you know the value of any one of $\sin\theta$, $\cos\theta$, or $\tan\theta$, you can always find the values of the other two.

EXAMPLE 3: If you are given that $\sin\theta = \dfrac{15}{17}$ and want to know the value of $\cos\theta$ or $\tan\theta$, draw right triangle ABC and label BC, the side opposite θ, as 15 and the hypotenuse as 17.

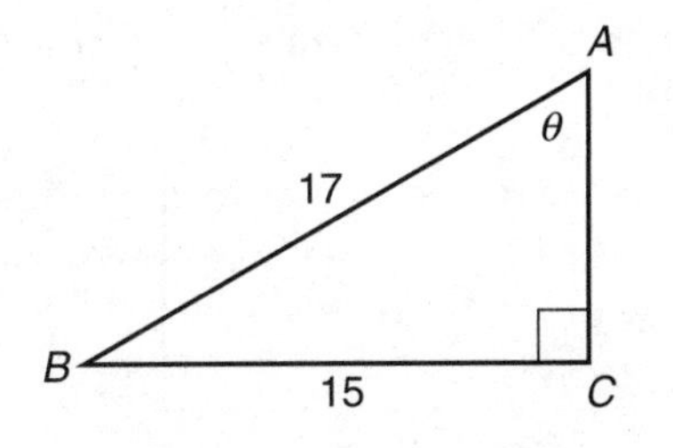

Now use the Pythagorean theorem to find AC:

$$(AC)^2 + 15^2 = 17^2 \Rightarrow AC^2 + 225 = 289 \Rightarrow$$
$$(AC)^2 = 64 \Rightarrow AC = 8.$$

So $\cos\theta = \dfrac{8}{17}$ and $\tan\theta = \dfrac{15}{8}$

If you were given that $\sin\theta = 0.835$, you would proceed exactly the same way. Draw a triangle and use the Pythagorean theorem to get

$x^2 + (0.835)^2 = 1^2 \Rightarrow x^2 + 0.697 = 1 \Rightarrow x^2 = 0.303 \Rightarrow x = 0.55$.

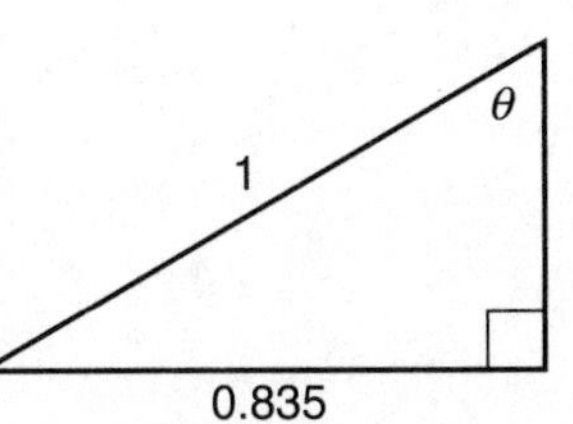

Now use SOHCAHTOA:

$\cos\theta = \frac{0.55}{1} = 0.55$ and $\tan\theta = \frac{0.835}{0.55} = 1.52.$

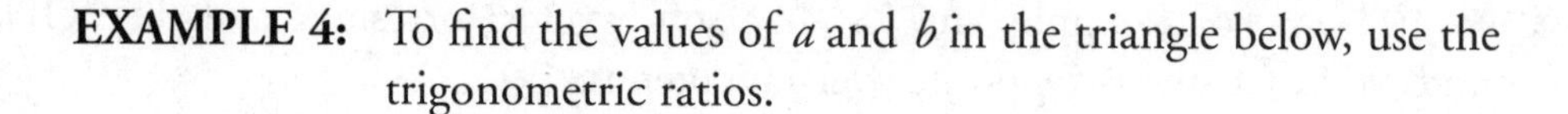

EXAMPLE 4: To find the values of a and b in the triangle below, use the trigonometric ratios.

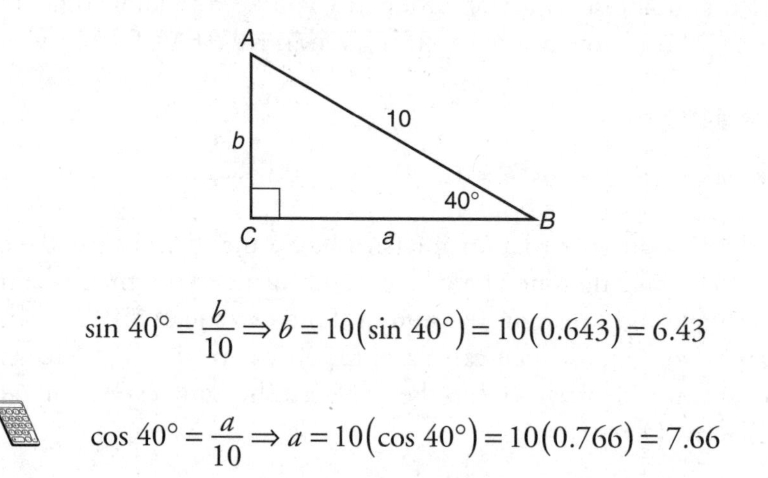

$$\sin 40° = \frac{b}{10} \Rightarrow b = 10(\sin 40°) = 10(0.643) = 6.43$$

$$\cos 40° = \frac{a}{10} \Rightarrow a = 10(\cos 40°) = 10(0.766) = 7.66$$

You could also be asked to find the measure of an angle θ if you are given the value of the sine, cosine, or tangent of that angle. For example, if you are given that $\sin\theta = 0.8$, to find the value of θ, you use your calculator to evaluate $\sin^{-1}(0.8)$. On most calculators, SIN^{-1} is the second function button above the SIN button. So depending on your calculator, you would either press [2nd] [SIN] [.8] [enter] or [.8] [2nd] [SIN].

EXAMPLE 5: What is the measure of the smallest angle in a 3-4-5 triangle?

Draw and label the triangle, which you should immediately recognize as a right triangle.

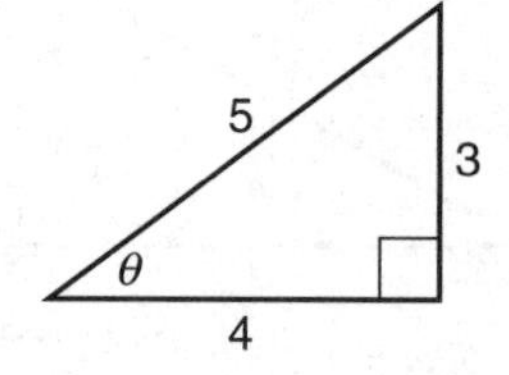

By KEY FACT H3, the smallest angle is opposite the smallest side. So $\sin\theta = \frac{3}{5}$ and $\theta = \sin^{-1}\left(\frac{3}{5}\right) = 36.87°$. Note that you can convert $\frac{3}{5}$ to 0.6 and take $\sin^{-1}(0.6)$, but it isn't necessary.

EXAMPLE 6: A 20-foot ladder is leaning against a vertical wall. If the base of the ladder is 13 feet from the wall, what is the angle formed by the ladder and the ground?

Of course, you start by drawing a diagram.

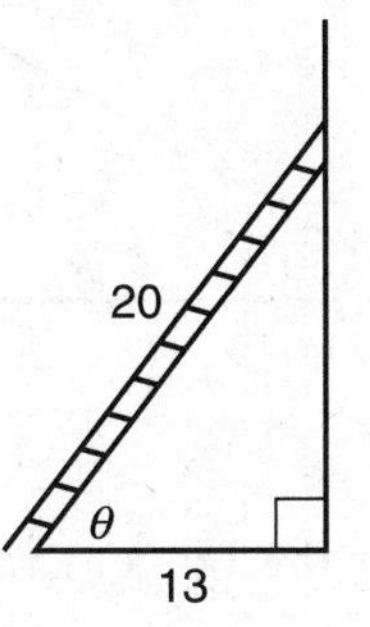

Then:

$$\cos\theta = \frac{13}{20} \Rightarrow \theta = \cos^{-1}\left(\frac{13}{20}\right) = 49.46°$$

WHAT YOU *DON'T* NEED TO KNOW

You have just reviewed the ONLY trigonometry you need to know for the Math 1 test. The College Board is very specific about this. Here is a list of several topics in trigonometry that you may have studied in school and that may be tested on the Math 2 test but are NOT included on the Math 1 test:

- Radian measure (**so keep your calculator in degree mode**)
- Angles whose measures are greater than 90° or less than 0°
- Reference angles
- The reciprocal trigonometric functions: secant, cosecant, and cotangent
- Graphs of the trigonometric functions
- Amplitude and period of the trigonometric functions
- Inverse trigonometric functions
- The law of sines and the law of cosines
- The trigonometric formula for the areas of a triangle and parallelogram
- Double-angle formulas and half-angle formulas
- Trigonometric identities (other than the two previously mentioned)

There is no question on the Math 1 test whose correct solution requires you to use any topic in the preceding list. However, you may always use anything you know.

For example, in the figure below, the area of ΔABC is $\frac{1}{2}(AC)(4)$. So in order to calculate the area, you need to know the value of AC. Since ΔABC is a 30-60-90 right triangle, by KEY FACT H8, $AC = 4\sqrt{3}$.

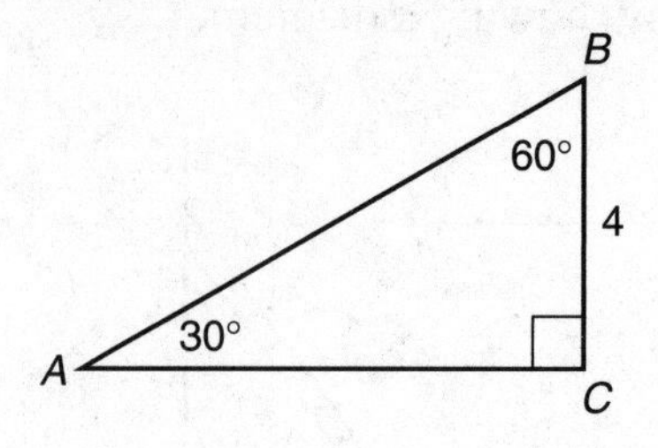

You can also find AC by using the tangent ratio: $\tan 30° = \frac{4}{AC}$ and solving for AC. However, if you know the law of sines and for any reason you prefer to use it, of course you may:

$$\frac{4}{\sin 30°} = \frac{AC}{\sin 60°} \Rightarrow AC = \frac{4 \sin 60°}{\sin 30°} = 6.93$$

Exercises

1. In the triangle below, what is the sum of a and b?

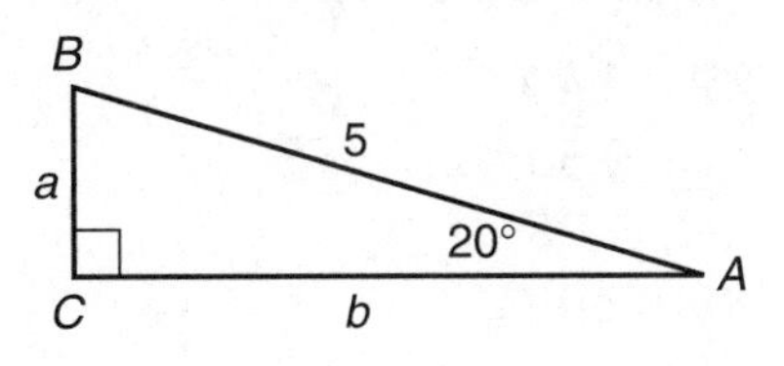

(A) 1.28
(B) 5
(C) 6.41
(D) 7.17
(E) 8.45

2. If $0° < \theta < 90°$, then which of the following is an expression for $\tan^2\theta\cos^2\theta + \cos^2\theta$?

(A) 0
(B) 1
(C) $\sin^2\theta$
(D) $\tan^2\theta + 1$
(E) $(\tan\theta\cos\theta + \cos\theta)^2$

3. A ladder is leaning against a wall, forming an angle of 65° with the ground. If the foot of the ladder is 8 feet from the wall, what is the length of the ladder, in feet?

(A) 3.38
(B) 7.25
(C) 8.83
(D) 17.15
(E) 18.93

4. If $0° < \theta < 90°$ and $\tan\theta = 5$, what is $\sin\theta + \cos\theta$?

(A) 0.19
(B) 0.85
(C) 1
(D) 1.18
(E) 5.2

5. If $0° < \theta < 90°$, then which of the following is equivalent to $\frac{\cos\theta}{\tan\theta} + \sin\theta$?

(A) 1
(B) $\frac{1}{\sin\theta}$
(C) $1 + \tan\theta$
(D) $\frac{\cos^2\theta}{\sin^2\theta}$
(E) $\cos\theta + \tan\theta\sin\theta$

6. What is the measure of the smallest angle in a 5-12-13 right triangle?

(A) 15°
(B) 21°
(C) 22.6°
(D) 45°
(E) 67.4°

7. If ΔABC is isosceles and $m\angle C = 90°$, which of the following must be true?

I. $\tan A = 1$
II. $\sin A = \cos A$
III. $\sin A = \cos B$

(A) I only
(B) II only
(C) III only
(D) I and II only
(E) I, II, and III

8. Which of the following is equal to $(3\sin 2\theta)(4\sin 2\theta) + (2\cos 2\theta)(6\cos 2\theta)$?

(A) 1
(B) 2
(C) 6
(D) 12
(E) 24

9. A kite string is tied to a peg in the ground. If the angle formed by the string and the ground is 70° and if there is 100 feet of string out, to the nearest foot, how high above the ground is the kite?

   (A) 34
   (B) 64
   (C) 74
   (D) 94
   (E) 274

10. If $0° < \theta < 90°$, which of the following is equivalent to $\frac{\sin^4\theta - \cos^4\theta}{\cos^2\theta - \sin^2\theta}$?

   (A) −1
   (B) 1
   (C) $\sin^2\theta - \cos^2\theta$
   (D) $\cos^2\theta - \sin^2\theta$
   (E) $\tan^2\theta - \cos^2\theta$

11. If the longer leg of a right triangle is twice as long as the shorter leg, what is the ratio of the measure of the larger acute angle to the measure of the smaller acute angle?

   (A) 0.42
   (B) 0.50
   (C) 1.78
   (D) 2.00
   (E) 2.39

12. A car is parked 120 feet from a building that is 350 feet tall. What is the measure of the angle of depression from the top of the building to the car?

   (A) 18.9°
   (B) 37.8°
   (C) 64.2°
   (D) 71.1°
   (E) 78.6°

ANSWERS EXPLAINED

Answer Key

1. **(C)**	3. **(E)**	5. **(B)**	7. **(E)**	9. **(D)**	11. **(E)**
2. **(B)**	4. **(D)**	6. **(C)**	8. **(D)**	10. **(A)**	12. **(D)**

SOLUTIONS

Each of the problems in this set of exercises is typical of a question you could see on a Math 1 test. When you take the model tests in this book and, in particular, when you take the actual Math 1 test, if you get stuck on questions such as these, you do not have to leave them out—you can almost always answer them by using one or more of the strategies discussed in the "Tactics" chapter. The solutions given here do *not* depend on those strategies; they are the correct mathematical ones.

1. **(C)** Use the sine and cosine ratios:

$$\sin 20° = \frac{a}{5} \Rightarrow a = 5(\sin 20°) = 5(0.342) = 1.710$$

$$\cos 20° = \frac{b}{5} \Rightarrow b = 5(\cos 20°) = 5(0.940) = 4.700$$

So $a + b = 1.710 + 4.700 = 6.41$

2. **(B)** $\tan^2\theta\cos^2\theta + \cos^2\theta = \frac{\sin^2\theta}{\cancel{\cos^2\theta}} \cdot \cancel{\cos^2\theta} + \cos^2\theta = \sin^2\theta + \cos^2\theta = 1$

3. **(E)** First draw a diagram.

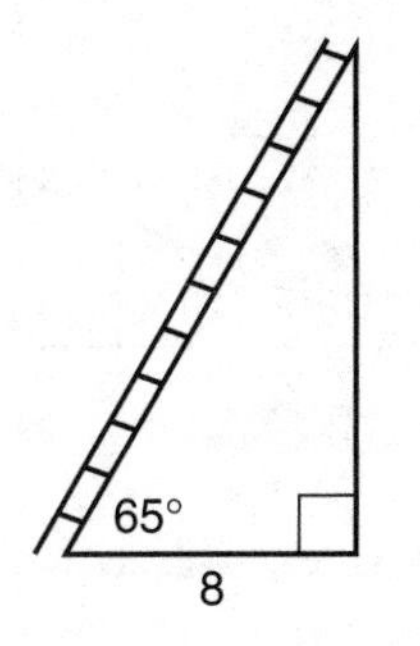

By using the cosine ratio, we get that $\cos 65° = \frac{8}{L}$, where L represents the length of the ladder.

$$L \cos 65° = 8 \Rightarrow$$
$$L = 8 \div (\cos 65°) = 18.93$$

4. **(D)** Draw a right triangle. Since $\tan\theta = \frac{\text{opposite}}{\text{adjacent}}$, label the side opposite θ as 5 and the side adjacent to θ as 1.

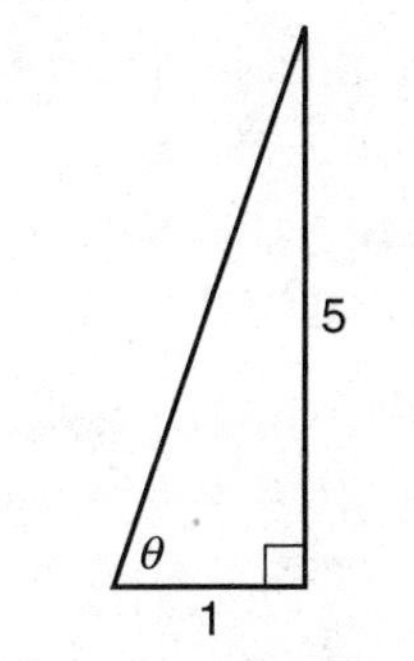

Then use the Pythagorean theorem to find the hypotenuse:

$$1^2 + 5^2 = c^2 \Rightarrow c^2 = 1 + 25 = 26 \Rightarrow c = \sqrt{26}$$

So $\sin\theta = \frac{5}{\sqrt{26}}$ and $\cos\theta = \frac{1}{\sqrt{26}}$ Then:

$$\sin\theta + \cos\theta = \frac{5}{\sqrt{26}} + \frac{1}{\sqrt{26}} = \frac{6}{\sqrt{26}} = 1.18$$

Alternatively, you could have found that $\theta = \tan^{-1}(5) = 78.69°$ and then added $\sin 78.69 + \cos 78.69 = 1.18$

5. **(B)** $\frac{\cos\theta}{\tan\theta} + \sin\theta = \frac{\cos\theta}{\frac{\sin\theta}{\cos\theta}} + \sin\theta = \frac{\cos^2\theta}{\sin\theta} + \sin\theta = \frac{\cos^2\theta}{\sin\theta} + \frac{\sin^2\theta}{\sin\theta} = \frac{1}{\sin\theta}$

6. **(C)** Draw and label a 5-12-13 right triangle. The smallest angle is B, the angle opposite the smallest side. Since $\sin B = \frac{5}{13}$, we have $B = \sin^{-1}\left(\frac{5}{13}\right) = 22.6°$.

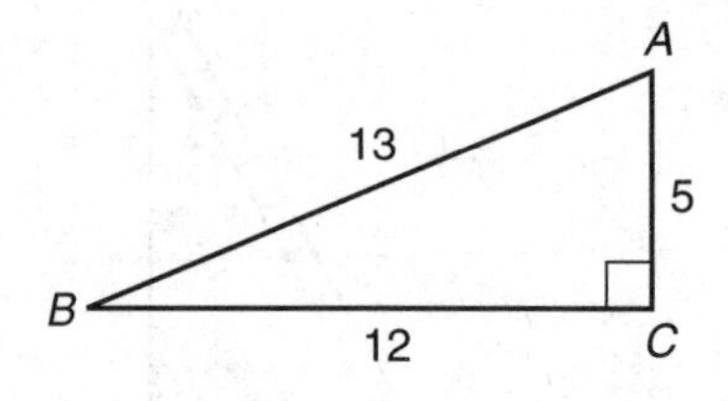

7. **(E)** Sketch isosceles right ΔABC with $AC = BC$ and $m\angle A = m\angle B = 45°$.

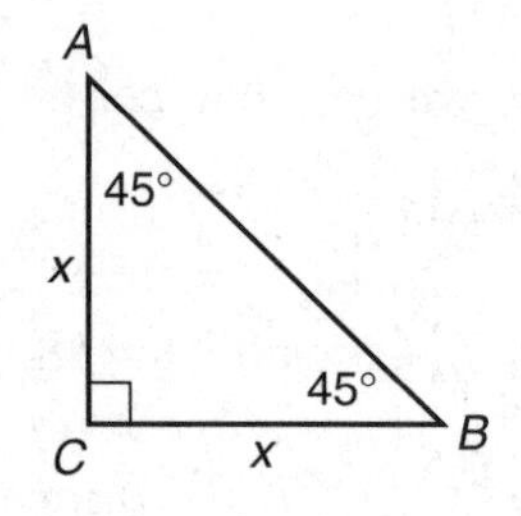

Then $\tan A = \frac{x}{x} = 1$ (I is true). $\sin A$, $\cos A$, $\sin B$, and $\cos B$ are all equal to $\frac{x}{AB}$ (II and III are true). Note that $\sin 45° = \cos 45° = 0.707$.

8. **(D)** $(3 \sin 2\theta)(4 \sin 2\theta) = 12 \sin^2 2\theta$
$(2 \cos 2\theta)(6 \cos 2\theta) = 12 \cos^2 2\theta$

$$\text{So, } (3 \sin 2\theta)(4 \sin 2\theta) + (2 \cos 2\theta)(6 \cos 2\theta) =$$
$$12 \sin^2 2\theta + 12 \cos^2 2\theta = 12(\sin^2 2\theta + \cos^2 2\theta) = 12(1) = 12$$

9. **(D)** If x represents the height of the kite, then

$$\sin 70° = \frac{x}{100} \Rightarrow x = 100 \sin 70° = 100(0.9396) = 93.96 \approx 94.$$

10. **(A)** Since $a^4 - b^4 = (a^2 - b^2)(a^2 + b^2)$:

$$\begin{aligned} \sin^4\theta - \cos^4\theta &= (\sin^2\theta - \cos^2\theta)(\sin^2\theta + \cos^2\theta) \\ &= (\sin^2\theta - \cos^2\theta)(1) = \sin^2\theta - \cos^2\theta \end{aligned}$$

$$\begin{aligned} \frac{\sin^4\theta - \cos^4\theta}{\cos^2\theta - \sin^2\theta} &= \frac{\sin^2\theta - \cos^2\theta}{\cos^2\theta - \sin^2\theta} \\ &= \frac{\sin^2\theta - \cos^2\theta}{-\left(\sin^2\theta - \cos^2\theta\right)} = -1 \end{aligned}$$

11. **(E)** Draw and label a diagram.
Then:

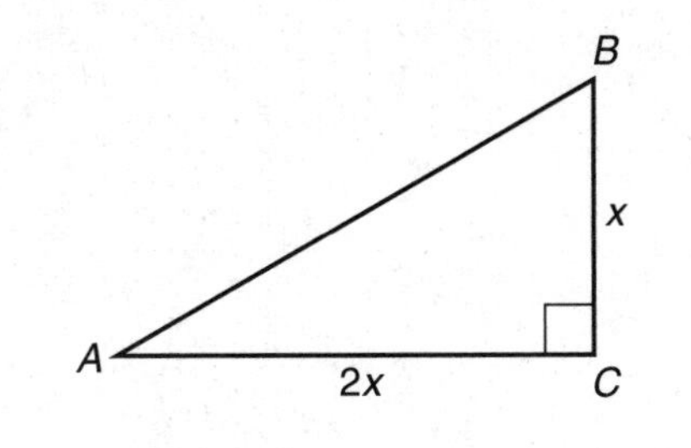

$$\tan B = \frac{2x}{x} = 2 \Rightarrow m\angle B = \tan^{-1}(2) = 63.435°$$

$$\tan A = \frac{x}{2x} = \frac{1}{2} \Rightarrow m\angle A = \tan^{-1}(0.5) = 26.565°$$

$$\text{So } \frac{m\angle B}{m\angle A} = \frac{63.435}{26.565} = 2.39$$

12. **(D)** Draw a diagram and label it.

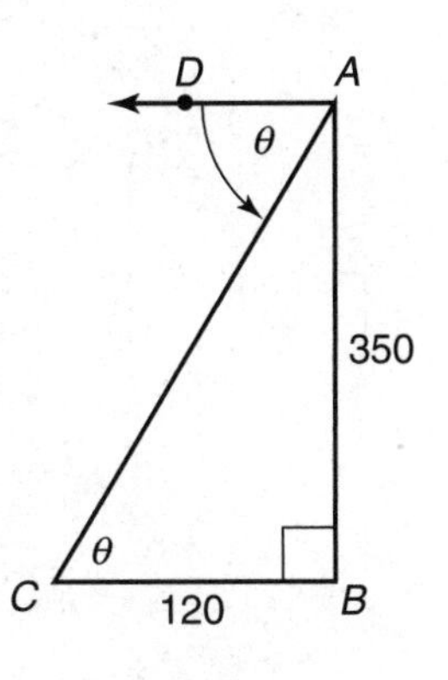

The angle of depression, $\angle DAC$, is the angle between the line of sight and the horizontal and by KEY FACT G6 is congruent to $\angle C$. Since $\tan C = \frac{350}{120}$, $m\angle C = \tan^{-1}\left(\frac{350}{120}\right) = 71.1°$.

FUNCTIONS

Functions and Their Graphs

- Relations
- Functions
- Combining Functions
- Composition of Functions
- Inverse Functions
- Exercises
- Answers Explained

The concept of a function is one of the most fundamental notions in mathematics. In this chapter, you will see two equivalent definitions of *function* and then review the most important facts you need to know about functions for the Math 1 test.

RELATIONS

Since a function is a special type of relation, we will first review the definition of a relation. A ***relation*** is a set of ordered pairs. The first and second coordinates of the ordered pairs can be anything whatsoever, although on the Math 1 test they are almost always numbers. The number of ordered pairs in a relation can be finite or infinite. Each of the following sets is a relation:

$R_1 = \{(1, 0), (1, 1), (3, 2)\}$

$R_2 = \{(0, 1), (1, 1), (2, 1)\} = \{(x, y) | \ x = 0, 1, \text{ or } 2 \text{ and } y = 1\}$

$R_3 = \{(0, 0), (1, 1), (2, 4)\} = \{(x, y) | \ y = x^2$
and $x = 0, 1, \text{ or } 2\}$

$R_4 = \{(x, y) | \ y = x^2 \text{ and } x \text{ is an integer}\}$

$R_5 = \{(x, y) | \ y = x^2 \text{ and } x \geq 0\}$

$R_6 = \{(x, y) | \ y = x^2\}$

$R_7 = \{(x, y) | \ x = y^2\}$

$R_8 = \{(x, y) | \ x^2 + y^2 = 25\}$

$R_9 = \{(x, y) | \ x$ is a state in the United States and y is the capital of $x\}$

$R_{10} = \{(x, y) | \ x$ is a word in the English language and y is the number of letters in $x\}$

R_1, R_2, R_3, R_9, and R_{10} are finite sets with 3, 3, 3, 50, and approximately 400,000 elements, respectively. R_4, R_5, R_6, R_7, and R_8 are all infinite sets.

Note that (2, 4) is in R_3, R_4, R_5, and R_6; (−1, 1) is in R_4 and R_6; $(\sqrt{2}, 2)$ is in R_5 and R_6; (4, 2) and (4, −2) are in R_7.

When the first and second coordinates of the ordered pairs are numbers, you can graph the pairs as explained in Chapter 13 on coordinate geometry. Here are the graphs of R_1, R_2, R_3, R_4, R_5, R_6, R_7, and R_8.

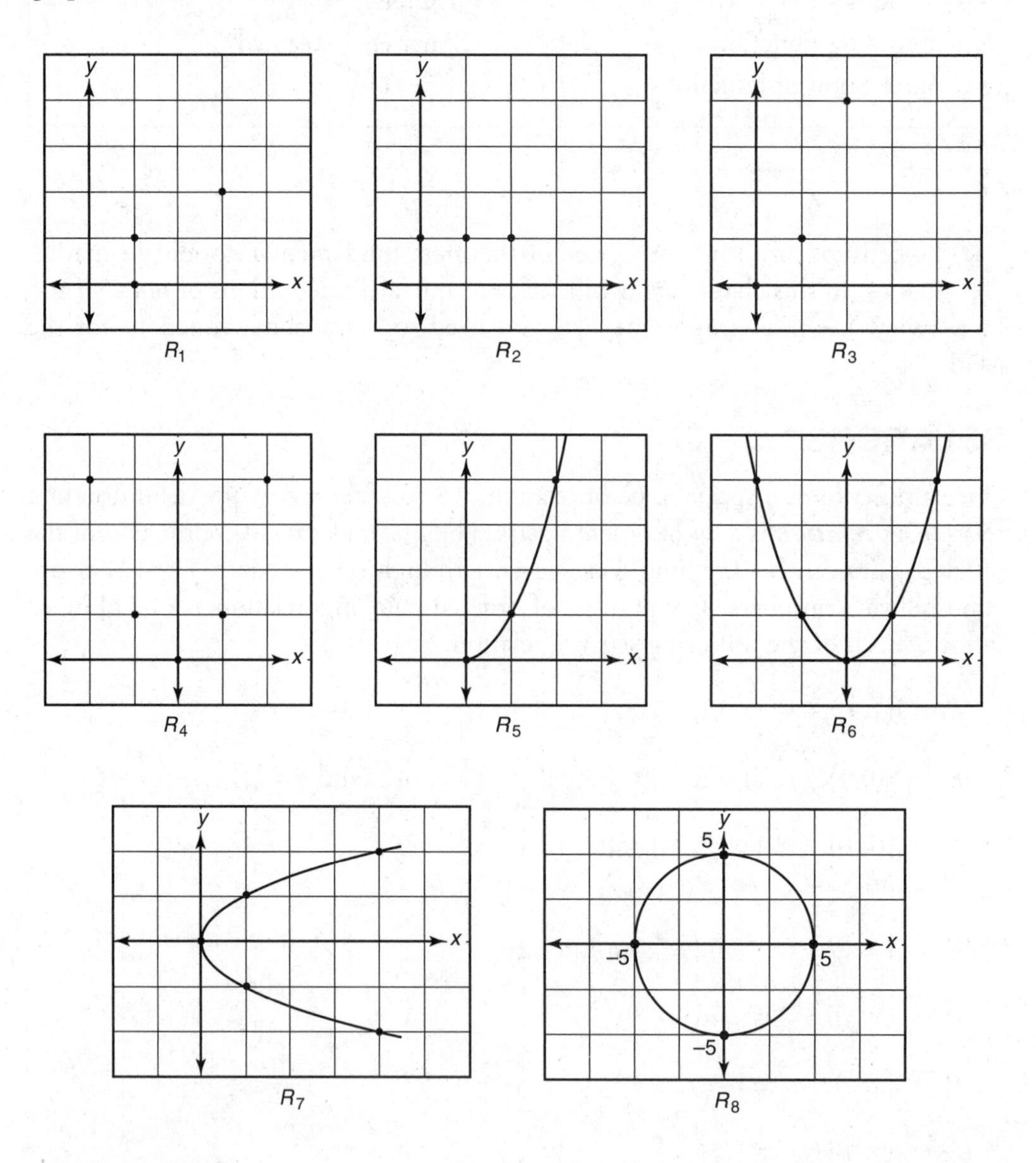

FUNCTIONS

A relation is a ***function*** if no two of the ordered pairs in the relation have the same first coordinate and different second coordinates. Some, but not all, relations are functions. R_1 is not a function because two of its ordered pairs, (1, 1) and (1, 0), have the same first coordinates but different second coordinates. Similarly, R_7 and R_8 are not functions: (1, 1) and (1, −1) are both in R_7 and (3, 4) and (3, −4) are both in R_8. All of the other relations in the list above *are* functions.

EXAMPLE 1: If *a* and *b* are real numbers, and if

{(1, 2), (3, 4), (*a*, 6), (7, *b*)} is a function, what do we know about *a* and *b*?

Since a function cannot have two different ordered pairs with the same first coordinate, *a* cannot be 1 or 3 and *a* can be 7 only if *b* is 6. Other than that, there are no restrictions on *a*. If *a* is 7, *b* must be 6; if *a* is not 7, then *b* can be any number whatsoever.

If you look at the graphs of R_1, R_7, and R_8, the relations that are not functions, you can see that in each of them it is possible to draw at least one vertical line that intersects the graph in more than one point.

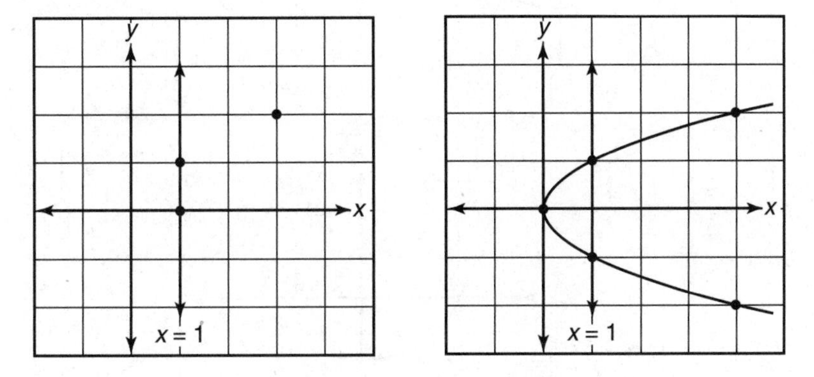

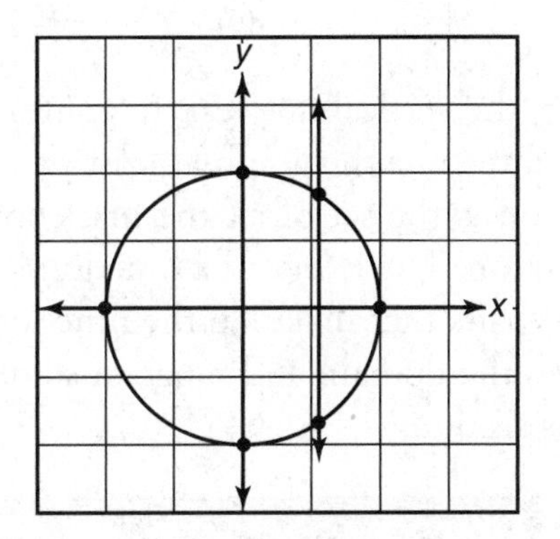

Key Fact N1

VERTICAL LINE TEST

A relation, *R*, is a function if and only if *no* vertical line can be drawn that intersects the graph of *R* more than once.

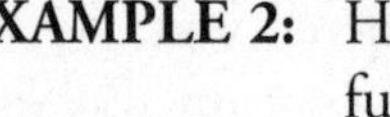

EXAMPLE 2: How many of the following graphs are graphs of functions?

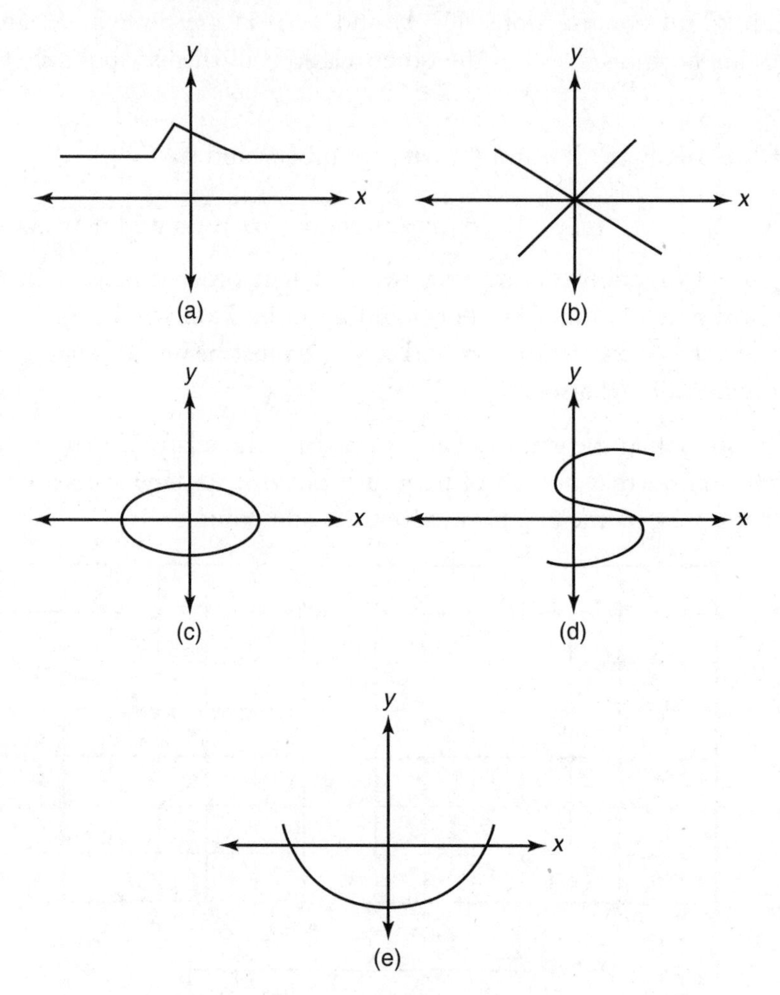

Graphs (a) and (e) satisfy the vertical line test. In contrast, in each of graphs (b), (c), and (d), the line $x = 1$ intersects the graph at least twice.

The ***domain*** of a function is the set of all the first components of the ordered pairs that make up the function. The ***range*** of a function is the set of all the second components of the ordered pairs that make up the function.

The following chart gives the domain and range of some of the functions in the list at the beginning of this section.

Function	Domain	Range
R_2	{0, 1, 2}	{1}
R_3	{0, 1, 2}	{0, 1, 4}
R_4	The set of all integers	The set of all perfect squares
R_6	The set of all real numbers	The set of all numbers ≥ 0
R_9	The set of all states in the United States	The set of all state capitals

A *function* can also be defined as a rule that assigns to each element in one set an element in a second set.

For example, you can think of R_6 as the rule that assigns to each real number its square and R_9 as the rule that assigns to each state its capital. When we think of a function as a rule rather than as a set of ordered pairs, we often represent it by the letter f, although any letter could be used. In discussing three functions, we would use f, g, and h rather than f_1, f_2, and f_3.

When we treat a function as a rule, the domain is the set of all numbers to which the function assigns a second number, and the range is the set of those second numbers. For example, if f is the squaring function, since f assigns 25 to -5, -5 is in the domain of f and 25 is in the range of f.

As you have just seen, there is a function that takes any real number and squares it. In fact, your calculator has a button that performs that function. If you enter 123 on your calculator and then press the $\boxed{x^2}$ button, your calculator instantly assigns to 123 its square, 15,129, and displays it. Again, you can think of the squaring function as the set of ordered pairs $R_6 = \{(x, y) | y = x^2\}$ or as the rule, f, that assigns to each real number x its square, x^2, in which case you write $f(x) = x^2$. Note that

- $f(2) = 2^2 = 4$
- $(2, 4)$ is in R_6
- $(2, 4)$ a point on the graph of $y = f(x) = x^2$

It is critically important that you understand function notation. When we write $f(x) = x^2 + 3$, we mean that f takes *anything* and assigns to it *the square of that thing plus 3*:

- $f(\textit{anything}) = (\textit{that thing})^2 + 3$
- $f(10) = 10^2 + 3 = 103$
- $f(0) = 0^2 + 3 = 3$
- $f(a) = a^2 + 3$
- $f(x + 1) = (x + 1)^2 + 3 = x^2 + 2x + 4$
- $f(x^2) = (x^2)^2 + 3 = x^4 + 3$
- $f(2^x) = (2^x)^2 + 3 = 2^{2x} + 3$
- $f(f(x)) = (f(x))^2 + 3 = (x^2 + 3)^2 + 3 = (x^4 + 6x^2 + 9) + 3 = x^4 + 6x^2 + 12$

Typically, there are about six questions on functions on the Math 1 test. Two of them occur among the first 25 questions and are relatively easy. The first, and easiest, will probably ask you to evaluate a function for one particular value of x, as in Example 3 on page 230. A slightly more difficult question might involve algebra, as in Example 4. The four other questions on functions will deal with more advanced topics that will be discussed later in this chapter.

EXAMPLE 3: If $f(x) = -2x^2 - 2x - 2$, what is $f(-3)$?

$$f(-3) = -2(-3)^2 - 2(-3) - 2 = -2(9) + 6 - 2$$
$$= -18 + 6 - 2 = -14$$

EXAMPLE 4: If $f(x) = x^2 + 3$, what is $f(a + b) - f(a - b)$?

$$f(a+b) - f(a-b) = ((a+b)^2 + 3) - ((a-b)^2 + 3)$$
$$= (a^2 + 2ab + b^2 + 3) - (a^2 - 2ab + b^2 + 3)$$
$$= a^2 + 2ab + b^2 + 3 - a^2 + 2ab - b^2 - 3$$
$$= 4ab$$

As you have already seen, since every real number can be squared, the domain of R_6, the squaring function, is the set of all real numbers. The domain of a function, however, can always be restricted by limiting the permissible x values. For example, the domain of $R_3 = \{(x, y) | y = x^2 \text{ and } x = 0, 1, 2\}$ is just the set $\{0, 1, 2\}$, and the domain of $R_5 = \{(x, y) | y = x^2 \text{ and } x \geq 0\}$ is the set of non-negative numbers. Note that limiting the domain may or may not affect the range. The range of $R_3 = \{0, 1, 4\}$, but the range of R_5 is the same as the range of R_6: $\{y \mid y \geq 0\}$. In the absence of any stated restriction, the domain of a function, $y = f(x)$, described by a rule is *the set of all real numbers except those for which f(x) is undefined.* On the Math 1 test, there are only two ways for a function to be undefined for some number a: when trying to evaluate $f(a)$ would involve (i) dividing by 0 or (ii) taking the square root, fourth root, or any even root of a negative number.

Key Fact N2

Unless specifically restricted, the domain of a function, $y = f(x)$, consists of all real numbers x except those for which $f(x)$ is undefined. A real number a is excluded from the domain of $y = f(x)$ if evaluating $f(a)$ would require

- **Dividing by zero.**
- **Taking a square root (or any even root) of a negative number.**

EXAMPLE 5: The domain of $f(x) = \dfrac{\sqrt{4 - x}}{x - 2}$ consists of all real numbers *except* 2, which would cause the denominator to be 0, and numbers greater than 4, which would cause the expression under the square root sign to be negative.

The domain of $f(x)$ is $\{x | x \leq 4 \text{ and } x \neq 2\}$.

In general, determining the range of a function is much harder than determining the domain of the function. Fortunately, questions on the Math 1 test that deal with range are straightforward and present no serious problems.

If you think of a function as a machine, the domain consists of those numbers that the machine will accept as inputs (the x values). The range consists of those numbers that come out of the machine (the y values).

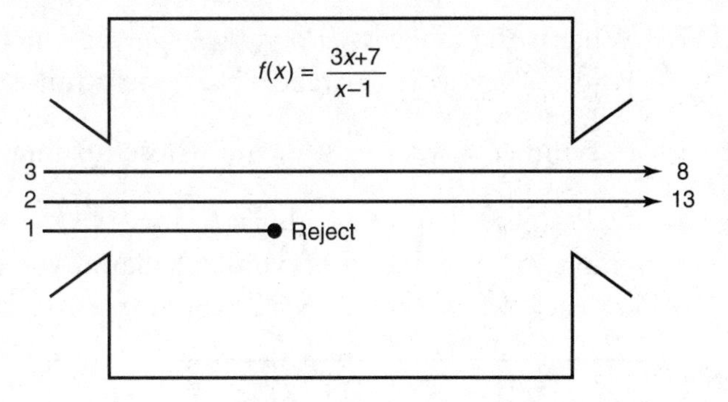

The machine above rejects 1 as an input since $f(1)$ is undefined. No other number, however, causes a problem. The domain of f is the set of all real numbers not equal to 1. From the diagram you can see that since $f(2) = 13$ and $f(3) = 8$, 13 and 8 are both in the range. Obviously, lots of other numbers are too. Since 100, π, and -83 are all in the domain, when 100 and π and -83 go in, some numbers come out. However, does $\frac{1}{2}$ ever come out? How about $-\frac{1}{2}$, $\sqrt{17}$, or 23? The answer is certainly not obvious.

In fact, all of those numbers *are* in the range. The only number that is not in the range is 3. To verify that 3 is not in the range, note that if 3 were in the range, there would be some number a such that $f(a) = 3$. However,

$$f(a) = 3 \Rightarrow \frac{3a+7}{a-1} = 3 \Rightarrow 3a+7 = 3(a-1) = 3a-3$$

However, if $3a+7 = 3a-3$, then, subtracting $3a$ from each side, we get that $7 = -3$, which clearly is wrong!

If you graph $y = \frac{3x+7}{x-1}$, you will see that the graph does not cross the horizontal line $y = 3$.

Key Fact N3

For any real number b, b is in the range of $f(x)$ if and only if the horizontal line $y = b$ intersects the graph of $y = f(x)$.

Graphing is actually a good way to determine the range, but on the Math 1 test, most range questions are easy enough to answer without graphing.

EXAMPLE 6: What is the range of $f(x) = 3x + 5$? The range of any linear function whose graph is not horizontal is the set of all real numbers. If b is any real number then

$$f\left(\frac{b-5}{3}\right) = \cancel{3}\left(\frac{b-5}{\cancel{3}}\right) + 5 = b.$$

EXAMPLE 7: What is the range of $f(x) = x^2 - 4x + 5$? The graph of $y = x^2 - 4x + 5$ is a parabola that opens upward, whose axis of symmetry is $x = \frac{4}{2} = 2$, and whose turning point is (2,1). (See Chapter 13.) So the range is $\{y \mid y \geq 1\}$. If you graph $y = x^2 - 4x + 5$ on your calculator, you can see that for each $b \geq 1$, the line $y = b$ would intersect the parabola.

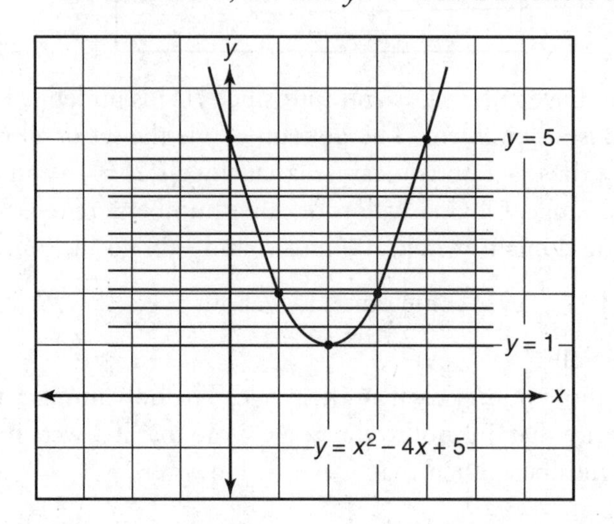

As you have already seen, the ***graph*** of a function, f, is the set of all ordered pairs (x, y) such that $y = f(x)$. On the Math 1 test, you won't have to draw graphs, but you may have to recognize and work with them.

Sometimes you are given a graph and asked which of the five answer choices is the equation of that graph.

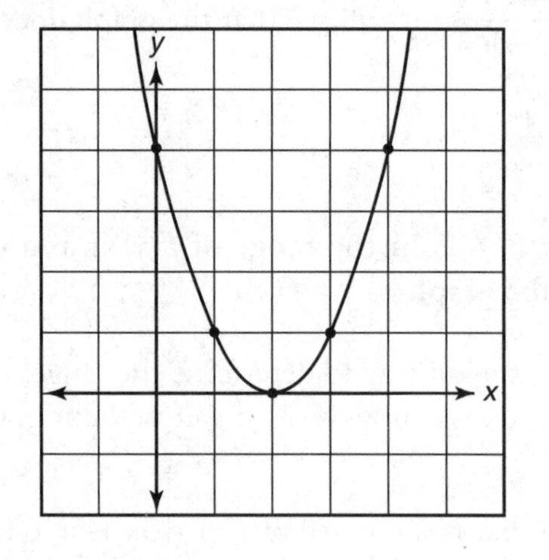

EXAMPLE 8: Which of the following could be the equation of the graph shown in the figure above?

(A) $y = -2x + 4$
(B) $y = 2x + 4$
(C) $y = x^2$
(D) $y = 2x^2 - 4$
(E) $y = x^2 - 4x + 4$

Solution. Since the graph passes through (2, 0), $x = 2$ and $y = 0$ must satisfy the equation. Test each of the five choices.

(A) Does $0 = -2(2) + 4$? Yes
(B) Does $0 = 2(2) + 4$? No
(C) Does $0 = 2^2$? No
(D) Does $0 = 2(2^2) - 4$? No
(E) Does $0 = 2^2 - 4(2) + 4$? Yes

So the answer is A or E. A, however, is the equation of a line, so the answer must be E. If you didn't notice that, then to break the tie, you would just try another point on the graph, say (0, 4). Unfortunately, in both equations A and E, when $x = 0$, $y = 4$; so testing (0, 4) didn't help. Try one more point: (1, 1). Now choice A does *not* work $1 \neq -2(1) + 4$, but choice E *does* work: $1 = 1^2 - 4(1) + 4$. The answer is E.

Another way the Math 1 test can test your understanding of the graphs of functions is to give you an equation and ask which of five graphs is the graph of the given equation.

EXAMPLE 9: Which of the following is the graph of $y = |x^2 - 4x + 2|$?

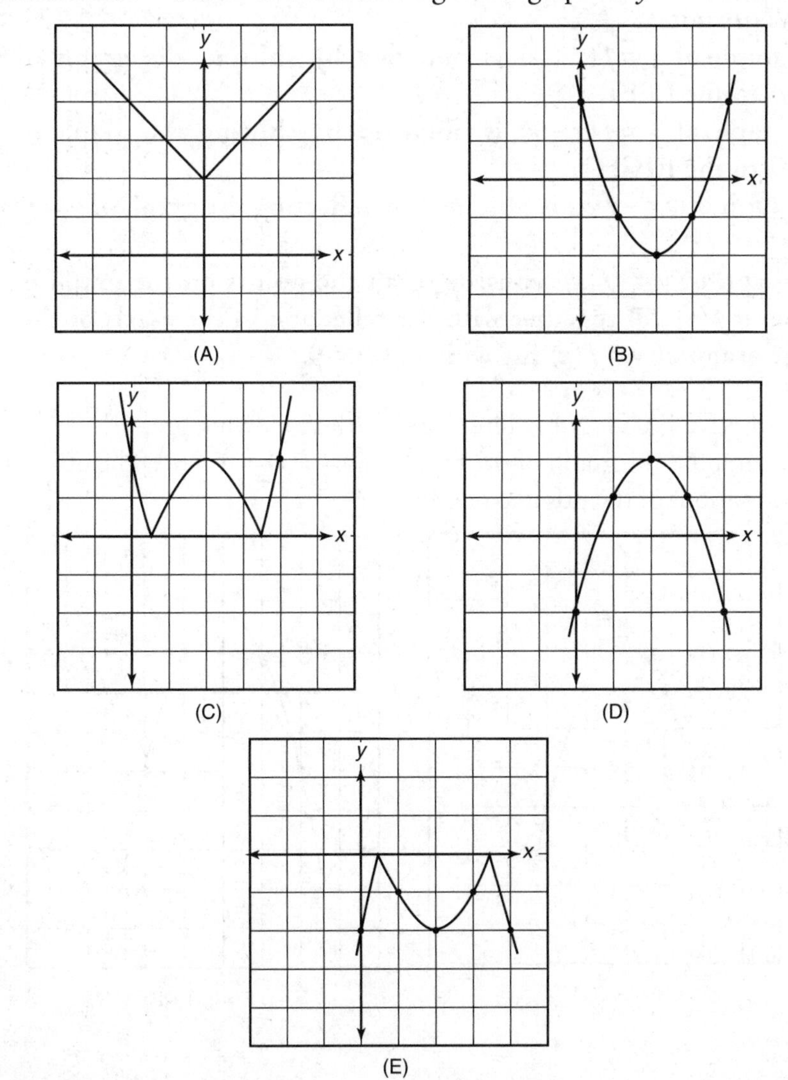

If you have a graphing calculator (and if you know how to enter absolute values), you can enter the equation and graph it. You should immediately be able to tell which of the answer choices is correct.

Finding the solution without using a calculator is just as easy. Since the absolute value of a number can never be negative, no point on the graph of $y = |x^2 - 4x + 2|$ can be below the *x*-axis. Eliminate choices B, D, and E. Then just test any number. For example, if $x = 1$, then $y = |1 - 4 + 2| = |-1| = 1$, so (1, 1) must be on the graph. (1, 1) *is* on graph C but *not* on graph A.

On the Math 1 test, there may be one question that shows you a graph and asks you which of five other graphs is related to the original one in a certain way. To answer such a question, you can either test points or use the six facts listed in the following KEY FACT.

Key Fact N4

If $f(x)$ is a function and r is a positive number:

(1) The graph of $y = f(x) + r$ is obtained by shifting the graph of $y = f(x)$ UP r units.

(2) The graph of $y = f(x) - r$ is obtained by shifting the graph of $y = f(x)$ DOWN r units.

(3) The graph of $y = f(x + r)$ is obtained by shifting the graph of $y = f(x)$ r units to the LEFT.

(4) The graph of $y = f(x - r)$ is obtained by shifting the graph of $y = f(x)$ r units to the RIGHT.

(5) The graph of $y = -f(x)$ is obtained by reflecting the graph of $y = f(x)$ in the x-axis.

(6) The graph of $y = |f(x)|$ consists of all the points on the graph of $y = f(x)$ whenever $f(x) \geq 0$ together with the reflection in the x-axis of those points on the graph of $y = f(x)$ for which $f(x) < 0$.

Each part of KEY FACT N4 is illustrated in the following graphs.

The first graph is the graph of the function $y = f(x) = 4 - x^2$. The other six graphs are transformations of the original graph.

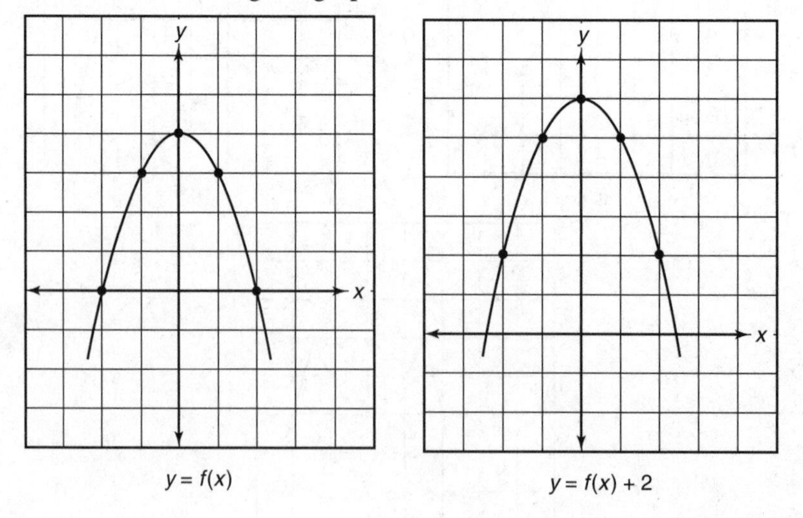

$y = f(x)$ $\qquad$ $y = f(x) + 2$

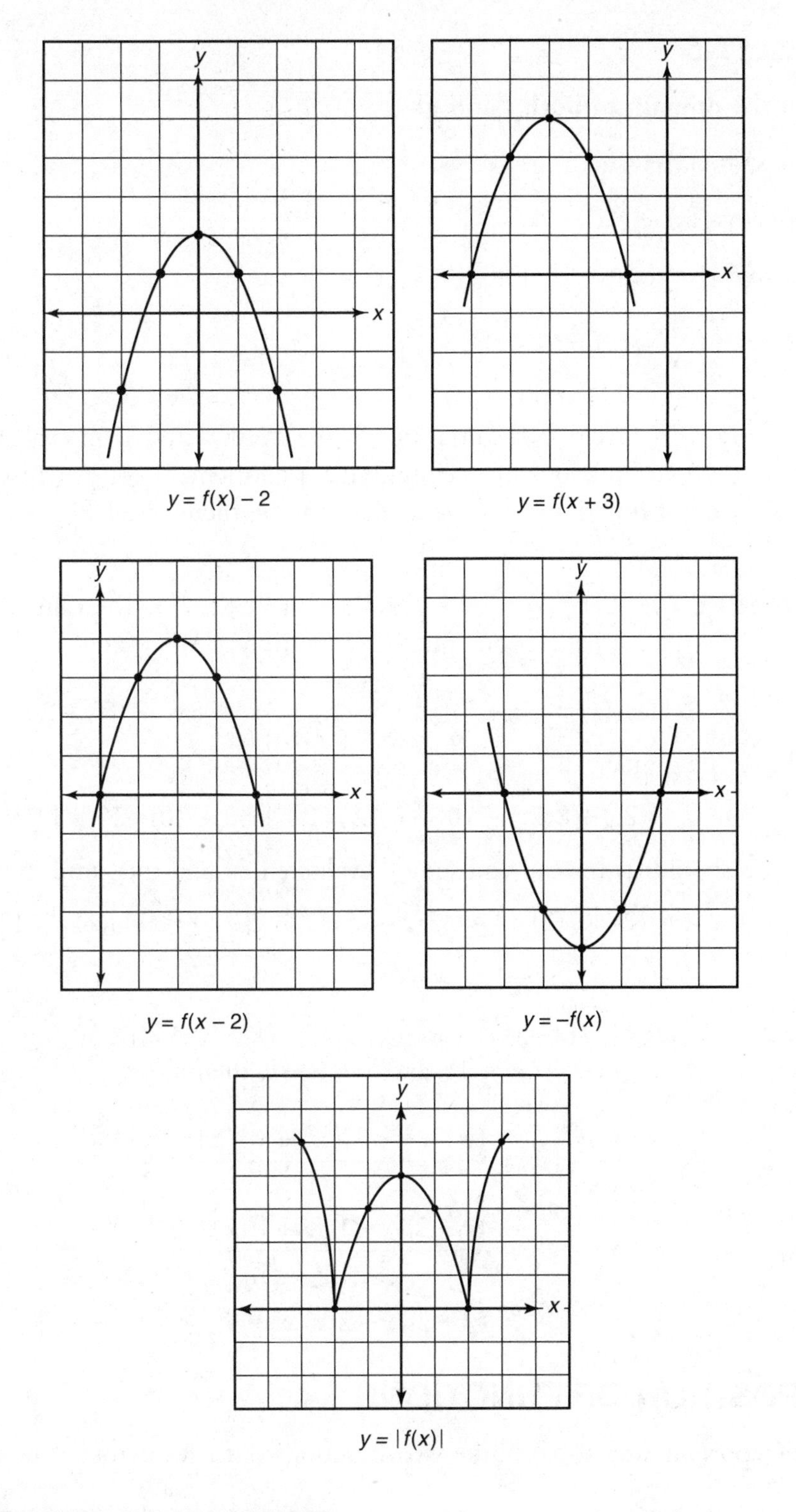

$y = f(x) - 2$ $y = f(x + 3)$

$y = f(x - 2)$ $y = -f(x)$

$y = |f(x)|$

COMBINING FUNCTIONS

If f and g are two functions with overlapping—possibly equal—domains, then for all numbers x that are in both domains, it is possible to define the ***sum***, ***difference***, ***product***, and ***quotient*** of f and g.

Key Fact N5

If x is in the domain of both f and g:

- $(f+g)(x)=f(x)+g(x)$
- $(f-g)(x)=f(x)-g(x)$
- $(fg)(x)=f(x)\cdot g(x)$
- $\left(\frac{f}{g}\right)(x)=\frac{f(x)}{g(x)},\ g(x)\neq 0$

The only way KEY FACT N5 comes up on the Math 1 test is in evaluating the combination of two functions or in simplifying a quotient. If you need to evaluate the combination of functions, don't actually combine them; use KEY FACT N5.

EXAMPLE 10: Let $f(x)=3x+5$ and $g(x)=x^2+x-5$. To evaluate $(fg)(2)$, you could first multiply $f(x)\cdot g(x)$:

$$f(x)g(x)=(3x+5)(x^2+x-5)=3x^3+8x^2-10x-25$$

and then plug in $x=2$:

$$3(2^3)+8(2^2)-10(2)-25=24+32-20-25=11$$

but you shouldn't. Instead you should evaluate first and then multiply:

$$(fg)(2)=f(2)g(2)=(3(2)+5)(2^2+2-5)=(11)(1)=11.$$

EXAMPLE 11: If $f(x)=(x^2+x-2)$, $g(x)=(x^2-x-2)$, $h(x)=(x^2-1)$, and $x\neq 1, -1$, then:

$$\left(\frac{fg}{h}\right)(x)=\frac{(x^2+x-2)(x^2-x-2)}{x^2-1}$$
$$=\frac{(x+2)\cancel{(x-1)}\,\cancel{(x+1)}(x-2)}{\cancel{(x+1)}\,\cancel{(x-1)}}$$
$$=(x+2)(x-2)=x^2-4.$$

COMPOSITION OF FUNCTIONS

A more important way to combine two functions is to form their ***composition***: $f\circ g$.

The definition of composition is given in KEY FACT N6.

Key Fact N6

The composition of two functions f and g is $(f\circ g)(x)=f(g(x))$.

The composition of f and g can be formed only if some of the numbers in the range of g are in the domain of f.

EXAMPLE 12: If $f(x)=\sqrt{x}$ and $g(x)=-x^2-2$, then $(f\circ g)(x)$ does not exist. For every real number x, $g(x)$ is negative, but no negative number is in the domain of f.

EXAMPLE 13: Let $f(x)=2x+3$ and $g(x)=x^2-1$.

To find $(f\circ g)(3)$ you have two choices:

(1) You can determine $f(g(x))$ and then plug in $x=3$:

$$(f\circ g)(x)=f(g(x))=2(g(x))+3=2(x^2-1)+3=2x^2-2+3=2x^2+1$$

$$\text{Then } (f\circ g)(3)=2(3)^2+1=2(9)+1=19$$

or

(2) You can calculate $(f\circ g)(3)$ directly:

$$(f\circ g)(3)=f(g(3))=f(8)=2(8)+3=19$$

EXAMPLE 14: Let f and g be the same functions as in Example 13: $f(x)=2x+3$ and $g(x)=x^2-1$. Then:

$$\begin{aligned}(g\circ f)(x)&=g(f(x))=g(2x+3)=(2x+3)^2-1\\&=(4x^2+12x+9)-1=4x^2+12x+8\\&=4(x^2+3x+2)=4(x+1)(x+2)\end{aligned}$$

$$\text{So } (g\circ f)(3)=4(3+1)(3+2)=4(4)(5)=80$$

Of course, $(g\circ f)(3)$ could have been calculated directly:

$$(g\circ f)(3)=g(f(3))=g(9)=9^2-1=80$$

Notice that, in general, $(f\circ g)\neq(g\circ f)$. In Examples 13 and 14, you saw:

$$(f\circ g)(3)=19 \text{ and } (g\circ f)(3)=80$$

It is possible, however, to have functions f and g for which $(f\circ g)=(g\circ f)$.

EXAMPLE 15: Let $f(x)=3x-2$ and $g(x)=\frac{x+2}{3}$. Then:

$$(f\circ g)(x)=f(g(x))=f\left(\frac{x+2}{3}\right)=3\left(\frac{x+2}{3}\right)-2=x$$

$$(g\circ f)(x)=g(f(x))=g(3x-2)=\frac{(3x-2)+2}{3}=x$$

INVERSE FUNCTIONS

If f and g are functions such that

(1) for every x in the domain of g, $f(g(x)) = x$ and
(2) for every x in the domain of f, $g(f(x)) = x$,

then we say that g is the ***inverse*** of f and write $g = f^{-1}$, which is read "f inverse." It is also true that f is the inverse of g: $f = g^{-1}$.

Key Fact N7

If for some function f, f^{-1} exists, then

$$f(f^{-1}(x)) = x \text{ and } f^{-1}(f(x)) = x.$$

The inverse, f^{-1}, of a function, f, undoes what f does. In Example 15, f multiplies a number by 3 and then subtracts 2 from it; g, which is f^{-1}, adds 2 to a number and then divides the result by 3.

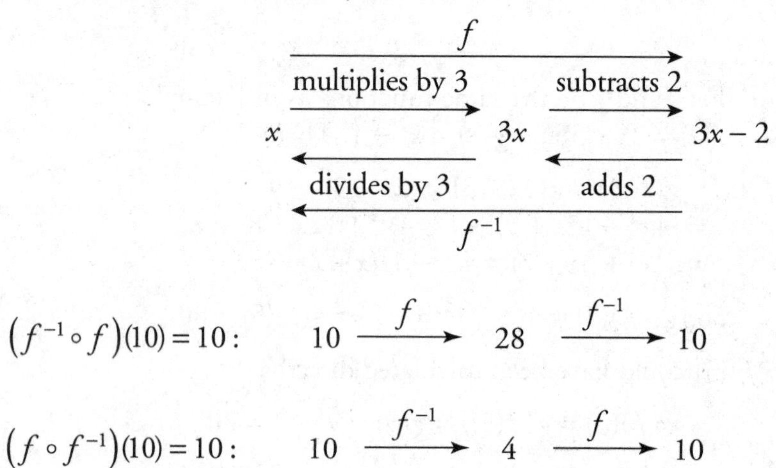

Not every function has an inverse, but many do. On the Math 1 test, you may be asked to find the inverse of a particular function. The procedure to do this is given in KEY FACT N8.

Key Fact N8

If f is a function of x, to find f^{-1}, first write $y = f(x)$. Then interchange x and y and solve for y.

EXAMPLE 16: If $f(x) = 3x - 2$, what is $f^{-1}(x)$?

Write $y = 3x - 2$. Then switch the x and y: $x = 3y - 2$. Now solve for y:

$$x = 3y - 2 \Rightarrow x + 2 = 3y \Rightarrow y = \frac{x+2}{3}$$

So $f^{-1}(x) = \frac{x+2}{3}$. Note that f^{-1}, which is function g in Example 15, simply undoes what f does: f multiplies a number by 3 and then subtracts 2; f^{-1} adds 2 to a number and then divides the result by 3.

Exercises

1. If {(1, 5), (2, 13), (x, y)} is a function, which of the following must be true?

   I. $x > 2$
   II. $y \neq 13$
   III. $x \neq 1$

   (A) None
   (B) I only
   (C) II only
   (D) III only
   (E) I and II only

2. If $f(x) = -3x^3 - 3\sqrt[3]{x}$, what is the value of $f(-8)$?

   (A) 78
   (B) 1,530
   (C) 1,542
   (D) 13,818
   (E) 13,830

3. What is the domain of the function $f(x) = \dfrac{(x-2)^3}{(x+2)^2}$?

   (A) All real numbers
   (B) $\{x \mid -2 < x < 2\}$
   (C) $\{x \mid x \neq 2\}$
   (D) $\{x \mid x \neq -2\}$
   (E) $\{x \mid x < -2 \text{ or } x > 2\}$

4. If $f(x) = x + \dfrac{1}{x}$, what is $f(f(x))$?

   (A) 1
   (B) $x + \dfrac{1}{x}$
   (C) $x^2 + \dfrac{1}{x^2}$
   (D) $\dfrac{x^2+1}{x}$
   (E) $\dfrac{x^2+1}{x} + \dfrac{x}{x^2+1}$

5. What is the range of the function $f(x) = \sqrt{x^2 - 4}$?

   (A) All real numbers
   (B) All real numbers greater than or equal to 0
   (C) All real numbers greater than or equal to 2
   (D) All real numbers greater than or equal to 2 or less than or equal to −2
   (E) All real numbers between −2 and 2, inclusive.

6. Which of the following graphs is not the graph of a function?

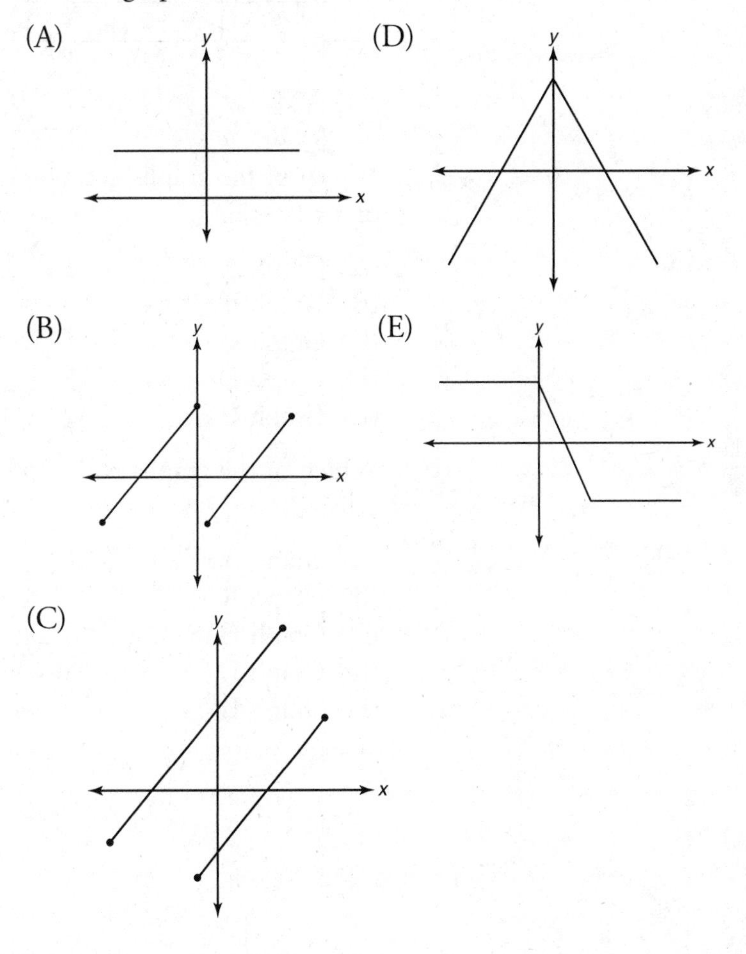

7. Which of the following is not the equation of a function?

(A) $y = 3$
(B) $x = 3$
(C) $y = x$
(D) $y = |x|$
(E) $y = \frac{|x|}{x}$

<u>Questions 8 and 9</u> refer to the following graphs. The first one is the graph of $y = |x|$.

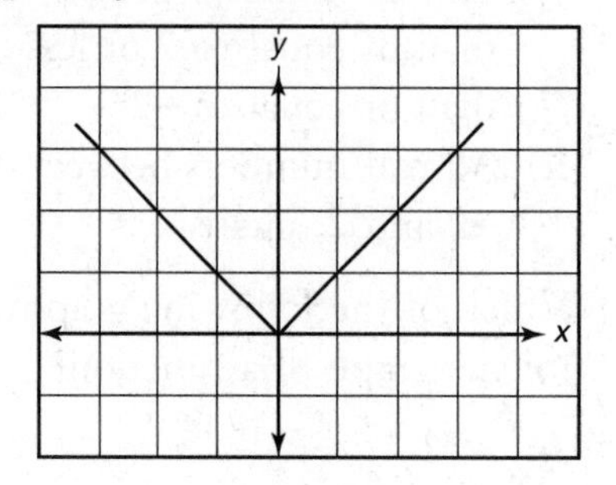

$y = f(x) = |x|$

8. Which of the graphs is the graph of $y = |x - 3|$?

(A) Graph A
(B) Graph B
(C) Graph C
(D) Graph D
(E) Graph E

9. Which of the graphs is that of $y = ||x| - 3|$?

(A) Graph A
(B) Graph B
(C) Graph C
(D) Graph D
(E) Graph E

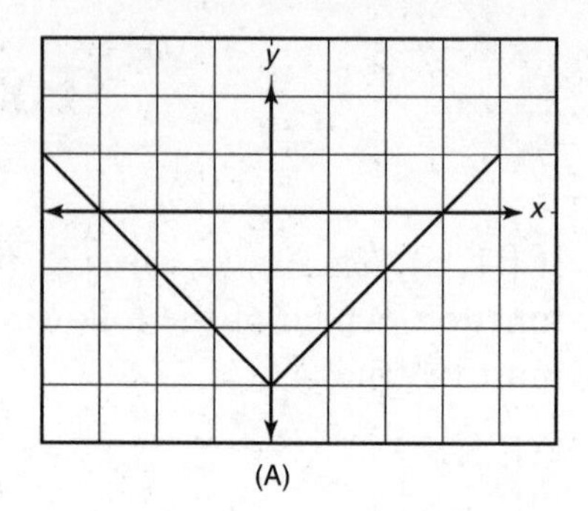

(A)

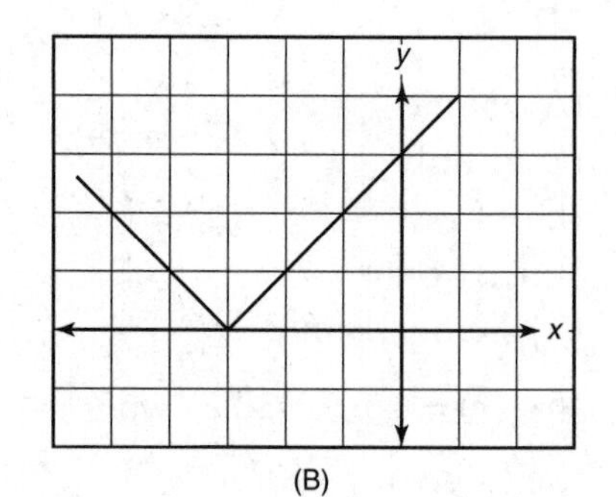

(B)

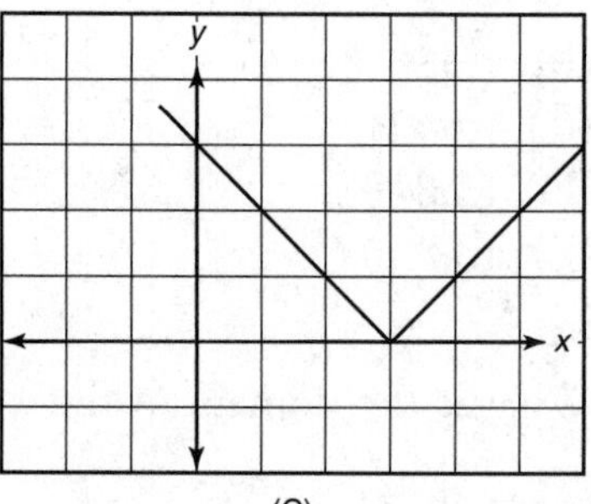

(C)

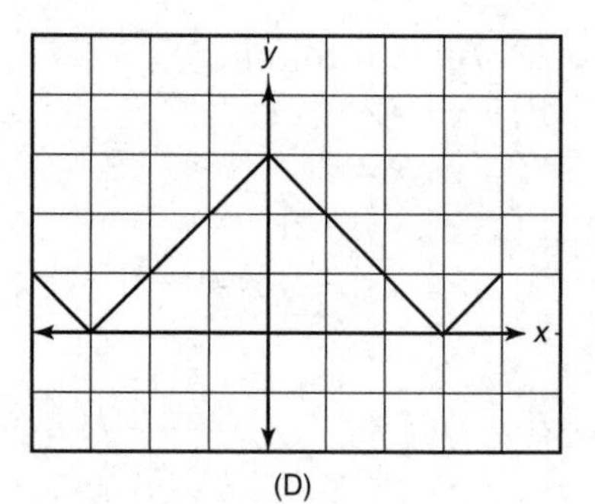

(D)

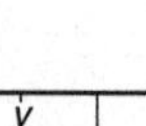

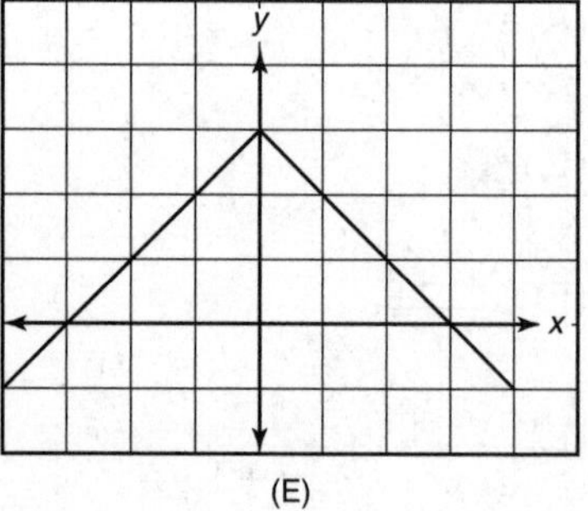

(E)

10. If $x \neq 1, 2$, $f(x) = x^2 - 1$, $g(x) = x^2 - 4$, and $h(x) = x^2 - 3x + 2$, then $\left(\frac{fg}{h}\right)(x) =$

    (A) $x^2 + 3x + 2$
    (B) $x^2 - 3x + 2$
    (C) $x^4 - 4$
    (D) $\frac{x^4 + 4}{x^2 - 3x + 2}$
    (E) $\frac{x^2 + 4}{-3x + 2}$

11. If $f(x) = x^2 - 1$ and $g(x) = x^2 + 1$, what is $(f \circ g)(x) - (g \circ f)(x)$?

    (A) -2
    (B) 0
    (C) $x^4 - 1$
    (D) $4x^2$
    (E) $4x^2 - 2$

12. If $f(x) = \frac{4 - 3x}{2}$, then $f^{-1}(x) =$

    (A) $\frac{4 - 2x}{3}$
    (B) $\frac{3x - 4}{2}$
    (C) $\frac{2x - 4}{3}$
    (D) $\frac{2 - 3x}{4}$
    (E) $\frac{3x - 2}{4}$

ANSWERS EXPLAINED

Answer Key

1. **(A)**	3. **(D)**	5. **(B)**	7. **(B)**	9. **(D)**	11. **(E)**
2. **(C)**	4. **(E)**	6. **(C)**	8. **(C)**	10. **(A)**	12. **(A)**

Solutions

Each of the problems in this set of exercises is typical of a question you could see on a Math 1 test. When you take the model tests in this book and, in particular, when you take the actual Math 1 test, if you get stuck on questions such as these, you do not have to leave them out—you can almost always answer them by using one or more of the strategies discussed in the "Tactics" chapter. The solutions given here do *not* depend on those strategies; they are the correct mathematical ones.

1. **(A)** Certainly, x can be 0 or -7 or 1.5, or any number less than 2. (I is false.) If $x = 2$, then y must be 13. So y can equal 13. (II is false.) If $y \neq 5$, then x cannot be 1; but it is perfectly all right to have $x = 1$ and $y = 5$. (III is false.) None of statements I, II, and III is true.

2. **(C)** $f(-8) = -3(-8)^3 - 3\sqrt[3]{-8}$
   $= (-3)(-512) - 3(-2)$
   $= 1{,}536 + 6 = 1{,}542$

3. **(D)** The only restriction is that the denominator of $f(x)$ cannot be 0. So $x \neq -2$.

4. **(E)** $f(f(x)) = f(x) + \dfrac{1}{f(x)} = \left(x + \dfrac{1}{x}\right) + \dfrac{1}{x + \dfrac{1}{x}} =$

$$\frac{x^2+1}{x} + \frac{1}{\dfrac{x^2+1}{x}} = \frac{x^2+1}{x} + \frac{x}{x^2+1}$$

5. **(B)** The square root of a number cannot be negative, so no negative number is in the range. If $b \geq 0$ and if $b = \sqrt{x^2 - 4}$, then $x^2 = b^2 + 4$ and $x = \sqrt{b^2 + 4}$. So every number greater than or equal to 0 is in the range.

6. **(C)** Only graph C fails the vertical line test. The line $x = 0$ crosses graph C twice.

7. **(B)** The graph of $x = 3$ is a vertical line that passes through (3, 0), (3, 1), (3, 2), . . . Since there is more than one ordered pair whose first component is 3, it is not the graph of a function. Also, the vertical line $x = 3$ intersects the line $x = 3$ more than once—infinitely many times, in fact. All the other choices are functions.

8. **(C)** By KEY FACT N4, the graph of $y = |x - 3|$ is obtained by shifting the graph of $y = |x|$ three units to the right.

9. **(D)** By KEY FACT N4, the graph of $y = |x| - 3$ is obtained by shifting the graph of $y = |x|$ three units down as shown in graph A. The graph of $y = ||x| - 3|$ consists of the reflection in the x-axis of all the points on graph A that are below the x-axis together with all the points on graph A that are on or above the x-axis. The correct graph is D.

10. **(A)** $$\left(\frac{fg}{h}\right)(x) = \frac{f(x)g(x)}{h(x)} = \frac{(x^2-1)(x^2-4)}{x^2-3x+2} =$$

$$\frac{\cancel{(x-1)}(x+1)\cancel{(x-2)}(x+2)}{\cancel{(x-2)}\,\cancel{(x-1)}} = (x+1)(x+2)$$

$$= x^2 + 3x + 2$$

11. **(E)** $$(f \circ g)(x) = f(g(x)) = f(x^2+1) = (x^2+1)^2 - 1$$
$$= (x^4 + 2x^2 + 1) - 1 = x^4 + 2x^2$$

$$(g \circ f)(x) = g(f(x)) = g(x^2 - 1) = (x^2 - 1)^2 + 1$$
$$= (x^4 - 2x^2 + 1) + 1 = x^4 - 2x^2 + 2$$

Finally,

$$(f \circ g)(x) - (g \circ f)(x)$$
$$= (x^4 + 2x^2) - (x^4 - 2x^2 + 2) = 4x^2 - 2$$

12. **(A)** Write $y = f(x) = \dfrac{4-3x}{2}$. Switch x and y, and solve for y:

$$x = \frac{4-3y}{2} \Rightarrow 2x = 4 - 3y \Rightarrow$$
$$3y = 4 - 2x \Rightarrow y = \frac{4-2x}{3}$$

STATISTICS, COUNTING, AND PROBABILITY

Basic Concepts of Statistics, Counting, and Probability

Very few questions on the Math 1 test are about statistics and probability. However, the ones that do appear are generally pretty easy if you know the appropriate definitions. Therefore, you should definitely review the material in this section.

STATISTICS

If you are analyzing a set of data, you must know four terms: ***mean*** (which on the Math 1 test is often referred to as the ***arithmetic mean***), ***median***, ***mode***, and ***range***.

The ***arithmetic mean*** of a set of numbers is what you have always called the ***average***. If A represents the average (arithmetic mean) of a set of n numbers, then:

$$A = \frac{\text{the sum of the } n \text{ numbers}}{n}$$

Key Fact O1

If A is the average (arithmetic mean) of a set of n numbers, then:

- $A = \dfrac{\text{sum}}{n}$
- $nA = \text{sum}$

When you take the Math 1 test, it is very unlikely that you will be asked to calculate the average of a list of numbers since that is just simple arithmetic. More likely, you will have to find the average of some algebraic expressions or have to apply the definition of average to a problem in geometry.

EXAMPLE 1: What is the average (arithmetic mean) of the measures of the five angles in a pentagon?

To find the average of five numbers you have to calculate their sum and divide by 5. By KEY FACT I2, the sum of the measures of the five angles in a pentagon is $(5 - 2) \times 180° = 3 \times 180° = 540°$. So the average is
$\frac{\text{sum}}{5} = \frac{540°}{5} = 108°$.

Tactic O1

Whenever you know the average, *A*, of a set of *n* numbers, multiply *A* by *n* to get their sum.

EXAMPLE 2: Lior has taken five Russian tests so far this year and his average is 88. What grade does he need on his sixth test to raise his average to 90?

Use TACTIC O1 twice. So far Lior has earned $5 \times 88 = 440$ points. If after six tests his average is to be 90, then he needs a total of $6 \times 90 = 540$ points. So on his sixth test he needs $540 - 440 = 100$ points.

EXAMPLE 3: A store has 30 employees—10 men and 20 women. The average salary of the men is $600 per week and the average salary of the women is $540 per week. What is the average weekly salary of all 30 employees?

If there were an equal number of men and women, the average would just be $570—the average of $600 and $540. However, since there are more women in the group than men, the women's salaries must be given greater weight. The 10 men earn a total of 10 × $600 = $6,000 per week, and the 20 women earn a total of 20 × $540 = $10,800 per week. So the total weekly earnings of all 30 employees is $6,000 + 10,800 = $16,800. Finally, $16,800 ÷ 30 = $560.

On the Math 1 test, the five answer choices for Example 3 would probably be $570, two amounts less than $570 (one of which would be $560, the correct answer), and two amounts greater than $570. As you saw above, the "obvious" choice of $570 is wrong. Because there are more women than men, the correct answer has to be closer to the women's figure than to the men's. Certainly, you would know that the correct answer is less than $570.

The following tactic gives the procedure for calculating a ***weighted average***.

Tactic O2

To calculate the weighted average of a set of numbers, multiply each number in the set by the number of times it appears. Then add all the products and divide by the total number of numbers in the set.

By using TACTIC O2, the solution to Example 3 above would look like this:

$$\frac{10(600)+20(540)}{10+20}=\frac{6{,}000+10{,}800}{30}=\frac{16{,}800}{30}=560$$

The ***median*** of a set of n numbers arranged in increasing order is the middle number (if n is odd) or the average of the two middle numbers (if n is even). The ***mode*** of a set of numbers is the number that occurs most often. Note that not every set of numbers has a mode. The ***range*** of a set of numbers is the difference between the largest and the smallest numbers in the set.

EXAMPLE 4: On a recent quiz, a group of 10 students earned the following scores: 2, 3, 9, 3, 5, 7, 7, 10, 7, 6. To find the median, first list the scores in increasing order: 2, 3, 3, 5, 6, 7, 7, 7, 9, 10. The median is 6.5, the average of the middle two scores. The mode is 7, the score that occurs most often in the list. The range is 8, the difference between the largest number (10) and the smallest number (2).

A Math 1 test could have one question that requires you to interpret the data presented in a table or a graph. The graph could be a simple bar graph, line graph, circle graph, or stem-and-leaf plot. Reading these graphs should cause absolutely no difficulty. The associated question would likely ask you to express some of the data as a percent of the total data; this should entail a straightforward use of percents, as reviewed in Chapter 3.

COUNTING

Occasionally, a question on the Math 1 test will begin, "How many . . . ?" This type of question is asking you to count something: the number of integers that satisfy a certain property, the number of ways there are to complete a particular task, the number of ways to rearrange a set of objects, and so on. There is often more than one way to accomplish this counting. However, on the Math 1 test, the best method is to use the counting principle.

Key Fact O2

COUNTING PRINCIPLE

If two jobs need to be completed and there are m ways to do the first job and n ways to do the second job, then there are $m \times n$ ways to do one job followed by the other. This principle can be extended to any number of jobs.

EXAMPLE 5: How many integers between 100 and 1,000 consist of three different odd digits?

You are looking for 3-digit numbers, such as 135, 571, and 975, all of whose digits are odd and all of which are different. The best way to answer this question is to use the counting principle. Think of writing a 3-digit number as three jobs that need to be done. The first job is to select one of the five odd digits (1, 3, 5, 7, 9) and use it as the digit in the hundreds place. This can be done in 5 ways. The second job is to select one of the four remaining odd digits for the tens place. This can be done in 4 ways. Finally, the third job is to select one of the three odd digits not yet used to be the digit in the ones place. This can be done in 3 ways. So, by using the counting principle, the total number of ways to write a 3-digit number with three different odd digits is $5 \times 4 \times 3 = 60$.

EXAMPLE 6: At the corporate headquarters of a large company, outside each office there is a keypad like the one in the figure below. To gain entrance to an office, an employee must punch in an access code and then press the enter button. If a code consists of the letter A or B followed by either two or three different digits, how many different access codes can be issued?

1	2	3
4	5	6
7	8	9
A	0	B
ENTER		

Again, we will use the counting principle. We have to calculate the number of two-digit codes and three-digit codes separately and then add them. For a two-digit code, there are 2 choices for the letter (A or B), 10 choices for the first digit (0 through 9), and 9 choices for the second digit (0 through 9, except for the digit already chosen). So there are $2 \times 10 \times 9 = 180$ two-digit codes. Similarly, there are $2 \times 10 \times 9 \times 8 = 1{,}440$ three-digit codes. The total number of access codes possible is $180 + 1{,}440 = 1{,}620$.

EXAMPLE 7: A school committee consists of 8 administrators and 7 teachers. In how many ways can the committee choose a chairperson, a vice-chairperson, and a secretary if the chairperson must be an administrator and the vice-chairperson must be a teacher? There are 8 ways to choose the chairperson and 7 ways to choose the vice-chairperson. The secretary can be any of the 15 committee members except the 2 already chosen. So there are 13 ways to choose the secretary. By the counting principle, there are $8 \times 7 \times 13 = 728$ ways to choose the three people.

TIP 

Combinations and permutations do *not* appear on the SAT Math 1 test. For example, you do not have to know how many ways there would be to choose 3 of the 8 administrators to serve on a subcommittee.

PROBABILITY

The ***probability*** that an ***event*** will occur is a number between 0 and 1, often written as a fraction, that indicates how likely it is that the event will happen. For example, if you spin the spinner below, there are six possible outcomes. Since each region is the same size, it is equally likely that the spinner will stop in any of the six regions. There is 1 chance in 6 that it will stop in the region marked 2. So we say that the probability of spinning a 2 is one-sixth and write $P(2) = \frac{1}{6}$. Since 2 is the only even number on the spinner, we could also say $P(\text{even}) = \frac{1}{6}$. There are 5 chances in 6 that the spinner will land in a region with an odd number in it, so $P(\text{odd}) = \frac{5}{6}$.

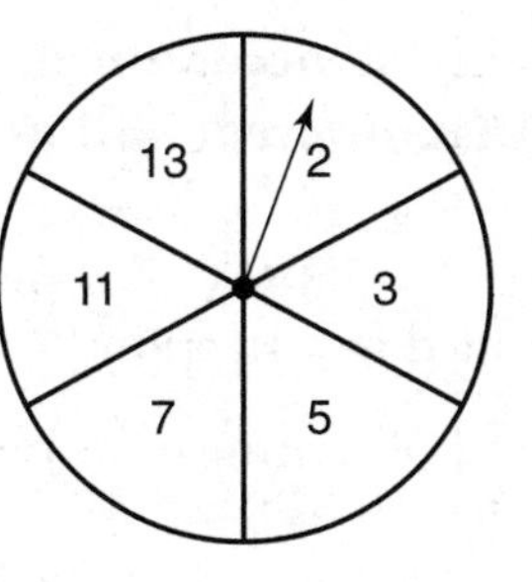

Key Fact O3

If E is any event, the probability that E will occur is given by

$$P(E) = \frac{\textbf{the number of ways event } E \textbf{ can occur}}{\textbf{the total number of possible outcomes}}$$

assuming that all of the possible outcomes are equally likely.

The examples cited below and in the next Key Fact all refer to the spinner shown above.

$$P(\text{number} > 15) = \frac{\text{the number of ways of getting a number} > 15}{\text{the total number of possible outcomes}} = \frac{0}{6} = 0$$

$$P(\text{prime number}) = \frac{\text{the number of ways of getting a prime number}}{\text{the total number of possible outcomes}} = \frac{6}{6} = 1$$

$$P(\text{number} < 4) = \frac{\text{the number of ways of getting a number} < 4}{\text{the total number of possible outcomes}} = \frac{2}{6} = \frac{1}{3}$$

Key Fact O4

Let E be an event and $P(E)$ the probability it will occur.

- **If E is *impossible* (such as getting a number greater than 15), $P(E) = 0$.**
- **If it is *certain* that E will occur (such as getting a prime number), $P(E) = 1$.**
- **In all cases $0 \le P(E) \le 1$.**
- **The probability that event E will not occur is $1 - P(E)$.**
- **If two or more events are mutually exclusive and constitute all the outcomes, the sum of their probabilities is 1. (For example, $P(\text{even}) + P(\text{odd}) = \frac{1}{6} + \frac{5}{6} = 1$.)**
- **The more likely that an event will occur, the higher its probability (the closer to 1 it is); the less likely that an event will occur, the lower its probability (the closer to 0 it is).**

Even though probability is defined as a fraction, we can write probabilities as decimals or percents, as well. Instead of writing $P(E) = \frac{1}{2}$, we can write $P(E) = .50$ or $P(E) = 50\%$.

EXAMPLE 8: An integer between 100 and 999, inclusive, is chosen at random. What is the probability that all three digits of the number are different odd numbers?

If E represents the event of picking a number with three different odd digits, then

$$P(E) = \frac{\text{number of 3-digit integers with different odd digits}}{\text{the total number of 3-digit integers}}.$$

In Example 4, we calculated that the numerator of this fraction is 60. The denominator is 900, which you could get by deleting the 99 integers from 1 to 99 from the 999 integers from 1 to 999 or by using the counting principle:

$$9 \times 10 \times 10 = 900$$

$$\text{So } P(E) = \frac{60}{900} = \frac{1}{15} = .0666$$

Key Fact O5

If an experiment is repeated two (or more) times, the probability that an event occurs and then a second event occurs is the product of the probabilities that each event occurs.

EXAMPLE 9: If the spinner discussed above is spun three times, what is the probability that it will stop in the region marked 2 each time?

Each time the spinner is spun, $P(2) = \frac{1}{6}$. Therefore,

$P(3 \text{ 2s in a row}) =$

$P(2 \text{ on the 1st spin}) \times P(2 \text{ on the 2nd spin}) \times P(2 \text{ on the 3rd spin}) =$

$\frac{1}{6} \times \frac{1}{6} \times \frac{1}{6} = \frac{1}{216} = .0046$

Exercises

1. If $a + b = 9$, $b + c = 13$, and $c + a = 14$, what is the average (arithmetic mean) of a, b, and c?

   (A) 6
   (B) 9
   (C) 12
   (D) 18
   (E) 36

2. Let m be the median and M the mode of the following set of numbers: 1, 7, 2, 4, 7, 9, 2, 7. What is the average (arithmetic mean) of m and M?

   (A) 5
   (B) 5.5
   (C) 6
   (D) 6.25
   (E) 6.5

3. Palindromes are numbers such as 44 and 383 and 1,441 that read the same from right to left as they do from left to right. If a 4-digit number is chosen at random, what is the probability that it will be a palindrome?

   (A) $\frac{1}{100}$
   (B) $\frac{1}{99}$
   (C) $\frac{1}{90}$
   (D) $\frac{1}{10}$
   (E) $\frac{1}{9}$

4. The following chart presents the number of employees at Acme Air-Conditioning in three age groups and the average monthly salary of the workers in each group.

Age Group	Number of Employees	Average Monthly Salary
Under 35	12	$3,100
35–50	24	$3,800
Over 50	14	$4,200

   What is the average (arithmetic mean) monthly salary, in dollars, for all the employees?

   (A) 3,652
   (B) 3,700
   (C) 3,744
   (D) 3,798
   (E) 3,800

5. If five fair coins are flipped, what is the probability that all five land on heads?

   (A) $\frac{1}{32}$
   (B) $\frac{1}{25}$
   (C) $\frac{1}{5}$
   (D) $\frac{1}{2}$
   (E) $\frac{5}{2}$

6. What is the average (arithmetic mean) of 3^6, 3^9, and 3^{18}?

   (A) 3^9
   (B) 3^{11}
   (C) 3^{23}
   (D) $3^2 + 3^3 + 3^6$
   (E) $3^5 + 3^8 + 3^{17}$

7. Each question on a multiple-choice test has five equally likely answers. If Nicole guesses randomly on four questions, what is the probability she will answer at least one correctly?

   (A) .25
   (B) .47
   (C) .5
   (D) .59
   (E) .8

8. How many 5-digit numbers greater than 16,666 are written with five different odd digits?

   (A) 72
   (B) 96
   (C) 108
   (D) 120
   (E) 125

9. If Ari, Ben, and Carol each choose a number at random between 1 and 5, inclusive, what is the probability that all of them choose different numbers?

   (A) $\frac{1}{125}$

   (B) $\frac{1}{25}$

   (C) $\frac{1}{5}$

   (D) $\frac{3}{5}$

   (E) $\frac{12}{25}$

10. The following stem-and-leaf plot displays the number of points scored in January by each member of the Central High School basketball team.

    1 | 0
    2 | 1 7 7
    3 |
    4 | 2 4 4 4 6 8 9
    5 | 3 7 7
    6 | 1 8

    Legend: 4 | 2 = 42 points

    What percent of the team's members scored between 35 and 55 points?

    (A) 25%
    (B) 43.75%
    (C) 50%
    (D) 56.25%
    (E) 75%

ANSWERS EXPLAINED

Answer Key

1. **(A)**	3. **(A)**	5. **(A)**	7. **(D)**	9. **(E)**
2. **(D)**	4. **(C)**	6. **(E)**	8. **(C)**	10. **(C)**

Solutions

Each of the problems in this set of exercises is typical of a question you could see on a Math 1 test. When you take the model tests in this book and, in particular, when you take the actual Math 1 test, if you get stuck on questions such as these, you do not have to leave them out—you can almost always answer them by using one or more of the strategies discussed in the "Tactics" chapter. The solutions given here do *not* depend on those strategies; they are the correct mathematical ones.

1. **(A)** Add the three equations:

$$\begin{aligned} a+b&=9 \\ b+c&=13 \\ \underline{c+a}&\underline{=14} \end{aligned}$$

$$2a+2b+2c=36 \Rightarrow a+b+c=18 \Rightarrow \frac{a+b+c}{3}=6$$

2. **(D)** First arrange the numbers in increasing order: 1, 2, 2, 4, 7, 7, 7, 9. Since there are an even number of numbers, the median is the average of the middle two: $m=\frac{4+7}{2}=5.5$. The mode, M, is 7, the number that appears most often. Finally, the average of m and M is $\frac{5.5+7}{2}=6.25$.

3. **(A)** First use the counting principle to calculate the number of 4-digit palindromes. There are 9 choices for the first digit (1–9) and 10 choices for the second digit (0–9). For the third digit, there is only 1 choice (it must be the same as the second digit), and for the fourth digit there is only 1 choice (it must be the same as the first digit). So in total there are $9\times10\times1\times1=90$ four-digit palindromes. Since there are 9,000 four-digit integers, the probability that a four-digit number chosen at random will be a palindrome is $\frac{90}{9{,}000}=\frac{1}{100}$.

4. **(C)** Use a weighted average:

$$\frac{12(3{,}100)+24(3{,}800)+14(4{,}200)}{(12+24+14)}=\frac{187{,}200}{50}=3{,}744$$

5. **(A)** Since there are two equally possible outcomes for each coin (a head or a tail), there are $2\times2\times2\times2\times2=32$ different equally possible outcomes. Only one of them consists of five heads, so the probability of getting five heads is $\frac{1}{32}$. Alternatively, you could recognize that for each coin, the probability of getting a head is $\frac{1}{2}$. Then the probability of getting five straight heads is $\frac{1}{2}\times\frac{1}{2}\times\frac{1}{2}\times\frac{1}{2}\times\frac{1}{2}=\frac{1}{32}$.

6. **(E)** The average is

$$\frac{3^6+3^9+3^{18}}{3}=\frac{3^6}{3}+\frac{3^9}{3}+\frac{3^{18}}{3}=3^5+3^8+3^{17}$$

7. **(D)** The only way to fail to get at least one right is to get all four wrong. On each question Nicole has a probability of $\frac{1}{5}$ of guessing the correct answer and $\frac{4}{5}$ of guessing the wrong answer. So the probability that Nicole gets all four wrong is $\frac{4}{5}\times\frac{4}{5}\times\frac{4}{5}\times\frac{4}{5}=.41$. The probability that she gets at least one correct is $1-.41=.59$.

8. **(C)** There are five odd digits: 1, 3, 5, 7, 9. First count how many numbers between 10,000 and 20,000 satisfy the conditions. If the first digit is a 1, for the number to be greater than 16,666, the second digit must be 7 or 9. So there is 1 choice for the first digit; 2 choices for the second digit; 3 choices for the third digit (anything except 1 and the digit chosen for the second digit); 2 choices for the fourth digit; and 1 choice for the fifth digit. So, between 10,000 and 20,000 there are $1\times2\times3\times2\times1=12$ integers greater than 16,666 that satisfy the given conditions. For numbers greater than 20,000, the only restriction (other than that the digits be different and odd) is that the first digit not be 1. So there are $4\times4\times3\times2\times1=96$ more numbers that satisfy the conditions. In total, there are $12+96=108$ 5-digit numbers greater than 16,666 that can be written with five different odd digits.

9. **(E)** Ari can choose any number. For Ben's number to be different from Ari's number, he must choose one of the 4 remaining numbers; the probability he will do that is $\frac{4}{5}$. Similarly, the probability that Carol's number will be different from the two numbers already chosen is $\frac{3}{5}$. Therefore, the probability that the three numbers chosen are all different is $\frac{4}{5}\times\frac{3}{5}=\frac{12}{25}$.

10. **(C)** Of the 16 members on the team, 8 scored between 35 and 55 points—all 7 players whose scores were in the 40s and the 1 player who scored 53 points. So $\frac{8}{16}=\frac{1}{2}=50\%$ of the members scored between 35 and 55 points.

MISCELLANEOUS TOPICS

Imaginary and Complex Numbers

- Imaginary Numbers
- Complex Numbers
- Exercises
- Answers Explained

Approximately 47 of the 50 questions on the Math 1 test are on topics covered in Chapters 2–16. However, as indicated on the chart on page 6, there are generally three questions on miscellaneous topics not covered in those chapters. In Chapters 17–19, you will read about the three topics that occur most often among the miscellaneous questions: imaginary numbers, sequences, and logic.

If your study time is limited, you may want to skip this material and concentrate on the chapters about algebra and geometry, which are the sources of the overwhelming majority of the questions on the Math 1 test. If, however, you have the time, you should study this chapter carefully. If you review the material in Chapters 17, 18, and 19, you should be able to answer any question that comes up on these topics.

IMAGINARY NUMBERS

By KEY FACT A1, if x is a real number, then x is positive, negative, or zero. If $x = 0$, then $x^2 = 0$; and if x is either positive or negative, then x^2 is positive. So if x is a real number, x^2 CANNOT be negative.

In the set of real numbers, the equation $x^2 = -1$ has no solution. In order to solve such an equation, mathematicians defined a new number, i, called the ***imaginary unit***, with the property that $i^2 = -1$. This number is often referred to as the square root of -1: $i = \sqrt{-1}$. Note that i is *not* a real number and does *not* correspond to any point on the number line.

All of the normal operations of mathematics—addition, subtraction, multiplication, and division—can be applied to the number i.

Addition:	$i + i = 2i$	$2i + 5i = 7i$
Subtraction:	$i - i = 0$	$2i - 5i = -3i$
Multiplication:	$(i)(i) = i^2 = -1$	$(2i)(5i) = 10i^2 = -10$
Division:	$\frac{i}{i} = 1$	$\frac{2i}{5i} = \frac{2}{5}$

Key Fact P1

If x is a positive number, then $\sqrt{-x} = \sqrt{-1}\sqrt{x} = i\sqrt{x}$.

Don't Get Fooled

$(\sqrt{-4})(\sqrt{-4}) = (2i)(2i) = 4i^2 = -4$

$(\sqrt{-4})(\sqrt{-4})$ is *not* equal to $\sqrt{(-4)(-4)} = \sqrt{16} = 4$.

EXAMPLE 1:

$$\sqrt{-16} = \sqrt{16}\sqrt{-1} = 4i \text{ and}$$
$$\sqrt{-12} = \sqrt{-1}\sqrt{12} = \sqrt{-1}\sqrt{4}\sqrt{3} = i \cdot 2 \cdot \sqrt{3} = 2i\sqrt{3}$$

EXAMPLE 2: What is $5\sqrt{-25} - 3\sqrt{-64}$?

$$5\sqrt{-25} = 5\sqrt{-1}\sqrt{25} = (5i)(5) = 25i \text{ and}$$
$$3\sqrt{-64} = 3\sqrt{-1}\sqrt{64} = (3i)(8) = 24i$$
$$\text{Therefore, } 5\sqrt{-25} - 3\sqrt{-64} = 25i - 24i = i$$

For the Math 1 test, you must be able to raise i to any positive integer power. In particular, note:

$i^1 = i$	($a^1 = a$ for *any* number)
$i^2 = -1$	(by definition)
$i^3 = -i$	$i \cdot i \cdot i = (i \cdot i)(i) = i^2 \cdot i = -1(i) = -i$
$i^4 = 1$	$i^4 = i \cdot i \cdot i \cdot i = (i \cdot i)(i \cdot i) = (-1)(-1) = 1$
$i^5 = i$	$i^5 = i^4 \cdot i = 1 \cdot i = i$
$i^6 = -1$	$i^6 = i^5 \cdot i = i \cdot i = i^2 = -1$
$i^7 = -i$	$i^7 = i^6 \cdot i = (-1)i = -i$
$i^8 = 1$	$i^8 = i^7 \cdot i = (-i)(i) = -i^2 = -(-1) = 1$

Note that the powers of i form a repeating sequence in which the four terms, $i, -1, -i, 1$ repeat in that order indefinitely.

As you will see in KEY FACT P2, this means that to find the value of i^n for any positive integer n, you should divide n by 4 and calculate the remainder.

Key Fact P2

For any positive integer n:

- **If n is a multiple of 4, $i^n = 1$**
- **If n is not a multiple of 4, $i^n = i^r$, where r is the remainder when n is divided by 4.**

EXAMPLE 3: To evaluate i^{375}, use your calculator to divide 375 by 4:

$$375 \div 4 = 93.75 \Rightarrow \text{the quotient is } 93$$

Then multiply 93 by 4:

$$93 \times 4 = 372 \Rightarrow \text{the remainder is } 375 - 372 = 3$$

So $i^{375} = i^3 = -i$.

The concepts of *positive* and *negative* apply only to real numbers. If a is positive, a is to the right of 0 on the number line. Since imaginary numbers do not lie on the number line, you cannot compare them. It is meaningless even to ask whether i is positive or negative or whether $12i$ is greater than or less than $7i$.

COMPLEX NUMBERS

The imaginary unit can be added to and multiplied by real numbers to form ***complex numbers***. Every complex number can be written in the form $a + bi$, where a and b are real numbers. a is called the ***real part*** and bi the ***imaginary part*** of the complex number $a + bi$. Two complex numbers are equal if, and only if, their real parts are equal and their imaginary parts are equal.

Key Fact P3

If $a + bi = c + di$, then $a = c$ and $b = d$.

EXAMPLE 4: If $2(3 + yi) = x + 8i$, what are the values of x and y?

$$x + 8i = 2(3 + yi) \Rightarrow x + 8i = 6 + 2yi$$

$$\text{So, } x = 6 \text{ and } 8 = 2y \Rightarrow x = 6 \text{ and } y = 4$$

The arithmetic of complex numbers follows all the rules you are familiar with for real numbers.

Key Fact P4

- **To add complex numbers, add their real parts and add their imaginary parts. For example:**

$$(3 + 5i) + (2 + 3i) = 5 + 8i$$

- **To subtract complex numbers, subtract their real parts and subtract their imaginary parts. For example:**

$$(3 + 5i) - (2 + 3i) = 1 + 2i$$

- **To multiply complex numbers, "FOIL" them as if they were binomials and replace i^2 by -1. For example:**

$$\begin{aligned}(3 + 5i)(2 + 3i) &= 6 + 9i + 10i + 15i^2 \\ &= 6 + 19i + 15(-1) \\ &= 6 + 19i - 15 \\ &= -9 + 19i\end{aligned}$$

Complex numbers can also be divided, but you do not need to know how to do this for the Math 1 test.

Exercises

1. If a and b are real numbers and if $ai^2 = bi^4$, which of the following must be true?

   I. $a = b$
   II. $a = -b$
   III. $|a| = |b|$

   (A) I only
   (B) II only
   (C) III only
   (D) II and III only
   (E) I, II, and III

2. Which of the following is a negative number?

   (A) i^{25}
   (B) i^{50}
   (C) i^{75}
   (D) i^{100}
   (E) i^{200}

3. Which of the following is equal to $(1 + i)^2$?

   (A) -2
   (B) 0
   (C) 2
   (D) $-2i$
   (E) $2i$

4. If a and b are real numbers, which of the following is equal to $(a + bi)(a - bi)$?

   (A) $a^2 - b^2$
   (B) $a^2 + b^2$
   (C) $a^2 + 2abi - b^2$
   (D) $a^2 - 2abi - b^2$
   (E) $a^2 - 2abi + b^2$

5. If a, b, c, and d are consecutive positive integers, what is the value of $i^a + i^b + i^c + i^d$?

   (A) 0
   (B) -1
   (C) 1
   (D) i
   (E) It cannot be determined from the information given.

6. If $i^m = i^n$, which of the following must be true?

   I. $m = n$
   II. $m + n$ is even
   III. $n - m$ is a multiple of 4.

   (A) I only
   (B) III only
   (C) I and III only
   (D) II and III only
   (E) I, II, and III

ANSWERS EXPLAINED

Answer Key

1. **(D)** 3. **(E)** 5. **(A)**
2. **(B)** 4. **(B)** 6. **(D)**

Solutions

Each of the problems in this set of exercises is typical of a question you could see on a Math 1 test. When you take the model tests in this book and, in particular, when you take the actual Math 1 test, if you get stuck on questions such as these, you do not have to leave them out—you can almost always answer them by using one or more of the strategies discussed in the "Tactics" chapter. The solutions given here do *not* depend on those strategies; they are the correct mathematical ones.

1. **(D)** Since $i^2 = -1$ and $i^4 = 1$, $ai^2 = bi^4 \Rightarrow a(-1) = b(1) \Rightarrow -a = b$. Hence $a = -b$ and $|a| = |-b| = |b|$. (II and III are true.) Statement I *could* be true—if a and b were each 0—but does not have to be true: $-2i^2 = 2i^4$, but $-2 \neq 2$. Only statements II and III are true.

2. **(B)** Since 100 and 200 are multiples of 4, $i^{100} = i^{200} = 1$. The remainders when 25, 50, and 75 are divided by 4 are 1, 2, and 3, respectively. Therefore, $i^{25} = i$, $i^{50} = -1$, and $i^{75} = -i$. Of these, only i^{50} is a negative number.

3. **(E)** $(1 + i)^2 = (1 + i)(1 + i) = 1 + 2i + i^2 = 1 + 2i + (-1) = 2i$

4. **(B)** $(a + bi)(a - bi) = a^2 - abi + abi - b^2i^2 = a^2 - b^2(-1) = a^2 + b^2$

5. **(A)** Since the values of the power of i repeat indefinitely in groups of four, i^a, i^b, i^c, i^d are in some order: i, -1, $-i$, and 1. So their sum is $i + (-1) + (-i) + 1 = 0$.

6. **(D)** Since $i^4 = i^8 = 1$, I is false. If m is odd and n is even, then i^m equals i or $-i$ and i^n equals 1 or -1, so $i^m \neq i^n$. Similarly, it can't be that m is even and n is odd. Therefore, m and n are both even or both odd. In either case, $m + n$ is even (II is true). In fact, $i^n = i^m \Rightarrow \frac{i^n}{i^m} = 1 \Rightarrow i^{n-m} = 1 \Rightarrow n - m$ is a multiple of 4 (III is true). Only statements II and III are true.

Sequences

On the Math 1 test, there is often one and occasionally two questions about sequences. In this section, you will read about the three different types of sequences that could be the source of a Math 1 test question.

A ***sequence*** is a list of objects separated by commas. Most often, the objects are numbers, but they don't have to be. A sequence can be finite, such as 2, 5, 10, 17, 26, or can be infinite, such as 2, 4, 6, 8, 10, . . . The numbers in the list are called the ***terms*** of the sequence.

The terms of a sequence don't have to follow any regular pattern or rule. Suppose John tossed a die 100 times and recorded the outcomes. If you saw the first 10 terms of that sequence—4, 6, 6, 2, 5, 3, 5, 4, 4, 2—you would have no way of knowing what the 11th term is.

On the Math 1 test, however, there is always a definite rule that determines the terms of a sequence. If Mary created a sequence of 100 terms by writing down the number 5 and then continually added 3 to each term, the first 10 terms would be 5, 8, 11, 14, 17, 20, 23, 26, 29, 32. Not only would you be able to know what the eleventh term is, as you will see, you can easily calculate the 88th, or any other, term.

On the Math 1 test, the most common type of question concerning sequences gives you a rule and asks you to find a particular term. The three types of sequences that occur most often are repeating sequences, arithmetic sequences, and geometric sequences.

REPEATING SEQUENCES

A sequence whose terms repeat in a cyclical pattern is called a ***repeating sequence***. For example, the following three sequences are repeating sequences:

a, b, c, a, b, c, a, b, c, a, b, c, . . .

0, 1, 0, 1, 0, 1, 0, 1, 0, 1, 0, 1, . . .

7, 1, 4, 2, 8, 5, 7, 1, 4, 2, 8, 5, . . .

Key Fact Q1

When a sequence consists of a group of k terms that repeat in the same order indefinitely, to find the nth term, find the remainder, r, when n is divided by k. The rth term and the nth term are equal.

The fraction $\frac{5}{7}$ is equivalent to the repeating decimal 0.714285714285 . . . in which the six digits, 7, 1, 4, 2, 8, 5, repeat indefinitely. Examples 1 and 2 refer to this sequence of digits.

EXAMPLE 1: To find the 1,000th digit to the right of the decimal point in the expansion of $\frac{5}{7}$, use KEY FACT Q1.

Since the repeating portion consists of 6 digits, divide 1,000 by 6.

$1{,}000 \div 6 = 166.666\ldots \Rightarrow$ the quotient is 166
$166 \times 6 = 996 \Rightarrow$ the remainder is $1{,}000 - 996 = 4$

Therefore, the 1,000th term is the same as the 4th term, namely 2.

EXAMPLE 2: What is the sum of the 101st through the 106th digits in the decimal expression of $\frac{5}{7}$?

You could repeat what was done in Example 1 six times, but, of course, you shouldn't. Any six consecutive terms of this sequence consist, in some order, of the same six digits—7, 1, 4, 2, 8, 5—whose sum is 27. In this case, the order is 8, 5, 7, 1, 4, 2, but you do not need to know that.

At some point in your study of math you may have seen the sequence 1, 1, 2, 3, 5, 8, 13, . . . This sequence, called the Fibonacci sequence, is defined by the following rule: the first two terms are both 1 and, starting with the third term, each term is the sum of the two preceding terms. For example, $2 = 1 + 1$; $13 = 5 + 8$; and since the sum of the 6th and 7th terms, 8 and 13, is 21, the 8th term is 21. Clearly this is *not* a repeating sequence, and it would be totally unreasonable to ask you to find the 100th term. However, there are some questions, such as the one in Example 3, that you could be asked.

EXAMPLE 3: Of the first 100 terms of the Fibonacci sequence, how many are odd? The sequence itself does not form a repeating sequence, but its pattern of odd (O) and even (E) terms does:

1	1	2	3	5	8	13	21	34
O	O	E	O	O	E	O	O	E

Note that the Os and Es form a repeating sequence:

$$\boxed{\text{O O E}},\ \boxed{\text{O O E}},\ \boxed{\text{O O E}},\ \ldots$$

The first 99 terms consist of 33 sets of $\boxed{\text{O O E}}$. Since each set contains two Os and one E, of the first 99 terms, 66 are odd and 33 are even. The 100th term is the first term in the next set and so is O. In all, there are 67 odd terms.

ARITHMETIC SEQUENCES

An ***arithmetic sequence*** is a sequence such as 5, 8, 11, 14, 17, . . . in which the difference between any two consecutive terms is the same. In this sequence, the difference is 3 ($8 - 5 = 3$; $11 - 8 = 3$; $14 - 11 = 3, \ldots$). An easy way to find the nth term of such a sequence is to start with the first term and add the common difference $n - 1$ times. Here, the 5th term is 17 which can be obtained by taking the first term, 5, and adding the common difference, 3, four times: $5 + 4(3) = 17$. In the same way, the 100th term is $5 + 99(3) = 5 + 297 = 302$.

Key Fact Q2

If $a_1, a_2, a_3, \ldots$ is an arithmetic sequence whose common difference is d, then $a_n = a_1 + (n - 1)d$.

EXAMPLE 4: If the 8th term of an arithmetic sequence is 10 and the 20th term is 58, what is the first term?

Use KEY FACT Q2 twice and subtract:

$$\begin{aligned} a_{20} &= a_1 + 19d = 58 \\ a_8 &= a_1 + \ \ 7d = 10 \\ \hline & \quad\quad 12d = 48 \Rightarrow d = 4 \end{aligned}$$

Then

$$10 = a_1 + 7d = a_1 + 7(4) = a_1 + 28 \Rightarrow a_1 = -18$$

GEOMETRIC SEQUENCES

A ***geometric sequence*** is a sequence such as 3, 6, 12, 24, 48, . . . in which the ratio between any two consecutive terms is the same. In this sequence, the ratio is

$$\frac{6}{3} = 2;\ \frac{12}{6} = 2;\ \frac{24}{12} = 2;\ \ldots.$$

An easy way to find the nth term of a geometric sequence is to start with the first term and multiply it by the common ratio $n - 1$ times. Here the 5th term is 48, which can be obtained by taking the first term, 3, and multiplying it by the common ratio, 2, four times: $3 \times 2 \times 2 \times 2 \times 2 = 3 \times 2^4 = 3 \times 16 = 48$. In the same way, the 100th term is 3×2^{99}.

Key Fact Q3

If a_1, a_2, a_3, . . . is a geometric sequence whose common ratio is r, then $a_n = a_1 r^{n-1}$.

EXAMPLE 5: What is the 12th term of the sequence 3, −6, 12, −24, 48, −96, . . . ?

This is a geometric sequence whose common ratio is −2. By KEY FACT Q3,

$$a_{12} = a_1(-2)^{11} = 3(-2)^{11} = 3(-2{,}048) = -6{,}144$$

Exercises

1. The first term of sequence I is 10, and each term after the first term is 3 more than the preceding term. The first term of sequence II is 100, and each term after the first term is 3 less than the preceding term. If x is the 16th term of sequence I and y is the 16th term of sequence II, what is $x + y$?

   (A) 45
   (B) 55
   (C) 90
   (D) 110
   (E) 200

2. What is the 500th digit to the right of the decimal point when $\frac{15}{37}$ is expressed as a decimal?

   (A) 0
   (B) 3
   (C) 4
   (D) 5
   (E) 7

3. July 4, 2009 falls on a Saturday. What day of the week will it be on July 4, 2011? (Note: 2009, 2010, and 2011 are all regular years with 365 days.)

   (A) Sunday
   (B) Monday
   (C) Tuesday
   (D) Friday
   (E) Saturday

4. The number of bacteria in a culture increases by 20% every 20 minutes. If there are 1,000 bacteria present at noon on a given day, to the nearest thousand, how many will be present at midnight of the same day?

   (A) 9,000
   (B) 43,000
   (C) 174,000
   (D) 591,000
   (E) 709,000

5. A gum ball dispenser is filled with exactly 1,000 pieces of gum. The gum balls always come out in the following order: 1 red, 2 blue, 3 green, 4 yellow, and 5 white. After the fifth white, the pattern repeats, starting with 1 red, and so on. What is the color of the last gum ball to come out of the machine?

   (A) red
   (B) blue
   (C) green
   (D) yellow
   (E) white

6. If the 21st term of an arithmetic sequence is 57 and the 89th term is 227, what is the 333rd term?

   (A) 367
   (B) 417
   (C) 587
   (D) 793
   (E) 837

ANSWERS EXPLAINED

Answer Key

1. **(D)** 3. **(B)** 5. **(D)**
2. **(A)** 4. **(E)** 6. **(E)**

Solutions

Each of the problems in this set of exercises is typical of a question you could see on a Math 1 test. When you take the model tests in this book and, in particular, when you take the actual Math 1 test, if you get stuck on questions such as these, you do not have to leave them out—you can almost always answer them by using one or more of the strategies discussed in the "Tactics" chapter. The solutions given here do *not* depend on those strategies; they are the correct mathematical ones.

1. **(D)** Sequence I is an arithmetic sequence whose common difference is 3, and sequence II is an arithmetic sequence whose common difference is −3.

$$\text{For sequence I: } a_{16} = 10 + 15(3) = 10 + 45 = 55$$
$$\text{For sequence II: } a_{16} = 100 + 15(-3) = 100 - 45 = 55$$

So $x = y = 55$, and $x + y = 110$.

2. **(A)** Use your calculator to divide: $15 \div 37 = 0.405405405$. So the question is equivalent to asking, "What is the 500th term in the repeating sequence 4, 0, 5, 4, 0, 5, 4, 0, 5, . . . ?" Since there are 3 terms in the repeating portion, divide 500 by 3:

$$500 \div 3 = 166.66 \Rightarrow \text{the quotient is } 166$$
$$166 \times 3 = 498 \Rightarrow \text{the remainder is } 500 - 498 = 2$$

So the 500th term is the same as the second term: 0.

3. **(B)** July 4, 2011 is exactly 2 years $= 2 \times 365 = 730$ days after July 4, 2009. The days of the week form a repeating sequence in which 7 terms repeat.

$$730 \div 7 = 104.2857\ldots \Rightarrow \text{the quotient is } 104$$
$$104 \times 7 = 728 \Rightarrow \text{the remainder is } 730 - 728 = 2$$

So 730 days from Saturday will be the same day as 2 days from Saturday, namely Monday.

4. **(E)** Every 20 minutes, the number of bacteria present is muliplied by a factor of 1.2. This creates a geometric sequence whose first term is 1,000 and whose common ratio is 1.2:

$$1{,}000,\ 1{,}200,\ 1{,}440,\ 1{,}728, \ldots$$

The 12 hours from noon to midnight consist of 36 20-minute intervals. So the number of bacteria at midnight is

$$1{,}000 \times (1.2)^{36} = 1{,}000(708.8) = 708{,}800 \approx 709{,}000.$$

5. **(D)** Since the pattern repeats itself after every 15 gum balls, divide 1,000 by 15. The quotient is 66, and the remainder is 10. Therefore, the 1,000th gum ball is the same color as the 10th gum ball, which is yellow.

6. **(E)** Since $a_n = a_1 + (n - 1)d$, we have

$$57 = a_{21} = a_1 + 20d$$
$$227 = a_{89} = a_1 + 88d$$

Subtracting the first equation from the second yields

$$170 = 68d \Rightarrow d = 2.5$$

So $57 = a_1 + 20(2.5) = a_1 + 50 \Rightarrow a_1 = 7$.
Finally, the 333rd term is

$$7 + 332(2.5) = 7 + 830 = 837$$

Logic

- Statements
- Negations
- Conditional Statements
- Exercises
- Answers Explained

Throughout the United States, math curricula vary considerably. Mathematical logic is a topic that is taught in some schools but not in many others. On the Math 1 test, there should be at most one question involving mathematical logic. If you have studied logic in school, you have learned much more logic than you need for this test. If you review the few facts presented in this short section, you should have no problem with the one logic question that may be on your Math 1 test.

If you never studied logic in school, don't worry. If you carefully read the small amount of material in this section, you should be able to handle the one logic question you may encounter.

STATEMENTS

In logic, a ***statement*** is a sentence for which it is meaningful to say that it is true or false. In logic, statements are usually referred to by letters such as p, q, and r. These letters will not appear on the Math 1 test, but using them here will simplify our discussion. Here are three statements:

p: $2 + 2 = 4$
q: Every prime number is odd.
r: There is a real number x such that $x^2 + 1 = 0$.

Note that statement p is true and that statements q and r are false.

NEGATIONS

The ***negation*** of a statement is the statement formed by putting the words "it is not true that" in front of the original statement. In logic the symbol ~ is used for negation. (Again, you won't see that symbol on the Math 1 test.)

Key Fact R1

A statement and its negation have opposite truth values:

- **If p is true, $\sim p$ is false.**
- **If p is false $\sim p$ is true.**

Here are the negations of statements p, q, and r given above:

$\sim p$: It is not true that $2 + 2 = 4$
or equivalently: $2 + 2 \neq 4$.
$\sim q$: It is not true that every prime number is odd
or equivalently: There is a prime number that is not odd.
$\sim r$: It is not true that there is a real number x such that $x^2 + 1 = 0$
or equivalently: There is no real number x for which $x^2 + 1 = 0$.

Note that $\sim p$ is false and $\sim q$ and $\sim r$ are true.

Key Fact R2

If a statement claims that all objects of a certain type have a particular property, the negation of that statement says that at least one of those objects does not have the property.

For example, statement q (falsely) asserts that all prime numbers are odd. Its negation, $\sim q$, (truthfully) states that there is at least one prime number that is not odd (2 is prime, but not odd).

Similarly, the negation of the statement "All roses are red" is the statement "There is at least one rose that is not red."

Key Fact R3

If a statement claims that some object of a certain type has a particular property, the negation of that statement says that none of those objects has the property.

For example, statement r (falsely) asserts that some real number is a solution of the equation $x^2 + 1 = 0$. Its negation, $\sim r$, (truthfully) states that no real number is a solution of the equation $x^2 + 1 = 0$.

Similarly, the negation of the statement "Some roses are black" is the statement "No roses are black."

CONDITIONAL STATEMENTS

A conditional statement is a sentence of the form "if p, then q" where p and q are given statements. In logic, this is written $p \rightarrow q$ and is read "p implies q."

For example:

If p is the statement "x is divisible by 4," and q is the statement "x is even," then $p \rightarrow q$ is the statement "If x is divisible by 4, then x is even."

There are three conditional statements related to the statement $p \rightarrow q$.

The ***converse*** of $p \rightarrow q$ is $q \rightarrow p$
(the order of the statements is reversed).
The ***inverse*** of $p \rightarrow q$ is $\sim p \rightarrow \sim q$
(the original statements are negated).

The ***contrapositive*** of $p \rightarrow q$ is $\sim q \rightarrow \sim p$
(the order of the statements is reversed and the statements are negated).

For example:

Conditional: $p \rightarrow q$:
If x is divisible by 4, then x is even.
Converse: $q \rightarrow p$:
If x is even, then x is divisible by 4.
Inverse: $\sim p \rightarrow \sim q$:
If x is not divisible by 4, then x is not even.
Contrapositive: $\sim q \rightarrow \sim p$:
If x is not even, then x is not divisible by 4.

Note that in the preceding example, the conditional and its contrapositives are both true. Whether a conditional statement is true or false, its contrapositive always has the same truth value.

Key Fact R4

- **A conditional statement is logically equivalent to its contrapositive. Either both statements are true or both are false.**
- **The converse and inverse of a conditional statement are logically equivalent. Either both statements are true or both are false.**

The logic question on a Math 1 test often asks which of five statements is equivalent to a given conditional statement. The correct answer is always the contrapositive of the given conditional.

EXAMPLE 1: A statement equivalent to the conditional, "If Adam lives in California, then he lives in the United States" is its contrapositive, "If Adam does not live in the United States, then he does not live in California."

Exercises

1. If the statement "Every perfect set is dense" is false, which of the following statements must be true?

   (A) Some perfect sets are dense.
   (B) No perfect sets are dense.
   (C) All perfect sets are not dense.
   (D) Some perfect sets are not dense.
   (E) No dense sets are perfect.

2. If the statement "Some sloths have three toes" is true, which of the following statements must be false?

   (A) All sloths have three toes.
   (B) No sloths have three toes.
   (C) Some sloths do not have three toes.
   (D) All three-toed animals are sloths.
   (E) Some three-toed animals are sloths.

3. Which of the following statements is equivalent to the statement "If it is sunny, I will go to the beach"?

   (A) If I go to the beach, it is sunny.
   (B) If I don't go to the beach, it isn't sunny.
   (C) If it isn't sunny, I won't go to the beach.
   (D) It is sunny, and I go to the beach.
   (E) It isn't sunny, and I don't go to the beach.

4. Which of the following statements is equivalent to "If $0 < x < 1$, then $\sqrt{x}$ is irrational"?

   (A) If $0 < x < 1$, then $\sqrt{x}$ is rational.
   (B) If $\sqrt{x}$ is irrational, then $0 < x < 1$.
   (C) If $\sqrt{x}$ is rational, then $0 < x < 1$.
   (D) If $\sqrt{x}$ is rational, then $x \leq 0$ or $x \geq 1$.
   (E) If $x \leq 0$ or $x \geq 1$, then $\sqrt{x}$ is rational.

5. John said, "Nicole and Caroline each passed today's test." If John's statement is false, which of the following statements *must* be true?

   (A) If Nicole passed, then Caroline passed.
   (B) If Nicole passed, then Caroline failed.
   (C) Nicole passed, but Caroline failed.
   (D) Nicole failed and Caroline failed.
   (E) Either Nicole or Caroline failed.

6. Which of the following is equivalent to the statement, "If John is eligible to vote, then John is a citizen"?

   (A) All citizens are eligible to vote.
   (B) If John is not eligible to vote, then John is not a citizen.
   (C) Some citizens are not eligible to vote.
   (D) If John is not a citizen, then John is not eligible to vote.
   (E) If John is a citizen, then John is eligible to vote.

ANSWERS EXPLAINED

Answer Key

1. **(D)**	3. **(B)**	5. **(E)**
2. **(B)**	4. **(D)**	6. **(D)**

Solutions

Each of the problems in this set of exercises is typical of a question you could see on a Math 1 test. When you take the model tests in this book and, in particular, when you take the actual Math 1 test, if you get stuck on questions such as these, you do not have to leave them out—you can almost always answer them by using one or more of the strategies discussed in the "Tactics" chapter. The solutions given here do *not* depend on those strategies; they are the correct mathematical ones.

1. **(D)** To answer the question, you do not need to know what is a perfect set or what is a dense set. You must simply use KEY FACT R2: if it is not true that all perfect sets are dense, then some (at least one) perfect sets are not dense.

2. **(B)** By KEY FACT R3, the negation of the statement "Some sloths have three toes" is "No sloth has three toes" (choice B). Since the negation of a true statement is false, B must be false. Note that E is true and that from the given statement, it is impossible to know whether A, C, and D are true or false.

3. **(B)** By KEY FACT R4, a conditional statement ($p \rightarrow q$) is equivalent to its contrapositive ($\sim q \rightarrow \sim p$). Therefore, "Sunny → Beach" is equivalent to "no Beach → not Sunny."

4. **(D)** Note that the original statement is false

$$\left(\sqrt{\frac{1}{4}} = \frac{1}{2}, \text{ which is not irrational}\right).$$

In fact, all five choices are false, but the only one that is equivalent to the original is its contrapositive.

If p is the statement "$0 < x < 1$" and q is the statement "$\sqrt{x}$ is irrational," then $\sim p$ is "$x \leq 0$ or $x \geq 1$" and $\sim q$ is "$\sqrt{x}$ is rational." So the contrapositive $(\sim q \rightarrow \sim p)$ is choice D:

$$\text{If } \sqrt{x} \text{ is rational, then } x \leq 0 \text{ or } x \geq 1.$$

5. **(E)** If you want to disprove the statement "Nicole passed *and* Caroline passed," you have to show that *either* Nicole failed *or* Caroline failed. Of course, John's statement would be false if they each failed, but John's statement is false even if only one of them failed.

6. **(D)** Any conditional statement $(p \rightarrow q)$ is equivalent to its contrapositive $(\sim q \rightarrow \sim p)$. The contrapositive of "If John is eligible to vote, then John is a citizen" is "If John is *not* a citizen, then John is *not* eligible to vote."

MODEL TESTS

Guidelines and Scoring

The three model tests in this section have the identical format as the actual Math 1 test you will take. There are 50 multiple-choice questions that generally proceed from easy to medium to difficult. On each test the percentage of questions on each topic in the Math 1 syllabus follows the breakdown in the chart on page 6.

To simulate as closely as possible the experience of taking a Math 1 test, you should follow these guidelines:

- Take the test in a quiet room in which there are no distractions. In particular, you should not be listening to music and your cell phone should be turned off.
- Give yourself exactly 60 minutes to complete the test. When you take the model tests, as well as when you take your actual Math 1 test, you should have an easy-to-read clock, stopwatch, or wrist watch on your desk.
- When you take the model tests, use the same calculator that you intend to use on the actual Math 1 test.
- Tear out the answer sheet and use it to bubble in your answers as you take each test.
- If necessary, before starting the test read over the directions that are printed on the first page. Familiarize yourself with them. On the day you take the actual Math 1 test, do not spend even one second reading the directions.
- You don't have to memorize the five formulas that are given to you in the reference box, but be sure that you know which formulas are there. If there are any formulas other than the five in the box that you want to review, do so now.

After you complete each model test:

- Use the Answer Key to calculate your raw score: give yourself 1 point for each correct answer and deduct $\frac{1}{4}$ point for each incorrect answer.
- Then use the conversion chart on page 8 to find your scaled score.
- Carefully read the solutions to each question that you left out or missed.
- If there is still anything you do not understand, look up those topics in the math review chapters and do the exercises at the end of each section covering a topic for which you needed extra help.

Model Test 1

50 Questions / 60 MINUTES

Directions: For each question, determine which of the answer choices is correct and fill in the oval on the answer sheet that corresponds to your choice.

Notes:

1. You will need to use a scientific or graphing calculator to answer some of the questions.
2. Be sure your calculator is in degree mode.
3. Each figure on this test is drawn as accurately as possible unless it is specifically indicated that the figure has not been drawn to scale.
4. The domain of any function f is the set of all real numbers x for which $f(x)$ is also a real number, unless the question indicates that the domain has been restricted in some way.
5. The box below contains five formulas that you may need to answer one or more of the questions.

Reference Information. This box contains formulas for the volumes of three solids and the areas of two of them.

For a sphere with radius r:

- $V = \frac{4}{3}\pi r^3$
- $A = 4\pi r^2$

For a right circular cone with radius r, circumference c, height h, and slant height l:

- $V = \frac{1}{3}\pi r^2 h$
- $A = \frac{1}{2}cl$

For a pyramid with base area B and height h:

- $V = \frac{1}{3}Bh$

Answer Sheet

MODEL TEST 1

1 Ⓐ Ⓑ Ⓒ Ⓓ Ⓔ	14 Ⓐ Ⓑ Ⓒ Ⓓ Ⓔ	27 Ⓐ Ⓑ Ⓒ Ⓓ Ⓔ	40 Ⓐ Ⓑ Ⓒ Ⓓ Ⓔ
2 Ⓐ Ⓑ Ⓒ Ⓓ Ⓔ	15 Ⓐ Ⓑ Ⓒ Ⓓ Ⓔ	28 Ⓐ Ⓑ Ⓒ Ⓓ Ⓔ	41 Ⓐ Ⓑ Ⓒ Ⓓ Ⓔ
3 Ⓐ Ⓑ Ⓒ Ⓓ Ⓔ	16 Ⓐ Ⓑ Ⓒ Ⓓ Ⓔ	29 Ⓐ Ⓑ Ⓒ Ⓓ Ⓔ	42 Ⓐ Ⓑ Ⓒ Ⓓ Ⓔ
4 Ⓐ Ⓑ Ⓒ Ⓓ Ⓔ	17 Ⓐ Ⓑ Ⓒ Ⓓ Ⓔ	30 Ⓐ Ⓑ Ⓒ Ⓓ Ⓔ	43 Ⓐ Ⓑ Ⓒ Ⓓ Ⓔ
5 Ⓐ Ⓑ Ⓒ Ⓓ Ⓔ	18 Ⓐ Ⓑ Ⓒ Ⓓ Ⓔ	31 Ⓐ Ⓑ Ⓒ Ⓓ Ⓔ	44 Ⓐ Ⓑ Ⓒ Ⓓ Ⓔ
6 Ⓐ Ⓑ Ⓒ Ⓓ Ⓔ	19 Ⓐ Ⓑ Ⓒ Ⓓ Ⓔ	32 Ⓐ Ⓑ Ⓒ Ⓓ Ⓔ	45 Ⓐ Ⓑ Ⓒ Ⓓ Ⓔ
7 Ⓐ Ⓑ Ⓒ Ⓓ Ⓔ	20 Ⓐ Ⓑ Ⓒ Ⓓ Ⓔ	33 Ⓐ Ⓑ Ⓒ Ⓓ Ⓔ	46 Ⓐ Ⓑ Ⓒ Ⓓ Ⓔ
8 Ⓐ Ⓑ Ⓒ Ⓓ Ⓔ	21 Ⓐ Ⓑ Ⓒ Ⓓ Ⓔ	34 Ⓐ Ⓑ Ⓒ Ⓓ Ⓔ	47 Ⓐ Ⓑ Ⓒ Ⓓ Ⓔ
9 Ⓐ Ⓑ Ⓒ Ⓓ Ⓔ	22 Ⓐ Ⓑ Ⓒ Ⓓ Ⓔ	35 Ⓐ Ⓑ Ⓒ Ⓓ Ⓔ	48 Ⓐ Ⓑ Ⓒ Ⓓ Ⓔ
10 Ⓐ Ⓑ Ⓒ Ⓓ Ⓔ	23 Ⓐ Ⓑ Ⓒ Ⓓ Ⓔ	36 Ⓐ Ⓑ Ⓒ Ⓓ Ⓔ	49 Ⓐ Ⓑ Ⓒ Ⓓ Ⓔ
11 Ⓐ Ⓑ Ⓒ Ⓓ Ⓔ	24 Ⓐ Ⓑ Ⓒ Ⓓ Ⓔ	37 Ⓐ Ⓑ Ⓒ Ⓓ Ⓔ	50 Ⓐ Ⓑ Ⓒ Ⓓ Ⓔ
12 Ⓐ Ⓑ Ⓒ Ⓓ Ⓔ	25 Ⓐ Ⓑ Ⓒ Ⓓ Ⓔ	38 Ⓐ Ⓑ Ⓒ Ⓓ Ⓔ	
13 Ⓐ Ⓑ Ⓒ Ⓓ Ⓔ	26 Ⓐ Ⓑ Ⓒ Ⓓ Ⓔ	39 Ⓐ Ⓑ Ⓒ Ⓓ Ⓔ	

1. If $\frac{x+3}{7} = \frac{x-3}{3}$, then $x =$

(A) -3
(B) 1.5
(C) 3
(D) 7.5
(E) 21

2. If $a = b^2$, $b = c\sqrt{2}$, and $a = 18$, which of the following could be the value of c?

(A) 3
(B) $3\sqrt{2}$
(C) 9
(D) 81
(E) $162\sqrt{2}$

3. What is the value of $||-5| - |-7||$?

(A) -12
(B) -2
(C) 2
(D) 12
(E) 35

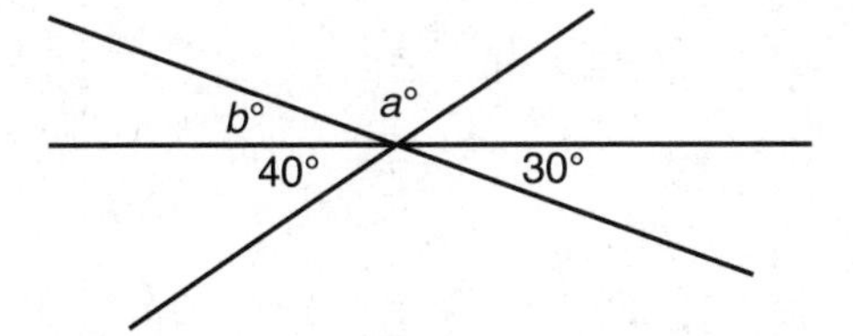

Note: Figure not drawn to scale.

4. In the figure above, what is the value of a?

(A) 70
(B) 100
(C) 110
(D) 120
(E) 290

USE THIS SPACE FOR SCRATCH WORK

USE THIS SPACE FOR SCRATCH WORK

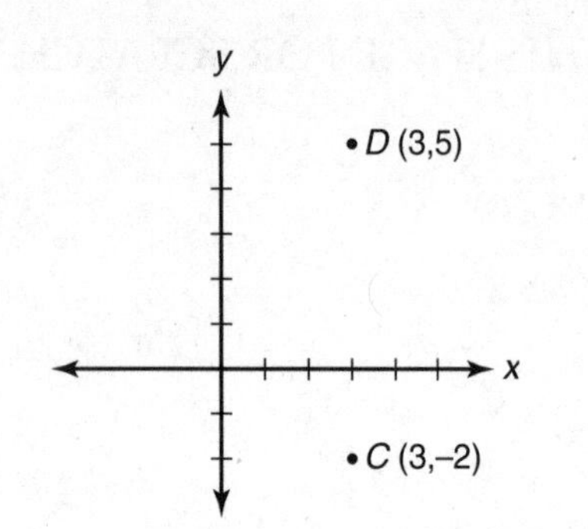

5. In the diagram above, C (3, −2) and D (3, 5) are two adjacent vertices of square $ABCD$. Which of the following could be the coordinates of A or B?

   (A) (−3, −2)
   (B) (−2, 3)
   (C) (−4, 5)
   (D) (3, 10)
   (E) (10, 3)

6. For all $x \neq -5, 3$, which of the following is equivalent to $\frac{x^2 - x - 6}{x+5} \cdot \frac{x^2 + 3x - 10}{2x - 6}$?

   (A) $x^2 - 2$
   (B) $x^2 - 4$
   (C) $\frac{x^2 - 4}{2}$
   (D) $\frac{(x-2)^2}{2}$
   (E) $\frac{(x+2)^2}{2}$

7. One week Susan earned \$1,000, which consisted of a base salary of \$475 and a 7% commission on her sales. What was the amount of her sales for that week?

   (A) \$36.75
   (B) \$75.25
   (C) \$700
   (D) \$6,786
   (E) \$7,500

USE THIS SPACE FOR SCRATCH WORK

8. If the sum of two numbers is 5 and the difference of the two numbers is 5, what is the product of the two numbers?

(A) 0
(B) 1
(C) 5
(D) 10
(E) 25

9. If $f(x) = -2x^2 - 3x$, $f(-5) =$

(A) −65
(B) −35
(C) 35
(D) 85
(E) 115

10. The circumference of circle O is a inches, and the area of circle O is b square inches. If $a = b$, what is the radius of circle O?

(A) 1
(B) 2
(C) 4
(D) 2π
(E) 4π

11. For what values of n is $\sqrt{n+2} = \frac{2}{3}$?

(A) $-\frac{14}{9}$ only

(B) $\frac{22}{9}$ only

(C) $-\frac{4}{3}$ only

(D) $-\frac{14}{9}$ and $-\frac{22}{9}$

(E) $-\frac{4}{3}$ and $-\frac{8}{3}$

12. For how many integer values of x can 6, 8, and x be the lengths of the three sides of a triangle?

(A) 1
(B) 2
(C) 10
(D) 11
(E) 13

USE THIS SPACE FOR SCRATCH WORK

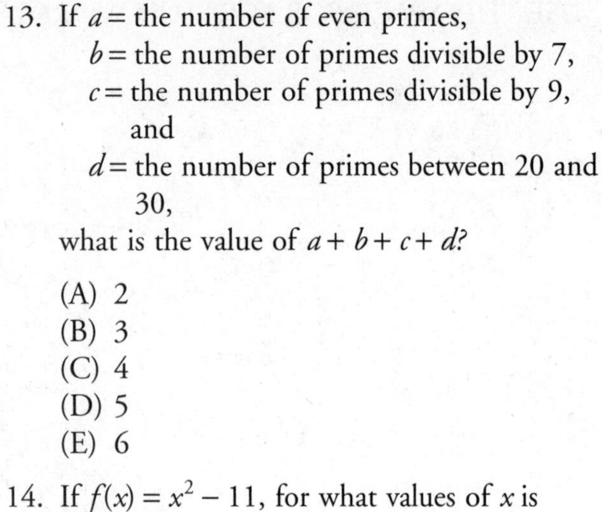

13. If a = the number of even primes,
    b = the number of primes divisible by 7,
    c = the number of primes divisible by 9, and
    d = the number of primes between 20 and 30,
    what is the value of $a + b + c + d$?

    (A) 2
    (B) 3
    (C) 4
    (D) 5
    (E) 6

14. If $f(x) = x^2 - 11$, for what values of x is $f(x) < 25$?

    (A) $x < 5$
    (B) $x < 6$
    (C) $x = 6$
    (D) $-5 < x < 5$
    (E) $-6 < x < 6$

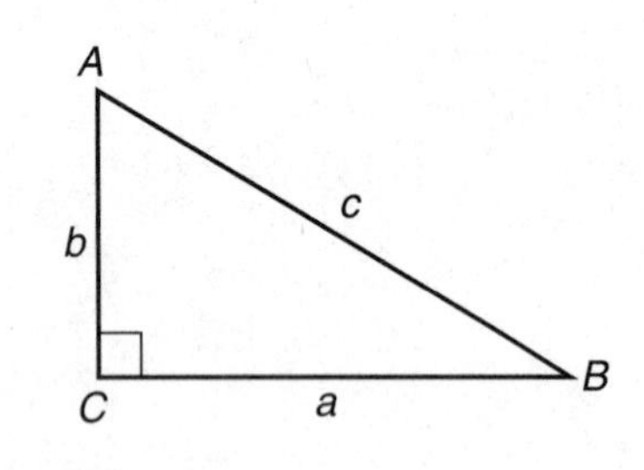

15. In ΔABC in the figure above, which of the following is equal to a?

    I. $\sqrt{c^2 - b^2}$
    II. $c \sin A$
    III. $b \tan A$

    (A) I only
    (B) II only
    (C) III only
    (D) I and II only
    (E) I, II, and III

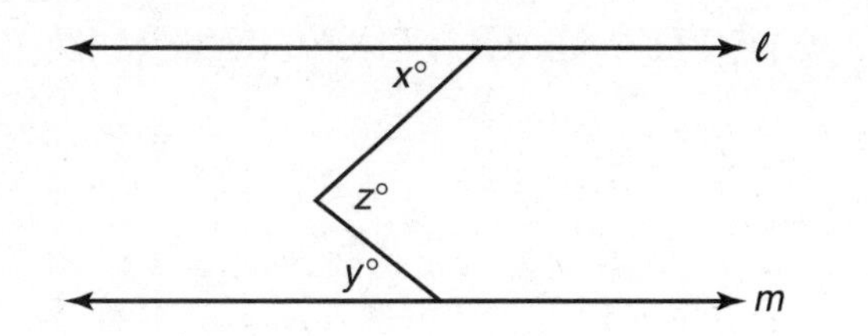

USE THIS SPACE FOR SCRATCH WORK

16. If in the figure above, $\ell \parallel m$, which of the following is a correct relationship between x, y, and z?

    (A) $y = x$
    (B) $y = x - z$
    (C) $x = z - y$
    (D) $x + y + z = 180$
    (E) There is not enough information to determine a relationship.

17. If a and b are positive integers and ab is divisible by 6, which of the following must be true?

    I. a or b is divisible by 2 and the other is divisible by 3.
    II. a or b is divisible by 6.
    III. ab is divisible by both 2 and 3.

    (A) None
    (B) I only
    (C) II only
    (D) III only
    (E) I and III only

18. What is the slope of the perpendicular bisector of line segment $\overline{AB}$ if $A = (0, -3)$ and $B = (4, 0)$?

    (A) $-\frac{4}{3}$
    (B) $-\frac{3}{4}$
    (C) 0
    (D) $\frac{3}{4}$
    (E) $\frac{4}{3}$

USE THIS SPACE FOR SCRATCH WORK

19. If $2^{3x-1} = 16$, then $x =$

(A) 0
(B) $\frac{1}{3}$
(C) $\frac{1}{2}$
(D) 1
(E) $\frac{5}{3}$

20. If the length and width of a rectangular solid are increased by 20% and the height is increased by 25%, by what percent is the volume of the solid increased?

(A) 1%
(B) 21.67%
(C) 65%
(D) 80%
(E) 180%

21. Which of the following is the equation of a parabola that passes through the points (0, 3) and (3, 0)?

(A) $y = -x + 3$
(B) $x^2 + y^2 = 9$
(C) $y = x^2 + 3$
(D) $y = -x^2 + 2x + 3$
(E) $y = x^3 - 3x^2 - x + 3$

22. If a 3-digit number is chosen at random, what is the probability that all three digits are prime numbers?

(A) $\frac{27}{900}$
(B) $\frac{125}{900}$
(C) $\frac{125}{999}$
(D) $\frac{64}{900}$
(E) $\frac{64}{999}$

23. If the lines whose equations are $y = ax + b$ and $x = cy + d$ are parallel, which statement must be true?

(A) $a = -\frac{1}{c}$

(B) $a = \frac{1}{c}$

(C) $c = a$

(D) $c = -a$

(E) $a + c = -1$

24. The domain of each of the following includes –5 EXCEPT

(A) $y = \sqrt{|x-1|}$

(B) $y = \left|\sqrt{x-1}\right|$

(C) $y = \frac{1}{x^2 + 25}$

(D) $y = \frac{1}{\sqrt{5-x}}$

(E) $y = \sqrt{25 - x^2}$

25. If in right ΔABC, $\sin A = \frac{3}{5}$, then what is the value of $\sin B$?

(A) $\frac{3}{5}$

(B) $\frac{3}{4}$

(C) $\frac{4}{5}$

(D) $\frac{5}{4}$

(E) $\frac{4}{3}$

USE THIS SPACE FOR SCRATCH WORK

USE THIS SPACE FOR SCRATCH WORK

5 | 8 9
6 | 0 4 6 8
7 | 3 5 5 5 6 7 7 9
8 | 2 2 3 5
9 | 0 1 4 4 4 8 9

Legend: 8 | 2 represents a grade of 82

26. The stem-and-leaf plot above gives the grades of the students in Mr. Smith's second period math class on the final exam. What percent of the students earned grades greater than 75?

(A) 15%
(B) 40%
(C) 50%
(D) 60%
(E) 75%

27. How many integers satisfy the inequality $\left|\frac{2x}{3} - 4\right| < 5$?

(A) 5
(B) 11
(C) 14
(D) 15
(E) Infinitely many

28. In 2004, Hamburger Heaven sold 20% more cheeseburgers than in 2003, and in 2004 the price of each cheeseburger they sold was 10% more than in 2003. The total income from the sale of cheeseburgers was what percent greater in 2004 than in 2003?

(A) 2%
(B) 20%
(C) 22%
(D) 30%
(E) 32%

USE THIS SPACE FOR SCRATCH WORK

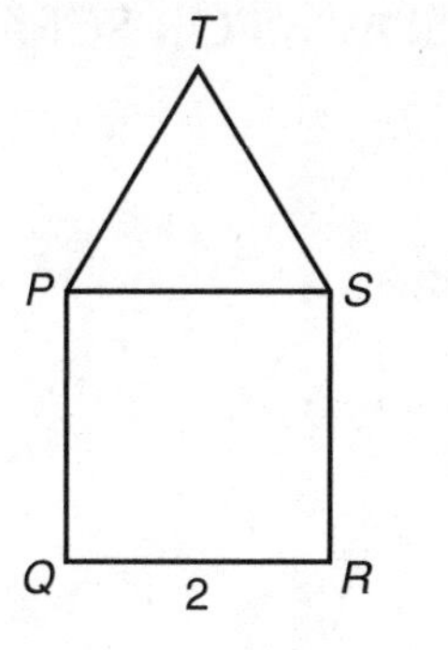

29. In the figure above, $PQRS$ is a square and PST is an equilateral triangle. If $QR = 2$, what is the length of QT (not shown)?

    (A) 3.48
    (B) 3.62
    (C) 3.86
    (D) 3.91
    (E) 4.12

30. What is the greatest integer k for which the equation $x^2 + 5x + k = 0$ has real solutions?

    (A) 4
    (B) 6
    (C) 7
    (D) 12
    (E) 16

31. What is the equation of the line that is tangent to the circle whose center is at (0, 2) and whose radius is 5 at the point (3, 6)?

    (A) $y = 3x + 6$
    (B) $y = \frac{4}{3}x + 2$
    (C) $y = -\frac{4}{3}x + 10$
    (D) $y = \frac{3}{4}x + \frac{15}{4}$
    (E) $y = -\frac{3}{4}x + \frac{33}{4}$

USE THIS SPACE FOR SCRATCH WORK

32. In how many ways can Al, Bob, Carol, Diane, and Ed stand in a straight line if Bob must be in the middle and Ed must be at one end?

(A) 6
(B) 12
(C) 24
(D) 48
(E) 120

33. For which of the following quadratic equations is the sum of its roots equal to the product of its roots?

(A) $x^2 + 5x + 5 = 0$
(B) $x^2 - 5x + 5 = 0$
(C) $2x^2 - 3x + 6 = 0$
(D) $2x^2 + 3x - 6 = 0$
(E) $2x^2 - 6x + 3 = 0$

34. If Adam can mow a large yard in 3 hours and Matthew can mow the same yard in 5 hours, when working together, how long will it take, in minutes, for the two boys to mow the yard?

(A) 92
(B) 112.5
(C) 120
(D) 132.5
(E) 240

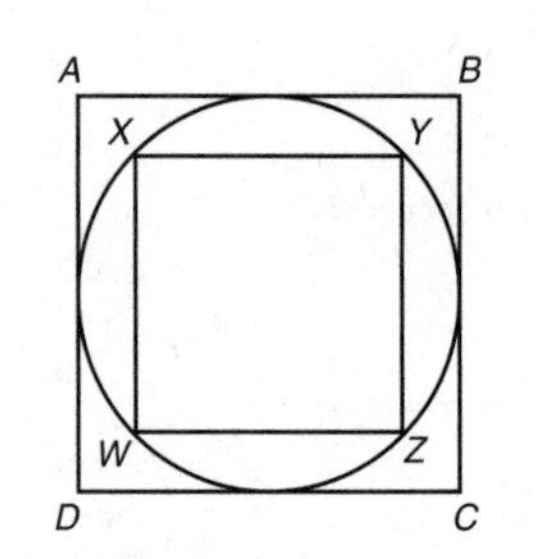

35. In the diagram above, a circle is inscribed in square *ABCD* and square *WXYZ* is inscribed in the circle. What is the ratio of the perimeter of square *ABCD* to the perimeter of square *WXYZ*?

(A) $2\sqrt{2}$
(B) 2
(C) $\sqrt{2}$
(D) $\frac{1}{\sqrt{2}}$
(E) $\frac{1}{2\sqrt{2}}$

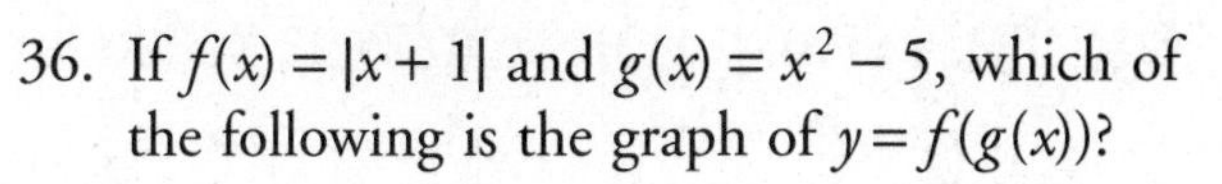

36. If $f(x) = |x + 1|$ and $g(x) = x^2 - 5$, which of the following is the graph of $y = f(g(x))$?

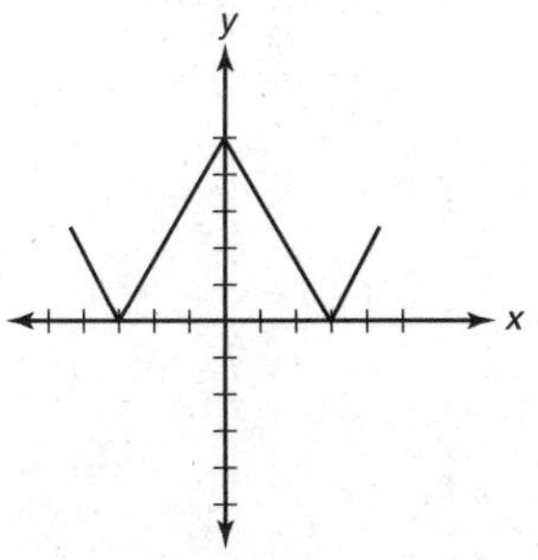

USE THIS SPACE FOR SCRATCH WORK

(A)

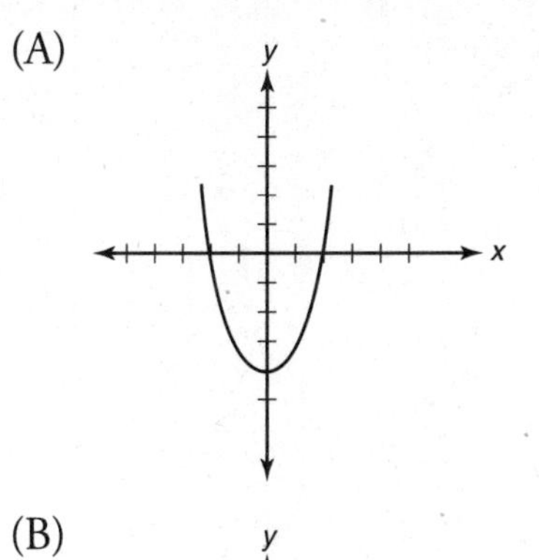

(B)

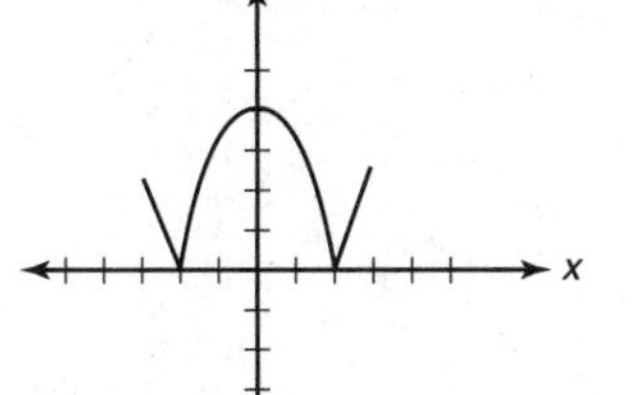

(C)

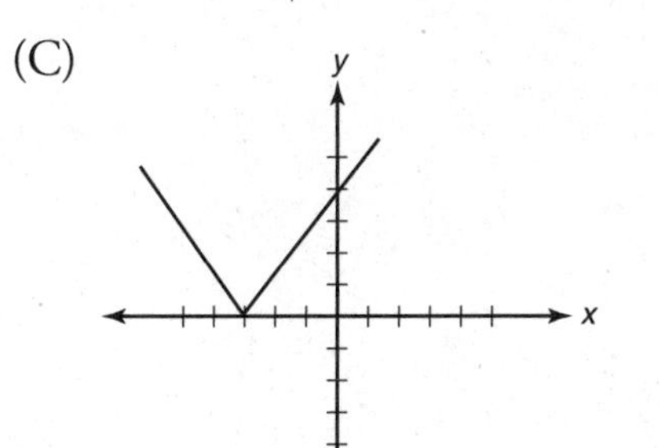

(D)

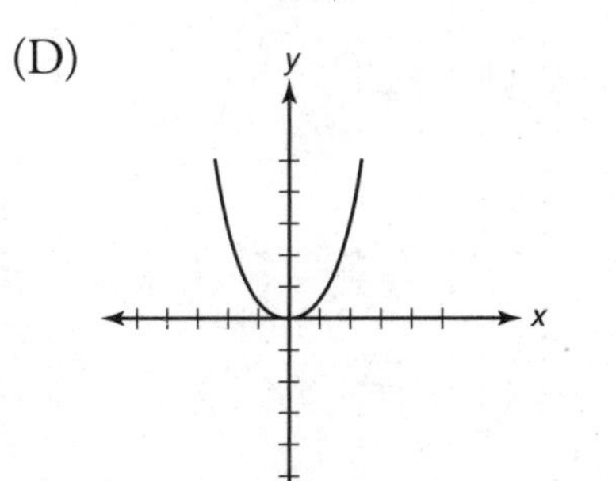

(E)

USE THIS SPACE FOR SCRATCH WORK

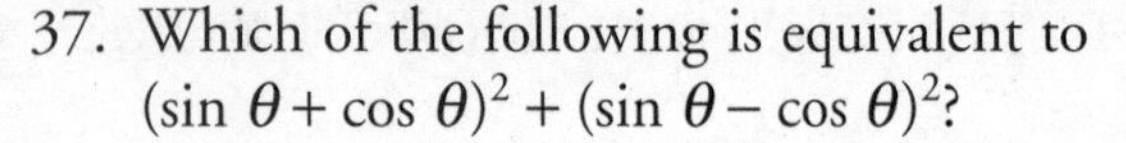

37. Which of the following is equivalent to $(\sin\theta + \cos\theta)^2 + (\sin\theta - \cos\theta)^2$?

    (A) 0
    (B) 2
    (C) $2\sin^2\theta$
    (D) $2\sin\theta\cos\theta$
    (E) $4\sin\theta\cos\theta$

38. If $i^2 = -1$, what is the value of $i^{75} - (-i)^{75}$?

    (A) $-2i$
    (B) $2i$
    (C) -2
    (D) 0
    (E) 2

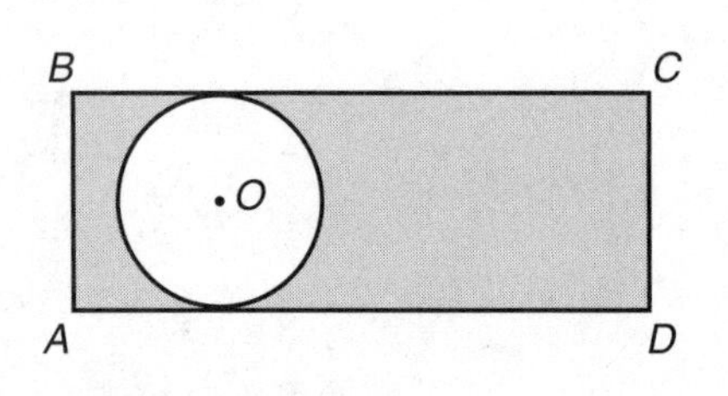

39. In the figure above, circle O is tangent to sides $\overline{AD}$ and $\overline{BC}$ of rectangle $ABCD$. If the area of the shaded region is 3 times the area of the circle, what is the ratio of the length of side $\overline{AD}$ to the circumference of the circle?

    (A) $\frac{3}{\pi}$
    (B) 1
    (C) $\frac{\pi}{3}$
    (D) 3
    (E) π

40. Which of the following is the x-coordinate of a point where the circle whose center is $(3, -4)$ and whose radius is 7 intersects the x-axis?

    I. -2.74
    II. 0
    III. 2.74

    (A) None
    (B) I only
    (C) II only
    (D) III only
    (E) I and II only

USE THIS SPACE FOR SCRATCH WORK

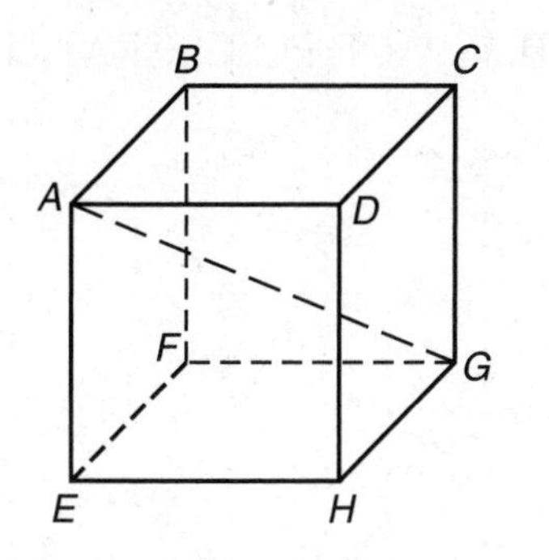

41. In cube $ABCDEFGH$, the length of diagonal $\overline{AG}$ is 9. What is the total surface area of the cube?

(A) 27
(B) 81
(C) 162
(D) $9\sqrt{3}$
(E) $81\sqrt{3}$

42. If $f(x)=\frac{x^2-16}{x^2+9}$, which of the following is a number that is NOT in the range of f?

(A) 1
(B) 0.5
(C) 0
(D) −0.5
(E) −1

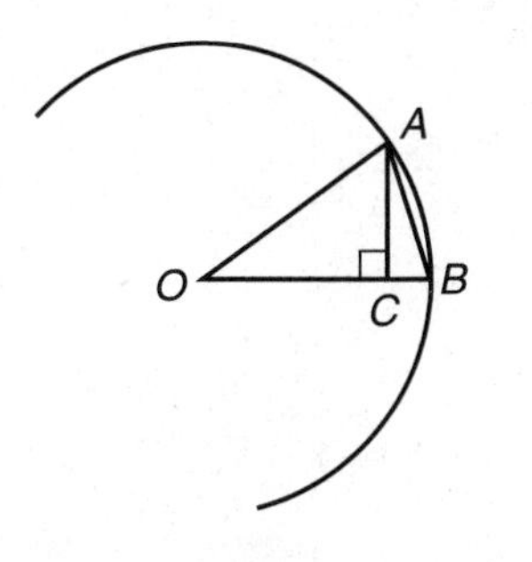

43. The figure above shows an arc of a circle whose center is at O. If $OA = 2$ and $AC = 1$, what is the area of ΔACB?

(A) $1-\frac{\sqrt{3}}{2}$
(B) $2-\sqrt{3}$
(C) 1
(D) 2
(E) $\frac{\pi}{3}$

44. If $f(x) = \sqrt[3]{x+7} - 7$, what is the value of $f^{-1}(7)$?

    (A) −7
    (B) 7
    (C) 42
    (D) 189
    (E) 2,737

45. At a national educational conference, all of the participants are teachers or administrators. If there are 584 teachers at the conference and 27% of the participants are administrators, how many administrators are attending the conference?

    (A) 158
    (B) 216
    (C) 312
    (D) 800
    (E) 2,163

46. What is the area of rhombus *ABCD*, if $AB = 10$ and $m\angle A = 45°$?

    (A) 50
    (B) 64.6
    (C) 70.7
    (D) 78.2
    (E) 100

47. If the statement "All practical numbers are prime" is false, which of the following statements must be true?

    (A) No practical numbers are prime.
    (B) No prime numbers are practical.
    (C) Some practical numbers are prime.
    (D) Some practical numbers are not prime.
    (E) Some prime numbers are not practical.

USE THIS SPACE FOR SCRATCH WORK

48. The base of pyramid 1 is a 3-4-5 triangle, and the base of pyramid 2 is a square whose sides are 3. If the volumes of the pyramid are equal, what is the ratio of the height of pyramid 1 to the height of pyramid 2?

(A) $\frac{4}{9}$

(B) $\frac{2}{3}$

(C) $\frac{4}{3}$

(D) $\frac{3}{2}$

(E) $\frac{9}{4}$

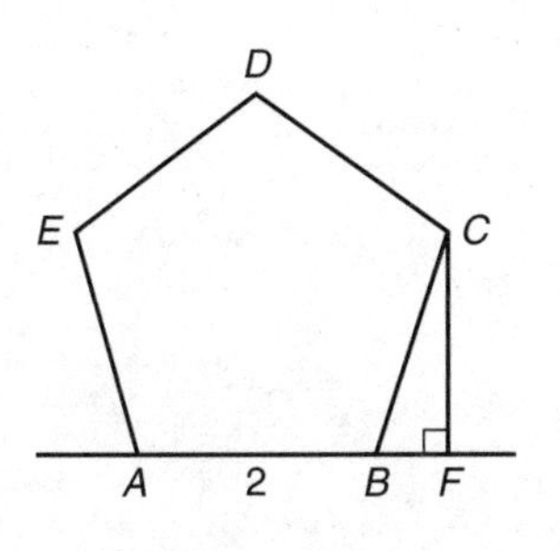

49. In the figure above, $ABCDE$ is a regular pentagon, B lies on $\overline{AF}$ and $\overline{CF} \perp \overline{BF}$. If $AB = 2$, what is the area of ΔBFC?

(A) 0.59
(B) 1.00
(C) 1.18
(D) 1.90
(E) 2.00

USE THIS SPACE FOR SCRATCH WORK

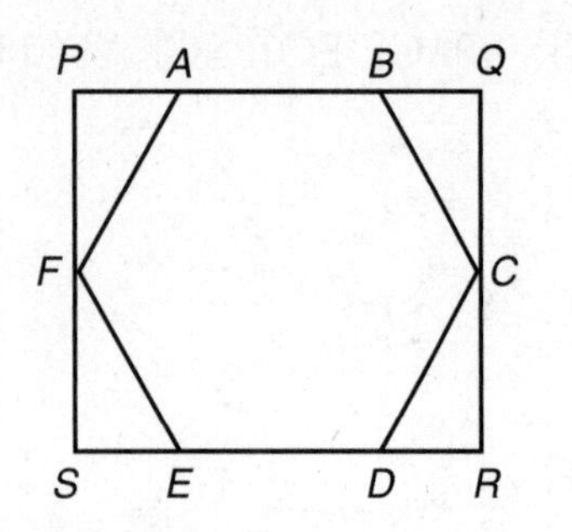

USE THIS SPACE FOR SCRATCH WORK

50. In the figure above, regular hexagon *ABCDEF* is inscribed in rectangle *PQRS*. If each side of the hexagon is 1, what is the area of the rectangle?

    (A) 2
    (B) 4
    (C) 6
    (D) $\sqrt{3}$
    (E) $2\sqrt{3}$

Answer Key

MODEL TEST 1

1. D	14. E	27. D	40. B
2. A	15. E	28. E	41. C
3. C	16. C	29. C	42. A
4. C	17. D	30. B	43. A
5. C	18. A	31. E	44. E
6. C	19. E	32. B	45. B
7. E	20. D	33. B	46. C
8. A	21. D	34. B	47. D
9. B	22. D	35. C	48. D
10. B	23. B	36. B	49. A
11. A	24. B	37. B	50. E
12. D	25. C	38. A	
13. C	26. D	39. B	

ANSWERS EXPLAINED

For many of the questions in this model test, an alternative solution, indicated by two asterisks (**), follows the first solution. Almost always the first solution is the direct mathematical one and the other is based on one of the tactics discussed in the "Tactics" chapter.

1. **(D)** Use TACTIC C2: always solve a proportion by cross multiplying:

$$\frac{x+3}{7} = \frac{x-3}{3} \Rightarrow 3(x+3) = 7(x-3) \Rightarrow$$
$$3x + 9 = 7x - 21 \Rightarrow 4x = 30 \Rightarrow x = 7.5$$

**Whenever a question asks for the value of a single variable, you can backsolve (TACTIC 2). However, you should not have to on a question as easy as this one.

2. **(A)** $a = b^2 \Rightarrow 18 = b^2 \Rightarrow b = \pm\sqrt{18} \Rightarrow c\sqrt{2} = \pm\sqrt{18}$. So,

$c = \frac{\pm\sqrt{18}}{\sqrt{2}} = \pm\sqrt{9} = \pm 3$. So $c = 3$ or $c = -3$, but only 3 is an answer choice.

**You don't have to rationalize or simplify $\frac{\sqrt{18}}{\sqrt{2}}$. You can just use your calculator.

3. **(C)** $||-5| - |-7|| = |5 - 7| = |-2| = 2$

4. **(C)** Since vertical angles are congruent (KEY FACT G4), $b = 30$. Then by KEY FACT G2, $a + b + 40 = 180$:

$$a + 30 + 40 = 180 \Rightarrow a + 70 = 180 \Rightarrow a = 110$$

5. **(C)** Since $CD = 5 - (-2) = 7$, each side of square $ABCD$ must be 7. Sketch the two possibilities for the square and label the four new vertices. Of the five choices, only (–4, 5) could be one of the vertices.

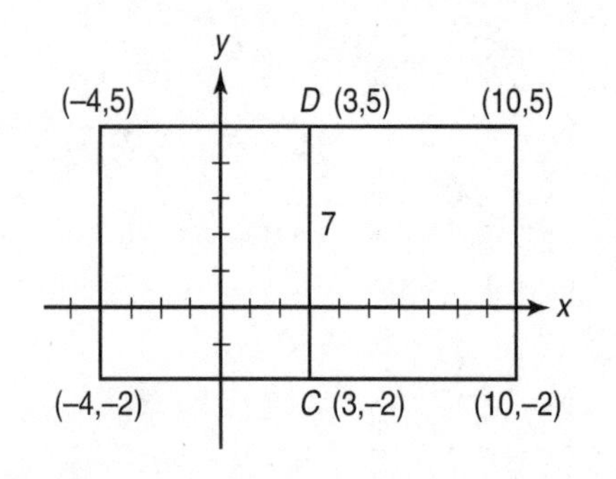

6. **(C)** Factor both numerators and reduce:

$$\frac{x^2 - x - 6}{x+5} \cdot \frac{x^2 + 3x - 10}{2x - 6} =$$
$$\frac{\cancel{(x-3)}(x+2)}{\cancel{x+5}} \cdot \frac{(x-2)\cancel{(x+5)}}{2\cancel{(x-3)}} = \frac{x^2 - 4}{2}$$

**If you get stuck trying to factor, just plug in a number. If $x = 1$:

$$\frac{x^2 - x - 6}{x+5} \bullet \frac{x^2 + 3x - 10}{2x - 6} = \frac{-6}{6} \bullet \frac{-6}{-4} = -\frac{6}{4} = -\frac{3}{2}$$

Only choice C equals $-\frac{3}{2}$ when $x = 1$.

7. **(E)** In the given week, Susan earned \$1,000 – \$475 = \$525 in commissions, which represented 7% of her sales. So

$$(0.07)(\text{Sales}) = \$525 \Rightarrow \text{Sales} = \frac{\$525}{0.07} = \$7{,}500$$

**Use TACTIC 2: backsolve starting with choice C. 7% of \$700 = \$49, which is much too small. Eliminate choices A, B, and C and test D or E.

8. **(A)** Represent the numbers by x and y. Then $x + y = 5$ and $x - y = 5$. Add the two equations:

$$\begin{aligned} x + y &= 5 \\ \underline{x - y} &\underline{= 5} \\ 2x \quad &= 10 \Rightarrow x = 5 \end{aligned}$$

Then $5 + y = 5 \Rightarrow y = 0$. So the product $xy = (5)(0) = 0$.

9. **(B)** $f(-5) = -2(-5)^2 - 3(-5) = -2(25) + 15 = -50 + 15 = -35$

10. **(B)** If r is the radius of circle O, then by KEY FACTS J4 and J8, $a = 2\pi r$ and $b = \pi r^2$. So:

$$a = b \Rightarrow 2\pi r = \pi r^2 \Rightarrow 2r = r^2 \Rightarrow r = 2$$

**Use TACTIC 2: backsolve. Calculate the area and circumference for each possible value of the radius.

11. **(A)** $\sqrt{n+2} = \frac{2}{3} \Rightarrow \left(\sqrt{n+2}\right)^2 = \left(\frac{2}{3}\right)^2 \Rightarrow n + 2 = \frac{4}{9} \Rightarrow n = \frac{4}{9} - 2 = -\frac{14}{9}$

**Use TACTIC 2: backsolve. Use your calculator to test the answers.

12. **(D)** By the triangle inequality (KEY FACT H9), x must be greater than 2 and less than 14. Therefore, there are 11 possible integer values of x: 3, 4, 5, . . . , 13.

13. **(C)** The only even prime is 2, so $a = 1$. The only prime that is divisible by 7 is 7, so $b = 1$. No prime is divisible by 9, so $c = 0$. The primes between 20 and 30 are 23 and 29, so $d = 2$. Finally, $a + b + c + d = 1 + 1 + 0 + 2 = 4$.

14. **(E)** $f(x) < 25 \Rightarrow x^2 - 11 < 25 \Rightarrow x^2 < 36 \Rightarrow -6 < x < 6$.

15. **(E)** By the Pythagorean theorem (KEY FACT H5),

$a^2 + b^2 = c^2 \Rightarrow a^2 = c^2 - b^2 \Rightarrow a = \sqrt{c^2 - b^2}$ (I is true).
Now use the trigonometric ratios SOHCAHTOA (KEY FACT M1):

$$\sin A = \frac{a}{c} \Rightarrow a = c \sin A \text{ and } \tan A = \frac{a}{b} \Rightarrow a = b \tan A$$

(II and III are true). So statements I, II, and III are true.

**Use TACTIC 3: plug in numbers for *a*, *b*, and *c*, say 3, 4, and 5.

16. **(C)** Extend a line segment to create a transversal.

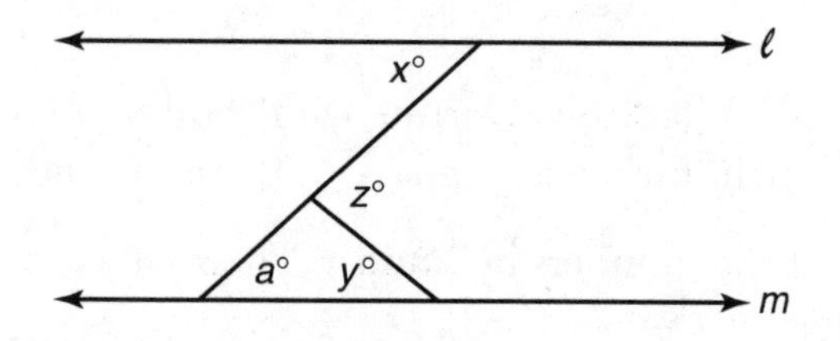

Then by KEY FACT G6, since ℓ and m are parallel, $a = x$. Since the measure of an exterior angle of a triangle is equal to the sum of the measures of the two opposite interior angles (KEY FACT H2):

$$z = a + y \Rightarrow z = x + y \Rightarrow x = z - y$$

**By TACTIC 5, since the given diagram is drawn to scale, you can trust it. The values of *x*, *y*, and *z* appear to be *about* 40, 45, and 80, respectively. Eliminate choices B and D. A and C are reasonable. However, since it is unlikely that the answer doesn't involve *z*, choice C is your best guess.

17. **(D)** If $a = 6$ and $b = 1$, ab is divisible by 6, but b is divisible by neither 2 nor 3. (Statement I is false.)

If $a = 2$ and $b = 3$, ab is divisible by 6, but neither a nor b is divisible by 6. (Statement II is false.)

If ab is a multiple of 6, it must be divisible by every divisor of 6, so ab is divisible by both 2 and 3. (Statement III is true.)

Only statement III is true.

18. **(A)** Use the slope formula (KEY FACT L5) to get the slope of $\overline{AB}$:

$$\frac{0-(-3)}{4-0} = \frac{3}{4}$$

By KEY FACT L7, the slope of *any* line that is perpendicular to $\overline{AB}$ is $-\frac{4}{3}$, the negative reciprocal of $\frac{3}{4}$.

**Use TACTIC 4: draw $\overline{AB}$ and line ℓ perpendicular to $\overline{AB}$. Clearly, the slope of ℓ is negative. Eliminate choices C, D, and E, and guess between A and B.

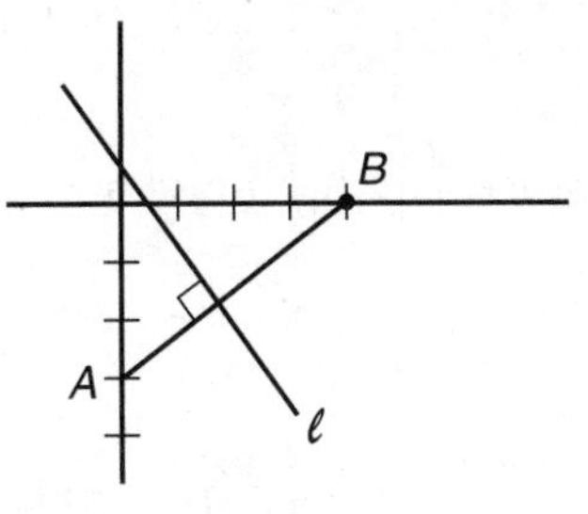

19. **(E)** $2^{3x-1} = 16 = 2^4 \Rightarrow 3x - 1 = 4 \Rightarrow 3x = 5 \Rightarrow x = \frac{5}{3}$

**Use TACTIC 2: backsolve starting with choice C. $2^{3\left(\frac{1}{2}\right)-1} = 2^{\frac{1}{2}} = 1.414$, which is too small. Eliminate choices A, B, and C and test choice D or E.

20. **(D)** If ℓ, w, and h are the dimensions of the original rectangular solid, then by KEY FACT K1, the original volume is ℓwh. The new volume is $(1.2\ell)(1.2w)(1.25h) = 1.8(\ell wh)$. So the new volume is 80% more than the original volume.

**Let the original solid be a cube whose edges are 100. Then the new solid has dimensions 120, 120, and 125. The volume of the cube is $100^3 = 1{,}000{,}000$ and the volume of the new rectangular solid is $(120)(120)(125) = 1{,}800{,}000$, which is an increase of 800,000. Finally, $\frac{800{,}000}{1{,}000{,}000} = 80\%$.

21. **(D)** By KEY FACT L11, only choices C and D are the equations of a parabola. Eliminate choices A, B, and E. $y = x^2 + 3$ (choice C) doesn't pass through (3, 0), whereas $y = -x^2 + 2x + 3$ (choice D) passes through both (0, 3) and (3, 0).

**You can test each answer choice by graphing each equation on a graphing calculator, but that clearly takes more time than the solution presented above.

22. **(D)** First determine how many 3-digit numbers can be written using only prime digits. Since there are 4 prime digits (2, 3, 5, and 7), if you want to write a 3-digit number using only prime digits, there are 4 choices for each digit. So, by the counting principle (KEY FACT O2), there are $4 \times 4 \times 4 = 64$ 3-digit numbers, each of whose digits is prime. Finally, since there are 900 3-digit numbers (the integers from 100–999), by KEY FACT O3, the probability that a 3-digit number chosen at random will have only prime digits is $\frac{64}{900}$.

23. **(B)** By KEY FACT L8, the slope of the line $y = ax + b$ is a. To find the slope of the line $x = cy + d$, solve for y:

$$x = cy + d \Rightarrow cy = x - d \Rightarrow y = \frac{1}{c}x - \frac{d}{c}$$

So the slope is $\frac{1}{c}$. Since parallel lines have equal slopes (KEY FACT L7), $a = \frac{1}{c}$.

24. **(B)** If $x = -5$, then $\left|\sqrt{-5-1}\right| = \left|\sqrt{-6}\right|$, which is undefined. In each of the other choices, if you replace x by -5, you will get a real number.

25. **(C)** First, draw and label triangle ABC.

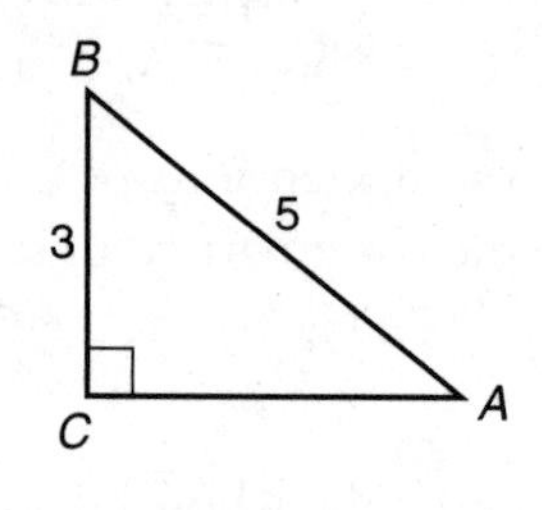

By KEY FACT M1, sine $= \frac{\text{opposite}}{\text{hypotenuse}}$, so let $BC = 3$ and $AB = 5$. Then, either by using the Pythagorean theorem (KEY FACT H5) or by recognizing that ΔABC is a 3-4-5 triangle, note that $AC = 4$. Therefore, $\sin B = \frac{4}{5}$.

**If $\sin A = \frac{3}{5}$, then $A = \sin^{-1}\left(\frac{3}{5}\right) = 36.87°$. So $B = 90° - 36.87° = 53.13°$, and $\sin B = \sin 53.13° = 0.8 = \frac{4}{5}$.

26. **(D)** From the given stem-and-leaf plot, you can see that the grades ranged from 58 to 99. Of the 25 grades in the chart, 15 of them are greater than 75 (4 between 76 and 79; 4 in the 80s; and 7 in the 90s). Finally, $\frac{15}{25} = 0.6 = 60\%$.

27. **(D)** $\left|\frac{2x}{3} - 4\right| < 5 \Rightarrow -5 < \left(\frac{2x}{3} - 4\right) < 5 \Rightarrow -1 < \frac{2x}{3} < 9 \Rightarrow$

$$-3 < 2x < 27 \Rightarrow -\frac{3}{2} < x < \frac{27}{2}$$

The integers that satisfy this inequality are $-1, 0, 1, 2, \ldots, 13$. There are 15 of them.

**Test some numbers: 0, 1, 2, 3 all satisfy the inequality. Keep going: $x = 6$ works; so does $x = 12$. That's already 13 integers that work. Eliminate A and B. Keep going: 13 works, but 14 does not:

$$\frac{2(14)}{3} - 4 = \frac{28}{3} - 4 = 9\frac{1}{3} - 4 = 5\frac{1}{3}$$

So the 14 integers from 0 to 13 all work. That is it unless there are any negative solutions. There is just one: −1, for a total of 15.

28. **(E)** Assume that in 2003 Hamburger Heaven sold x cheeseburgers at a price of y dollars each. Then in 2004, they sold $x + 0.20x = 1.2x$ cheeseburgers at a price of $y + 0.10y = 1.1y$ dollars each. So the total income in 2003 was xy, whereas in 2004 it was $(1.2x)(1.1y) = 1.32xy$, an increase of 32%.

    **Do exactly what was done above except replace x and y by 100. Assume that in 2003, Hamburger Heaven sold 100 cheeseburgers for \$100 each for a total of \$10,000. Then in 2004, Hamburger Heaven sold 120 cheeseburgers for \$110 each, for a total of \$13,200. Finally, \$13,200 is \$3,200 more than \$10,000, an increase of $\frac{3,200}{10,000} = 0.32 = 32\%$.

29. **(C)** Draw in $\overline{TQ}$. Also draw $\overline{TU}$ perpendicular to $\overline{QR}$, intersecting $\overline{PS}$ at V.

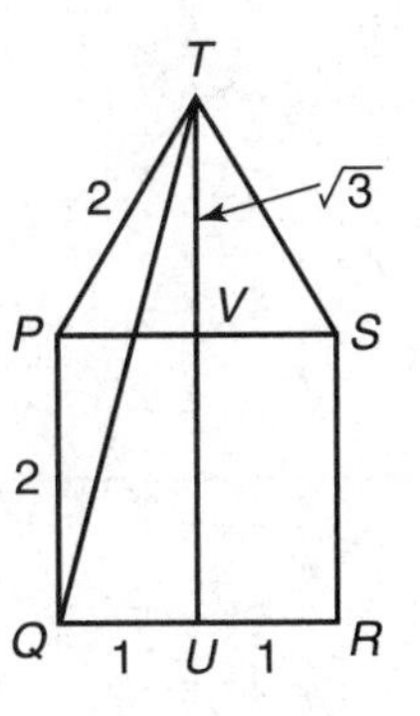

Since ΔPVT is a 30-60-90 right triangle, by KEY FACT H8, $PV = QU = 1$, and $TV = \sqrt{3}$. So $TU = 2 + \sqrt{3}$. Now use the Pythagorean theorem in ΔTUQ:

$$(TQ)^2 = (QU)^2 + (TU)^2 = 1^2 + (2 + \sqrt{3})^2 = 14.928 \Rightarrow TQ = \sqrt{14.928} = 3.86$$

30. **(B)** By KEY FACT E2, a quadratic equation has real solutions if and only if its discriminant is greater than or equal to 0. In the equation $x^2 + 5x + k = 0$, the discriminant, D, is $b^2 - 4ac$, where $a = 1$, $b = 5$, and $c = k$. So $D = 5^2 - 4(1)k = 25 - 4k$. Then:

$$25 - 4k \geq 0 \Rightarrow 25 \geq 4k \Rightarrow 6.25 \geq k$$

The largest integer value of k for which $k \leq 6.25$ is 6.

**You can use a graphing calculator to backsolve. For example, to test choice C, graph the parabola $y = x^2 + 5x + 7$. Since the graph doesn't cross the x-axis, the equation $x^2 + 5x + 7 = 0$ does not have real solutions. Eliminate choices C, D, and E and test choice B. It works.

31. **(E)** To write the equation of a line, you need to know a point on the line and the slope of the line. Here, the point is (3, 6). By the slope formula (KEY FACT L5), the radius drawn from the center (0, 2) to the point (3, 6) has slope $\frac{6-2}{3-0} = \frac{4}{3}$. By KEY FACT J11, the tangent line is perpendicular to the radius and so by KEY FACT L7, the slope of the tangent line is $-\frac{3}{4}$. Finally, using the point-slope formula for a line (KEY FACT L9), $y - y_1 = m(x - x_1)$, you get:

$$y - 6 = -\frac{3}{4}(x-3) \Rightarrow y - 6 = -\frac{3}{4}x + \frac{9}{4} \Rightarrow y = -\frac{3}{4}x + \frac{33}{4}$$

**Use TACTIC 4: make a quick sketch of the circle and the line, ℓ, tangent to the circle at (3, 6). Since the slope of ℓ is clearly negative, you can eliminate choices A, B, and D. If your sketch is good, you can see that the y-intercept is less than 10, so you can eliminate choice C as well.

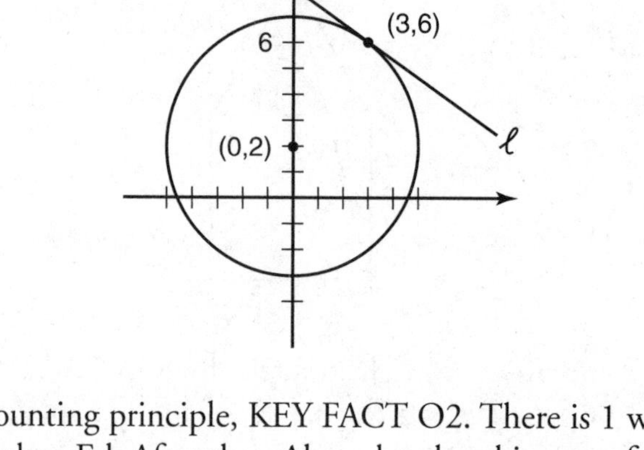

32. **(B)** Use the counting principle, KEY FACT O2. There is 1 way to place Bob and 2 ways to place Ed. After that, Al can be placed in one of the 3 remaining spots, followed by Carol, who can stand in one of the other 2 spots. Finally, there is only 1 spot left for Diane. The total number of ways is $1 \times 2 \times 3 \times 2 \times 1 = 12$.

**List the possibilities. First put Al in the middle and put Ed at the front of the line:

EBACD ECABD EDABC
EBADC ECADB EDACB

Clearly, there will be 6 more possibilities with Ed at the back of the line, so there are a total of 12 ways for the five people to line up.

33. **(B)** By KEY FACT E3, if $ax^2 + bx + c = 0$, the sum of the roots is $\frac{-b}{a}$ and the product of the roots is $\frac{c}{a}$. In choice B, $\frac{c}{a} = \frac{5}{1} = 5$ and $\frac{-b}{a} = \frac{-(-5)}{1} = 5$. None of the other choices works.

34. **(B)** Adam works at the rate of $\frac{1 \text{ yard}}{3 \text{ hours}} = \frac{1}{3}\frac{\text{yards}}{\text{hour}}$. Similarly, Matthew's rate of work is $\frac{1}{5}\frac{\text{yards}}{\text{hour}}$. Together they can mow $\left(\frac{1}{3} + \frac{1}{5}\right) = \frac{8}{15}$ yards in one hour.

Finally:

$$1 \text{ yard} = \left(\frac{8}{15}\frac{\text{yards}}{\text{hour}}\right)(t \text{ hours}) \Rightarrow 1 = \frac{8}{15}t \Rightarrow t = \frac{15}{8}$$

So when working together, the time required for Adam and Matthew to mow the lawn is:

$$\frac{15}{8} = 1\frac{7}{8} = 1.875 \text{ hours} = 1 \text{ hour and } 52.5 \text{ minutes}$$
$$= 112.5 \text{ minutes}$$

If you get stuck on a problem like this, don't omit it

**Use TACTIC 8: and at least eliminate the absurd choices. Since Adam can mow the lawn in 3 hours = 180 minutes, obviously working together will take less time. Eliminate choice E. If each boy could mow the lawn in 3 hours, together they would take 1.5 hours = 90 minutes. Since Matthew works much slower, it will take them more than 90 minutes. Eliminate choice A and guess among choices B, C, and D.

35. **(C)** Let $WZ = 1$. Then the perimeter of square $WXYZ$ is 4. Since ΔWZY is a right isosceles triangle, by KEY FACT H7, the length of hypotenuse $\overline{WY}$ is $\sqrt{2}$. However, $\overline{WY}$ is also a diameter of the circle, as is $\overline{EF}$ in the diagram below. Since $EF = AB$, the perimeter of square $ABCD$ is $4\sqrt{2}$. Finally the ratio of the perimeters of the two squares $\frac{4\sqrt{2}}{4} = \sqrt{2}$.

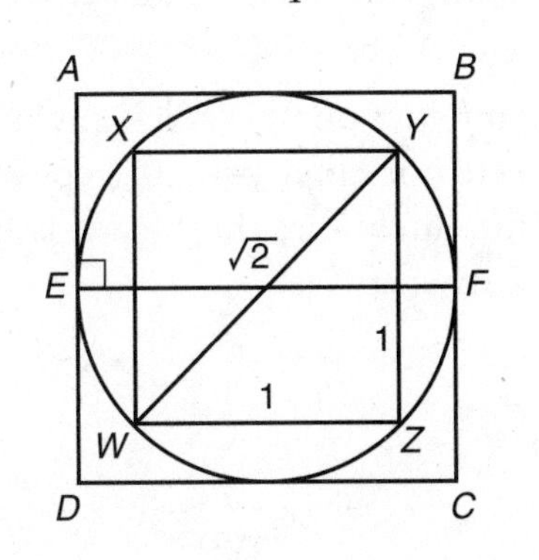

**Use TACTIC 8: eliminate absurd choices. Since the perimeter of square *ABCD* is greater than the perimeter of square *WXYZ*, the ratio must be greater than 1. So you can eliminate choices D and E. Also, $2\sqrt{2}$ is almost 3. From the diagram, $\overline{AB}$ is clearly less than 2 times as long as $\overline{XY}$. So the perimeter of *ABCD* is less than 2 times the perimeter of *WXYZ*. Eliminate choices A and B as well.

36. **(B)** $f(g(x)) = f(x^2 - 5) = |(x^2 - 5) + 1| = |x^2 - 4|$

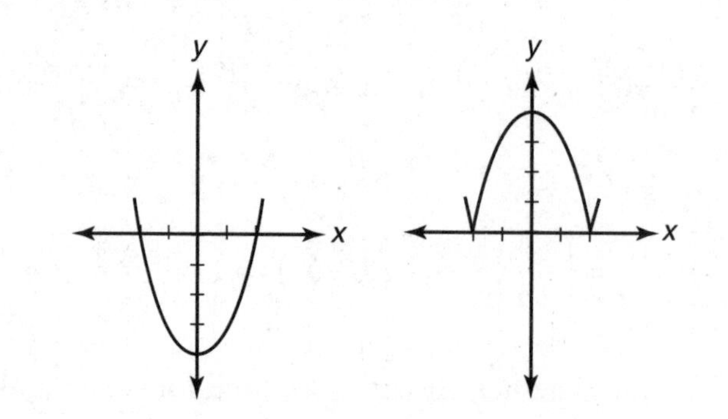

By KEY FACT N4, the graph of $y = x^2 - 4$ is the parabola above left, which is the parabola $y = x^2$ shifted 4 units down. The graph of $y = |x^2 - 4|$ (above right) results from reflecting the negative portion of the graph on the left in the *x*-axis.

**Test some values: if $y = |x^2 - 4|$, then when $x = 0$, $y = 4$. Only choices B and C pass through (0, 4). When $x = 2$, $y = 0$ and of choices B and C, only B passes through (2, 0).

37. **(B)** First, expand $(\sin\theta + \cos\theta)^2$:

$$(\sin\theta + \cos\theta)^2 = (\sin\theta + \cos\theta)(\sin\theta + \cos\theta)$$
$$= \sin^2\theta + 2\sin\theta\cos\theta + \cos^2\theta$$

Since by KEY FACT M2 $\sin^2\theta + \cos^2\theta = 1$:

$$(\sin\theta + \cos\theta)^2 = 1 + 2\sin\theta\cos\theta$$

Similarly, $(\sin\theta - \cos\theta)^2 = 1 - 2\sin\theta\cos\theta$

Finally, add the two expressions:

$$(\sin\theta + \cos\theta)^2 + (\sin\theta - \cos\theta)^2 =$$
$$(1 + \cancel{2\sin\theta\cos\theta}) + (1 - \cancel{2\sin\theta\cos\theta}) = 2$$

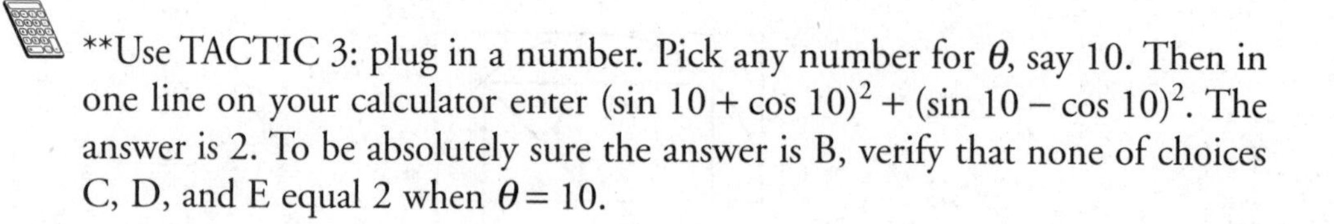

**Use TACTIC 3: plug in a number. Pick any number for θ, say 10. Then in one line on your calculator enter $(\sin 10 + \cos 10)^2 + (\sin 10 - \cos 10)^2$. The answer is 2. To be absolutely sure the answer is B, verify that none of choices C, D, and E equal 2 when $\theta = 10$.

38. **(A)** By KEY FACT P2, i^{75} is equal to i^r, where r is the remainder when 75 is divided by 4. Since $75 \div 4 = 18.75$, the quotient is 18. Then $18 \times 4 = 72$ and $75 - 72 = 3$. So the remainder is 3.

    Therefore, $i^{75} = i^3$ and $(-i)^{75} = (-1)^{75}(i)^{75} = (-1)(i)^3 = -i^3$
    Since $i^3 = -i$, then $-i^3 = i$
    Finally, $i^{75} - (-i)^{75} = i^3 - (-i^3) = -i - i = -2i$

39. **(B)** If r is the radius of the circle, then by KEY FACT J8, its area is πr^2. Therefore, the area of the shaded region is $3\pi r^2$, and the area of the entire rectangle is $\pi r^2 + 3\pi r^2 = 4\pi r^2$. $\overline{AB}$, the width of the rectangle, is equal to the diameter of the circle. So $AB = 2r$, and

    the area of $ABCD = 4\pi r^2 = (AB)(AD) = 2r(AD) \Rightarrow AD = \dfrac{4\pi r^2}{2r} = 2\pi r.$

    By KEY FACT J4, though, $2\pi r$ is exactly the circumference of the circle. So the ratio of AD to the circumference is 1.

    **Use TACTICS 5 and 8: trust the diagram and eliminate absurd choices. Clearly $\overline{AD}$ isn't 3 times the circumference, so eliminate choices D and E and guess.

40. **(B)** By KEY FACT L10, the equation of the circle whose center is (3, –4) and whose radius is 7 is $(x - 3)^2 + (y + 4)^2 = 49$. This circle intersects the x-axis at points where the y-coordinate is 0. So, replace y by 0 and solve:

    $$(x - 3)^2 + (0 + 4)^2 = 49 \Rightarrow (x - 3)^2 + 16 = 49 \Rightarrow$$
    $$(x - 3)^2 = 33 \Rightarrow x - 3 = \pm\sqrt{33} \Rightarrow x = 3 \pm \sqrt{33}$$

    So $x = 3 + \sqrt{33} \approx 3 + 5.74 = 8.74$ or
    $x = 3 - \sqrt{33} \approx 3 - 5.74 = -2.74$

    Only I is true.

    **Use TACTIC 4: make a quick sketch of the circle. You can immediately see that the circle does not intersect the x-axis at either 0 or 2.74. You cannot be sure that –2.74 is correct, so you have to guess between choices A and B. However, from the sketch it appears that there is an intersection point near –3, so it would make sense to guess B.

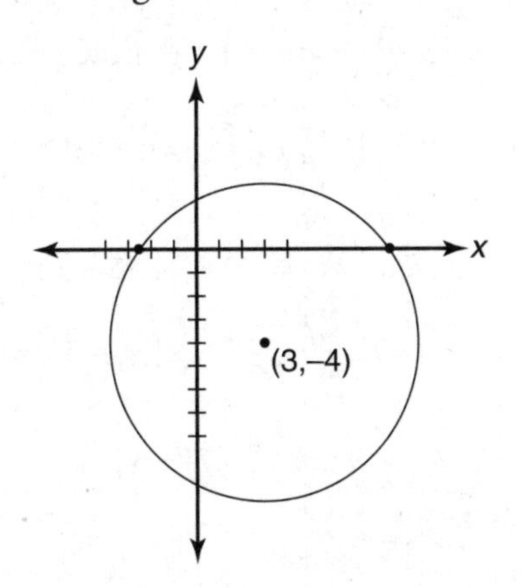

41. **(C)** If s is the side of the cube, then by KEY FACT H7, the length of $\overline{EG}$, a diagonal of the square base, is $s\sqrt{2}$. Then by the Pythagorean theorem (KEY FACT H5), in right ΔAEG:

$$s^2 + (s\sqrt{2})^2 = 9^2 \Rightarrow s^2 + 2s^2 = 81 \Rightarrow 3s^2 = 81 \Rightarrow s^2 = 27.$$

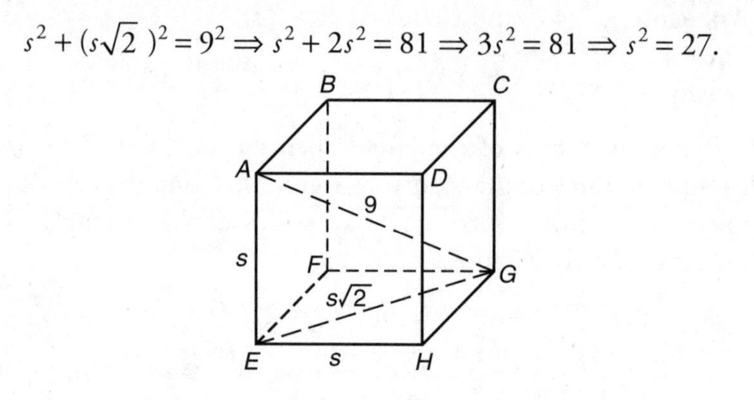

The formula for the total surface area of a cube is $A = 6s^2$ (KEY FACT K2), so $A = 6 \times 27 = 162$.

42. **(A)** If $\frac{x^2 - 16}{x^2 + 9} = 1$, then $x^2 - 16 = x^2 + 9$, which is impossible ($x^2 - 16$ is *always* less than $x^2 + 9$). So 1 is not in the range of $f(x)$. You can easily verify that each of the other numbers is in the range, but you don't need to.

43. **(A)** Since $\overline{OA}$ and $\overline{OB}$ are radii, $OB = OA = 2$. By the Pythagorean theorem (KEY FACT H5):

$$1^2 + (OC)^2 = 2^2 \Rightarrow (OC)^2 = 3 \Rightarrow OC = \sqrt{3}$$

$$\text{So } CB = OB - OC = 2 - \sqrt{3}$$

Then by KEY FACT H10, the area of ΔACB is:

$$\frac{1}{2}bh = \frac{1}{2}\left(2 - \sqrt{3}\right)(1) = 1 - \frac{\sqrt{3}}{2}$$

**Use TACTIC 5: trust the diagram. BC is less than one-half of AC, which is 1. So the area of ΔABC is less than $\frac{1}{2}(1)\left(\frac{1}{2}\right) = 0.25$. Clearly, the answer is not C, D, or E. Even choice B is greater than 0.25. Choose A.

44. **(E)** Use KEY FACT N8. To find $f^{-1}(x)$, write $y = \sqrt[3]{x+7} - 7$, switch x and y, and solve for y:

$$x = \sqrt[3]{y+7} - 7 \Rightarrow x + 7 = \sqrt[3]{y+7} \Rightarrow (x+7)^3 = y + 7 \Rightarrow y = (x+7)^3 - 7$$

$$\text{So } f^{-1}(x) = (x+7)^3 - 7 \Rightarrow f^{-1}(7) = (7+7)^3 - 7 = 14^3 - 7 = 2{,}737$$

**Since $f^{-1}(7)$ is the number x such that $f(x) = 7$, you can backsolve: Only $f(2{,}737) = 7$.

45. **(B)** If the administrators constitute 27% of the total, then the teachers are $100\% - 27\% = 73\%$ of the total. So if T is the total number of participants:

$$0.73T = 584 \Rightarrow T = 800$$

Therefore, there are $800 - 584 = 216$ administrators at the conference.

**If x represents the number of administrators, then the total number of participants is $x + 584$. So x is equal to 27% of $x + 584$:

$$x = 0.27(x + 584) = 0.27x + 157.68 \Rightarrow 0.73x = 157.68 \Rightarrow x = 216$$

**Use TACTIC 2: backsolve. If there are 312 administrators, then there are a total of $312 + 584 = 896$ participants. However, 27% of 896 is only 241.92, so 312 is too big. Eliminate choices C, D, and E and test 216. It works.

46. **(C)** A rhombus is a parallelogram in which all the sides are the same length. By KEY FACT I9, the formula for the area of any parallelogram is $A = bh$. Sketch the rhombus and draw in the height.

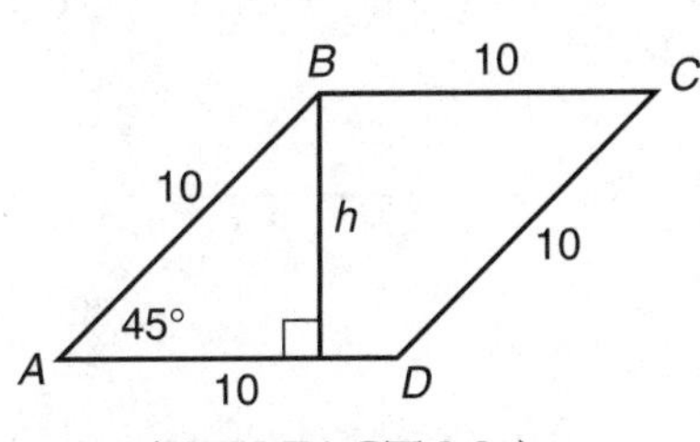

To find h, use the sine ratio (KEY FACT M1):

$$\sin 45° = \frac{h}{10} \Rightarrow h = 10 \sin 45° = 10(0.707) = 7.07$$

Alternatively, you can find h by using KEY FACT H7, which says that in a 45-45-90 right triangle, the length of each leg is equal to the length of the hypotenuse divided by $\sqrt{2}$: $h = \frac{10}{\sqrt{2}} = 7.07$.

Finally, $A = bh = 10(7.07) = 70.7$

47. **(D)** To answer this question, you do not have to know what a practical number is (or even what a prime number is for that matter). You just have to understand the logic. To disprove the statement "All roses are red," you have to produce some (one or more) roses that are not red. Similarly, to show the falseness of the statement "All practical numbers are prime," you need to show there is at least one practical number that is not prime.

48. **(D)** By KEY FACT K7, the formula for the volume of a pyramid is $V=\frac{1}{3}Bh$, where B is the area of the base and h is the height. (Remember that this fact is given to you on the first page of the Math 1 test.)

The base of pyramid 1 is a 3-4-5 right triangle, whose area is $\frac{1}{2}(3)(4)=6$. The base of pyramid 2 is a square of side 3 whose area is $3^2=9$. If h_1 and h_2 represent the two heights, then:

$$\frac{1}{3}(6)h_1=\frac{1}{3}(9)h_2 \Rightarrow 2h_1=3h_2 \Rightarrow \frac{h_1}{h_2}=\frac{3}{2}$$

49. **(A)** By KEY FACT I2, the sum of the degree measures of the interior angles of pentagon $ABCDE$ is $(5-2)180°=3(180°)=540°$. Since the pentagon is regular, the degree measure of each angle is $\frac{540°}{5}=108°$, and so m$\angle CBF$ is $180°-108°=72°$.

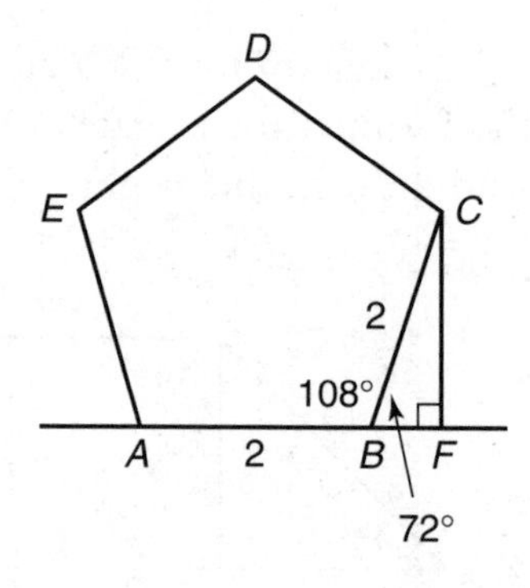

Then $\sin 72°=\frac{CF}{2} \Rightarrow CF=2\sin 72°=2(0.95)=1.90$ and

$\cos 72°=\frac{BF}{2} \Rightarrow BF=2\cos 72=2(0.31)=0.62$

Finally, the area of $\Delta BFC=\frac{1}{2}(BF)(CF)=\frac{1}{2}(0.62)(1.90)=0.59$

**If you don't see how to use trigonometry to answer this question, you still must answer it. It is clear that the length of leg $\overline{CF}$ is less than 2, which is the length of hypotenuse $\overline{BC}$. From the diagram, it is clear that $BF<1$. So $A<\frac{1}{2}(2)(1)=1$. Only choice A is less than 1.

50. **(E)** Draw in three diagonals that divide the hexagon into six congruent equilateral triangles.

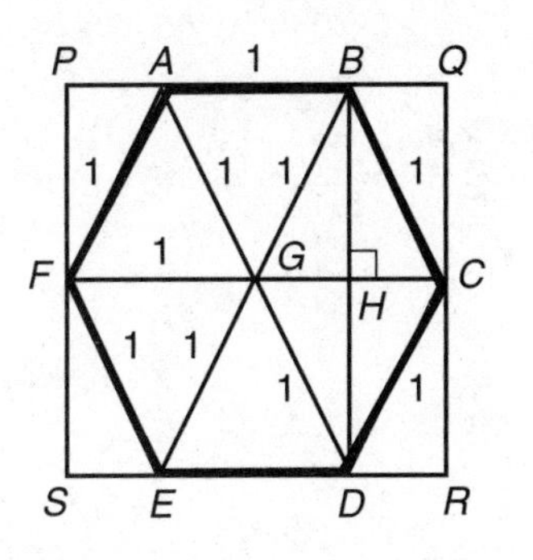

The length $\overline{PQ}$ of the rectangle is equal to $\overline{FC}$, which is 2. Altitude $\overline{BH}$ divides ΔBCG into two 30-60-90 triangles. By KEY FACT H8, $GH = \frac{1}{2}$ and $BH = \frac{1}{2}\sqrt{3}$. Similarly, $HD = \frac{1}{2}\sqrt{3}$.

So $QR = BD = 2\left(\frac{1}{2}\sqrt{3}\right) = \sqrt{3}$.

Finally, the area of the rectangle is $(PQ)(QR) = 2\sqrt{3}$.

**Use TACTIC 5: trust the diagram. Clearly *PQ*, which is 2, is greater than *QR*, so the area of the rectangle must be less than 4. Eliminate choices B and C. Also, *QR* is clearly greater than 1, so the area of the rectangle is greater than 2. Eliminate choices A and D.

Model Test 2

50 Questions / 60 MINUTES

Directions: For each question, determine which of the answer choices is correct and fill in the oval on the answer sheet that corresponds to your choice.

Notes:

1. You will need to use a scientific or graphing calculator to answer some of the questions.
2. Be sure your calculator is in degree mode.
3. Each figure on this test is drawn as accurately as possible unless it is specifically indicated that the figure has not been drawn to scale.
4. The domain of any function f is the set of all real numbers x for which $f(x)$ is also a real number, unless the question indicates that the domain has been restricted in some way.
5. The box below contains five formulas that you may need to answer one or more of the questions.

Reference Information. This box contains formulas for the volumes of three solids and the areas of two of them.

For a sphere with radius r:

- $V = \frac{4}{3}\pi r^3$
- $A = 4\pi r^2$

For a right circular cone with radius r, circumference c, height h, and slant height l:

- $V = \frac{1}{3}\pi r^2 h$
- $A = \frac{1}{2}cl$

For a pyramid with base area B and height h:

- $V = \frac{1}{3}Bh$

Answer Sheet

MODEL TEST 2

1 Ⓐ Ⓑ Ⓒ Ⓓ Ⓔ	14 Ⓐ Ⓑ Ⓒ Ⓓ Ⓔ	27 Ⓐ Ⓑ Ⓒ Ⓓ Ⓔ	40 Ⓐ Ⓑ Ⓒ Ⓓ Ⓔ
2 Ⓐ Ⓑ Ⓒ Ⓓ Ⓔ	15 Ⓐ Ⓑ Ⓒ Ⓓ Ⓔ	28 Ⓐ Ⓑ Ⓒ Ⓓ Ⓔ	41 Ⓐ Ⓑ Ⓒ Ⓓ Ⓔ
3 Ⓐ Ⓑ Ⓒ Ⓓ Ⓔ	16 Ⓐ Ⓑ Ⓒ Ⓓ Ⓔ	29 Ⓐ Ⓑ Ⓒ Ⓓ Ⓔ	42 Ⓐ Ⓑ Ⓒ Ⓓ Ⓔ
4 Ⓐ Ⓑ Ⓒ Ⓓ Ⓔ	17 Ⓐ Ⓑ Ⓒ Ⓓ Ⓔ	30 Ⓐ Ⓑ Ⓒ Ⓓ Ⓔ	43 Ⓐ Ⓑ Ⓒ Ⓓ Ⓔ
5 Ⓐ Ⓑ Ⓒ Ⓓ Ⓔ	18 Ⓐ Ⓑ Ⓒ Ⓓ Ⓔ	31 Ⓐ Ⓑ Ⓒ Ⓓ Ⓔ	44 Ⓐ Ⓑ Ⓒ Ⓓ Ⓔ
6 Ⓐ Ⓑ Ⓒ Ⓓ Ⓔ	19 Ⓐ Ⓑ Ⓒ Ⓓ Ⓔ	32 Ⓐ Ⓑ Ⓒ Ⓓ Ⓔ	45 Ⓐ Ⓑ Ⓒ Ⓓ Ⓔ
7 Ⓐ Ⓑ Ⓒ Ⓓ Ⓔ	20 Ⓐ Ⓑ Ⓒ Ⓓ Ⓔ	33 Ⓐ Ⓑ Ⓒ Ⓓ Ⓔ	46 Ⓐ Ⓑ Ⓒ Ⓓ Ⓔ
8 Ⓐ Ⓑ Ⓒ Ⓓ Ⓔ	21 Ⓐ Ⓑ Ⓒ Ⓓ Ⓔ	34 Ⓐ Ⓑ Ⓒ Ⓓ Ⓔ	47 Ⓐ Ⓑ Ⓒ Ⓓ Ⓔ
9 Ⓐ Ⓑ Ⓒ Ⓓ Ⓔ	22 Ⓐ Ⓑ Ⓒ Ⓓ Ⓔ	35 Ⓐ Ⓑ Ⓒ Ⓓ Ⓔ	48 Ⓐ Ⓑ Ⓒ Ⓓ Ⓔ
10 Ⓐ Ⓑ Ⓒ Ⓓ Ⓔ	23 Ⓐ Ⓑ Ⓒ Ⓓ Ⓔ	36 Ⓐ Ⓑ Ⓒ Ⓓ Ⓔ	49 Ⓐ Ⓑ Ⓒ Ⓓ Ⓔ
11 Ⓐ Ⓑ Ⓒ Ⓓ Ⓔ	24 Ⓐ Ⓑ Ⓒ Ⓓ Ⓔ	37 Ⓐ Ⓑ Ⓒ Ⓓ Ⓔ	50 Ⓐ Ⓑ Ⓒ Ⓓ Ⓔ
12 Ⓐ Ⓑ Ⓒ Ⓓ Ⓔ	25 Ⓐ Ⓑ Ⓒ Ⓓ Ⓔ	38 Ⓐ Ⓑ Ⓒ Ⓓ Ⓔ	
13 Ⓐ Ⓑ Ⓒ Ⓓ Ⓔ	26 Ⓐ Ⓑ Ⓒ Ⓓ Ⓔ	39 Ⓐ Ⓑ Ⓒ Ⓓ Ⓔ	

1. If $7x - 3x = 7x + 3x - 24$, then what is the value of x?

 (A) -4
 (B) -2
 (C) 0
 (D) 2
 (E) 4

2. If $w = 2$, then what is the value of $(w - 3)(w^2 - 3)$?

 (A) -12
 (B) -6
 (C) -1
 (D) 1
 (E) 17

3. If $a \neq 0$, then which of the following is equivalent to $\dfrac{2}{\left(\dfrac{4}{3a^3}\right)}$?

 (A) $\dfrac{3a^3}{2}$

 (B) $\dfrac{8a^3}{2}$

 (C) $\dfrac{8}{3a^3}$

 (D) $\dfrac{2}{3a^3}$

 (E) $24a^3$

USE THIS SPACE FOR SCRATCH WORK

USE THIS SPACE FOR SCRATCH WORK

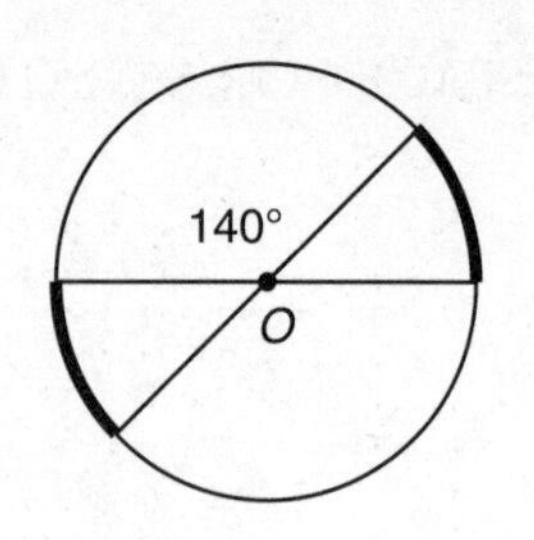

4. The figure above is a circle whose center is O. The sum of the lengths of the two darkened arcs is what fraction of the circumference of the circle?

(A) $\frac{1}{9}$

(B) $\frac{1}{8}$

(C) $\frac{2}{9}$

(D) $\frac{1}{4}$

(E) $\frac{1}{3}$

5. If $3a + 2b = 2a + 3b$, then $b =$

(A) -1
(B) 0
(C) 1
(D) a
(E) $5a$

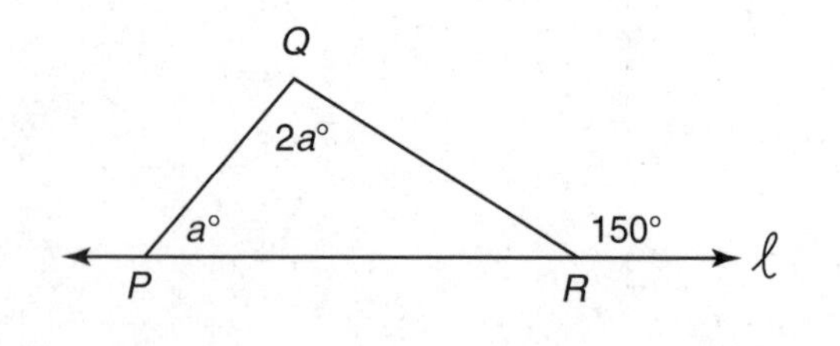

6. In the figure above, if P and R lie on line l, what is the value of a?

(A) 30
(B) 45
(C) 50
(D) 60
(E) 75

7. What is the slope of the line whose equation is $3x + 4y = 24$?

(A) $-\frac{4}{3}$

(B) $-\frac{3}{4}$

(C) $\frac{3}{4}$

(D) $\frac{4}{3}$

(E) 6

8. If $g(x) = \frac{100x + 12}{x^3}$, then $g(-0.2) =$

(A) –1,520
(B) –1,000
(C) –200
(D) 1,000
(E) 1,520

9. If $x^2 - y^2 = x - y$ and $x < y$, what is the average (arithmetic mean) of x and y?

(A) 0

(B) $\frac{1}{2}$

(C) 1

(D) $\frac{x}{2}$

(E) $\frac{y}{2}$

10. If $f(x) = 4x^4 - 4$, for what value of x is $f(x) = 4$?

(A) 0
(B) 1
(C) 1.19
(D) 1.41
(E) 1.68

USE THIS SPACE FOR SCRATCH WORK

11. On January 1, 2000, John had 2,000 baseball cards and Bob had 1,000 baseball cards. If John adds 150 cards per year to his collection and Bob adds 220 cards per year to his collection, what is the earliest year in which Bob will have more cards than John?

(A) 2010
(B) 2014
(C) 2015
(D) 2020
(E) 2070

12. The length of a rectangle is three times its width, and the perimeter of the rectangle is three times the perimeter of a square whose area is 100. What is the area of the rectangle?

(A) 150
(B) 300
(C) 360
(D) 450
(E) 675

13. Which of the following numbers is a counterexample to the statement, "All numbers that are divisible by 2 and 4 are divisible by 8"?

(A) 6
(B) 8
(C) 10
(D) 12
(E) 24

14. What is the distance between the points whose coordinates are (–2, 5) and (4, –3)?

(A) $\sqrt{8}$
(B) $\sqrt{40}$
(C) $\sqrt{68}$
(D) 6
(E) 10

USE THIS SPACE FOR SCRATCH WORK

15. On a map of the United States, 0.3 inches represents 75 miles. On the map, how many inches are there between two cities that are actually 420 miles from one another?

(A) 1.7
(B) 1.9
(C) 4.3
(D) 22.5
(E) 23.6

USE THIS SPACE FOR SCRATCH WORK

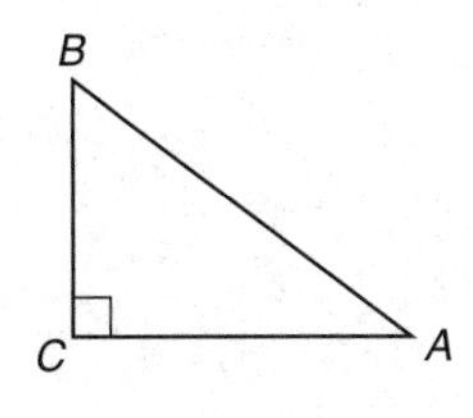

16. In right triangle ABC in the figure above, if $\sin A = 0.6$, what is $\sin B$?

(A) 0.36
(B) 0.40
(C) 0.60
(D) 0.64
(E) 0.80

17. On the final exam in Mrs. Johnson's math class, the average (arithmetic mean) grade of the 25 students was 90. If the average grade of the 15 girls in the class was 94, what was the average grade of the 10 boys?

(A) 84
(B) 85
(C) 86
(D) 88
(E) 92

USE THIS SPACE FOR SCRATCH WORK

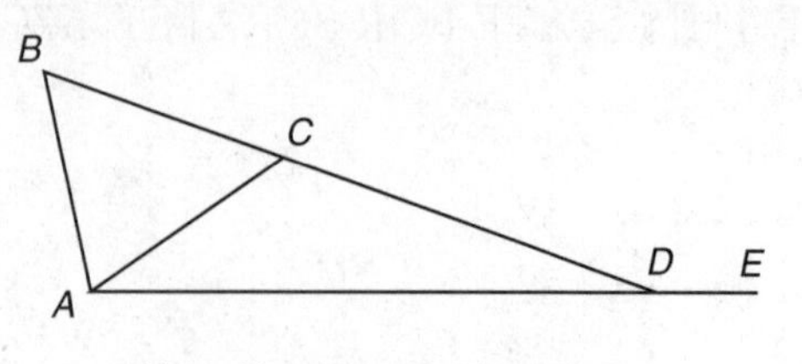

Note: Figure not drawn to scale.

18. If in the figure above, $AB = BC = CD = AC$, what is the measure of $\angle CDE$?

(A) 120°
(B) 135°
(C) 145°
(D) 150°
(E) 160°

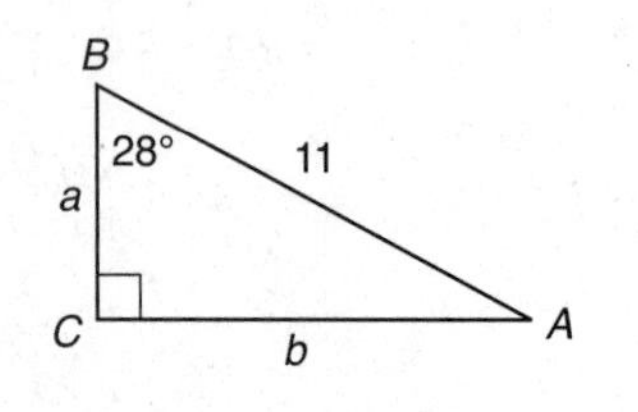

Note: Figure not drawn to scale.

19. In the figure above, what is the value of $\frac{b}{a}$?

(A) 0.47
(B) 0.53
(C) 0.67
(D) 0.88
(E) 1.89

20. If $f(x) = \frac{x^2 - 1}{3 - 2x}$, which of the following numbers is NOT in the domain of f?

(A) −1
(B) 0
(C) 0.66
(D) 1
(E) 1.5

21. What is the area of a triangle whose sides are 5, 5, and 6?

(A) 6
(B) 12
(C) 16
(D) 18
(E) 24

22. A and B have coordinates (–3, 1) and (6, 13), respectively. How many points on line $\overleftrightarrow{AB}$ are twice as far from A as from B?

(A) 0
(B) 1
(C) 2
(D) 3
(E) More than 3

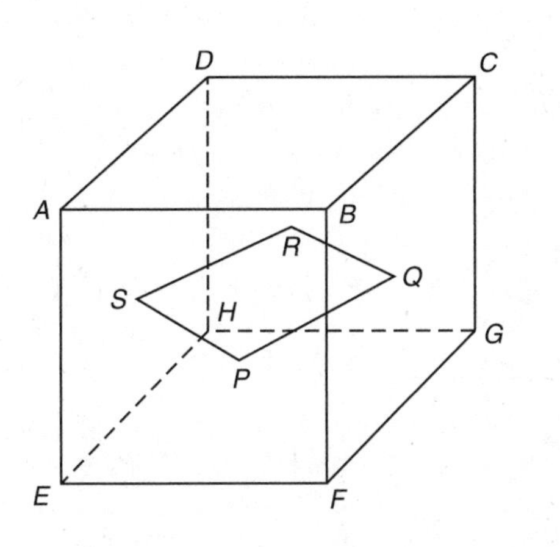

23. In the figure above, the length of each edge of the cube is 2. If P, Q, R, and S are the centers of faces $ABFE$, $BCGF$, $DCGH$, and $ADHE$, respectively, what is the perimeter of quadrilateral $PQRS$?

(A) 4
(B) $4\sqrt{2}$
(C) 6
(D) $4\sqrt{3}$
(E) 8

24. Five actors are auditioning for the 3 parts available in a play. Each part will be performed by one of these 5 actors, and no actor will perform more than one part. In how many ways can the part assignments be made?

(A) 6
(B) 15
(C) 27
(D) 60
(E) 120

USE THIS SPACE FOR SCRATCH WORK

25. If x and y are positive integers and $7x + 11y = z$, which of the following could be a factor of z?

    I. 5
    II. 48
    III. 77

    (A) None
    (B) I only
    (C) III only
    (D) I and III only
    (E) I, II, and III

USE THIS SPACE FOR SCRATCH WORK

26. In similar triangles ABC and DEF, $\overline{AB}$ and $\overline{DE}$ are corresponding sides and $AB = 5$ and $DE = 7$. If the area of triangle ABC is 10, what is the area of triangle DEF?

    (A) 5.1
    (B) 7.1
    (C) 12.0
    (D) 14.0
    (E) 19.6

27. If $3^{100}4^{100}5^{100} = 2^{20a}15^{50b}$, then $a + b =$

    (A) 7
    (B) 12
    (C) 50
    (D) 100
    (E) 200

28. How many four-digit numbers are there in which the tens digit is 8 and none of the digits is 9?

    (A) 210
    (B) 392
    (C) 504
    (D) 648
    (E) 729

29. In a trapezoid, which of the following could be true?

    I. The diagonals are congruent
    II. Two adjacent sides are congruent
    III. Two adjacent angles are congruent

    (A) I only
    (B) II only
    (C) I and II only
    (D) I and III only
    (E) I, II, and III

30. Which of the following is an equation of a line that has the same x-intercept as the line whose equation is $y = 3x - 6$?

(A) $y = 3x - 4$
(B) $y = 2x - 6$
(C) $y = 6x - 3$
(D) $y = x + 2$
(E) $y = 2x - 4$

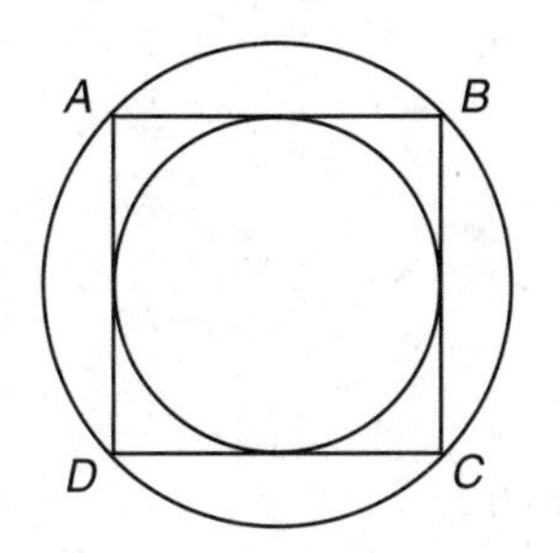

31. In the diagram above, square $ABCD$ is inscribed in a circle and a circle is inscribed in square $ABCD$. What is the ratio of the area of the outer circle to the area of the inner circle?

(A) $2\sqrt{2}$
(B) 2
(C) $\sqrt{2}$
(D) $\frac{1}{\sqrt{2}}$
(E) $\frac{1}{2\sqrt{2}}$

32. Which of the following is the equation of the circle whose center is at (2, –2) and whose radius is 2?

(A) $(x - 2)^2 + (y + 2)^2 = 2$
(B) $(x - 2)^2 + (y + 2)^2 = 4$
(C) $(x + 2)^2 - (y - 2)^2 = 2$
(D) $(x + 2)^2 + (y - 2)^2 = 4$
(E) $(x - 2)^2 - (y + 2)^2 = 4$

33. If $x > 0$ and $(2^{x^2 - 1})(4^{x + 2}) = 8^{x + 3}$, then $x =$

(A) 1
(B) 2
(C) 3
(D) 5
(E) 6

USE THIS SPACE FOR SCRATCH WORK

USE THIS SPACE FOR SCRATCH WORK

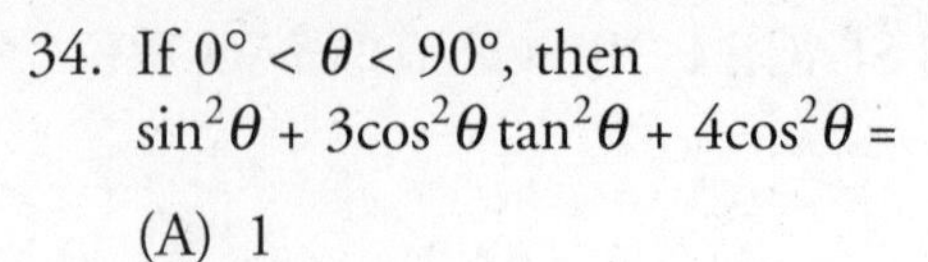

34. If $0° < \theta < 90°$, then
    $\sin^2\theta + 3\cos^2\theta\tan^2\theta + 4\cos^2\theta =$

    (A) 1
    (B) 2
    (C) 4
    (D) $3\tan^2\theta + 5$
    (E) $5\cos^2\theta + \sin^2\theta$

35. What is the value of $\log_8 16$?

    (A) $\frac{1}{2}$
    (B) $\frac{3}{4}$
    (C) $\frac{4}{3}$
    (D) 2
    (E) 4

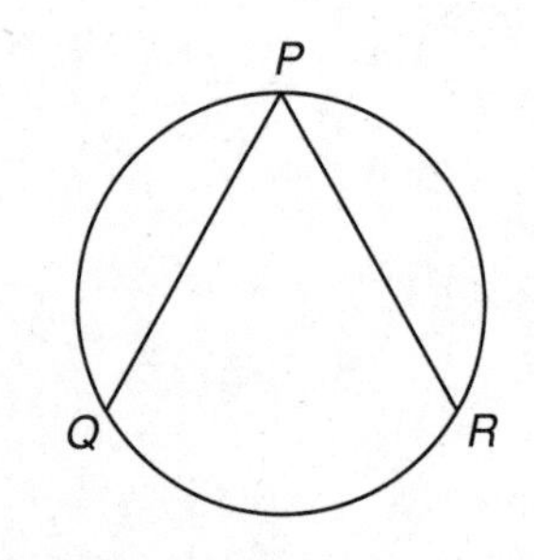

Note: Figure not drawn to scale.

36. In the circle above, the ratio of the measure of arc $\overset{\frown}{PQ}$ to the measure of arc $\overset{\frown}{QR}$ to the measure of arc $\overset{\frown}{RP}$ is 3 to 4 to 5. What is the measure of angle P?

    (A) 30°
    (B) 45°
    (C) 60°
    (D) 75°
    (E) 90°

37. If $f(x) = 3x + 2$ and $g(x) = 2x + 3$, then what is the y-intercept of the line whose equation is $y = g(f(x))$?

    (A) 1
    (B) 5
    (C) 6
    (D) 7
    (E) 11

38. If $b^2 - 4ac = 13$, how many times does the graph of $y = ax^2 + bx + c$ intersect the x-axis?

(A) 0
(B) 1
(C) 2
(D) 3
(E) More than 3

39. For what value of k will the graphs of $3x + 4y + 5 = 0$ and $kx + 6y + 7 = 0$ NOT intersect?

(A) –8
(B) 0
(C) 4.5
(D) 5
(E) 8

40. For which of the following equations is the sum of the roots equal to twice the product of the roots?

(A) $x^2 - 5x + 10 = 0$
(B) $x^2 + 5x - 10 = 0$
(C) $x^2 + 10x + 5 = 0$
(D) $x^2 - 10x - 5 = 0$
(E) $x^2 - 10x + 5 = 0$

41. If the length of a rectangle is three times its width, what is the sine of the angle that the diagonal makes with the longer side?

(A) 0.316
(B) 0.333
(C) 0.500
(D) 0.866
(E) 0.948

USE THIS SPACE FOR SCRATCH WORK

USE THIS SPACE FOR SCRATCH WORK

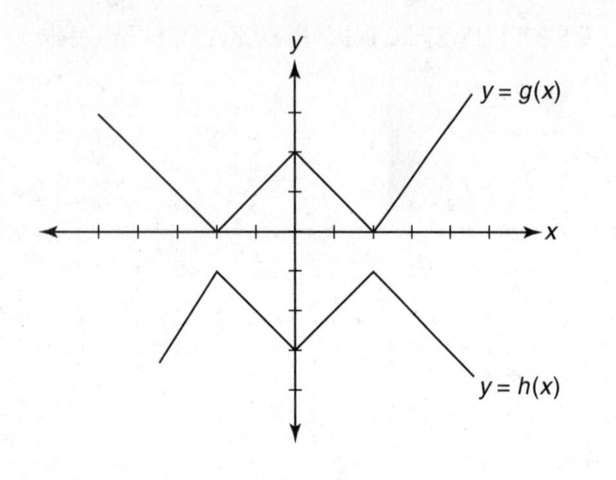

42. The figure above shows the graphs of $y = g(x)$ and $y = h(x)$. What is $h(g(4)) - g(h(4))$?

(A) –3
(B) –2
(C) –1
(D) 0
(E) 1

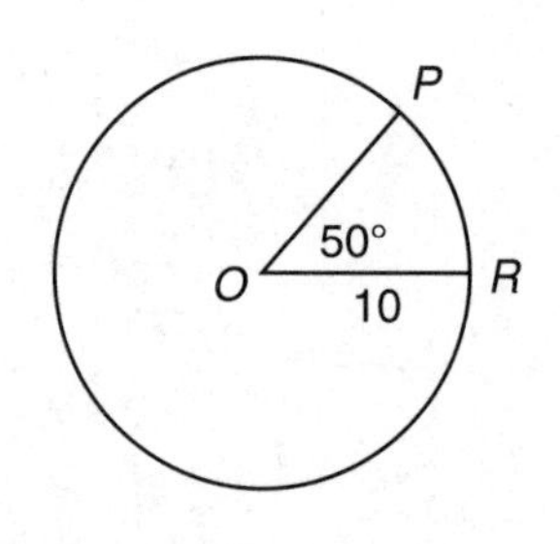

43. If in the figure above, O is the center of the circle, what is the area of sector OPR?

(A) 8.73
(B) 13.89
(C) 36.41
(D) 43.63
(E) 157.08

44. If $\log_b 2 = x$ and $\log_b 3 = y$, what is $\log_b 72$?

(A) $2x + 3y$
(B) $3x + 2y$
(C) $6xy$
(D) $9xy$
(E) x^3y^2

45. A jar contains only red and blue marbles, 40 of which are red and 75 of which are blue. How many red marbles must be added to the jar so that when one marble is drawn from the jar, the probability of drawing a red one will be $\frac{3}{4}$?

(A) 85
(B) 120
(C) 185
(D) 225
(E) 300

46. Spheres 1 and 2 have radii of 3 and 9, respectively. If the surface area of sphere 3 is the average (arithmetic mean) of the surface areas of spheres 1 and 2, what is the radius of sphere 3?

(A) 5.65
(B) 6.00
(C) 6.71
(D) 7.23
(E) 7.50

47. While speaking to Jane, Adam made the following true statement, "If I win the lottery, I will buy you a car." Which of the following statements could be true?

I. Adam won the lottery and did not buy Jane a car.
II. Adam did not win the lottery and bought Jane a car.
III. Adam won the lottery or Adam bought Jane a car.

(A) None
(B) III only
(C) I and III only
(D) II and III only
(E) I, II, and III

USE THIS SPACE FOR SCRATCH WORK

48. If b and c are real numbers and $1 + i$ is a root of the equation $x^2 + bx + c = 0$, what is the value of $b + c$?

(A) -2
(B) -1
(C) 0
(D) 1
(E) 2

49. If the volume of a sphere is equal to the volume of a cube, what is the ratio of the edge of the cube to the radius of the sphere?

(A) 1.16
(B) 1.33
(C) 1.53
(D) 1.61
(E) 2.05

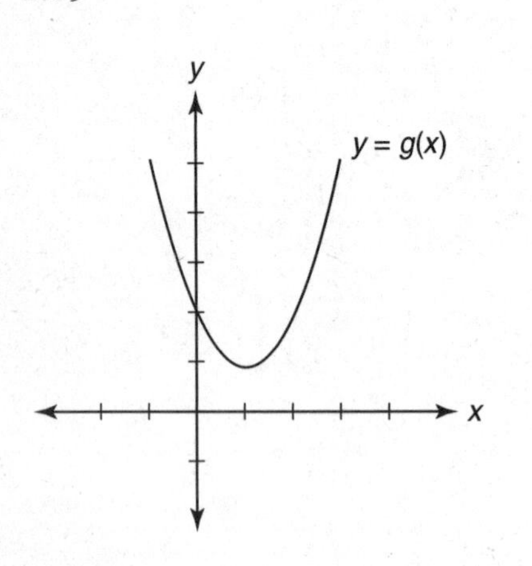

50. The graph of $y = g(x)$ shown above is the result of shifting the graph of $y = f(x)$ 3 units to the right and 2 units up. If the equation of $g(x)$ is $g(x) = x^2 - 2x + 2$, what is the equation of $f(x)$?

(A) $f(x) = x^2 + 4x + 3$
(B) $f(x) = x^2 + 4x + 15$
(C) $f(x) = x^2 + 6x + 19$
(D) $f(x) = x^2 - 8x + 15$
(E) $f(x) = x^2 - 8x + 19$

USE THIS SPACE FOR SCRATCH WORK

Answer Key

MODEL TEST 2

1. E	14. E	27. B	40. E
2. C	15. A	28. D	41. A
3. A	16. E	29. E	42. A
4. C	17. A	30. E	43. D
5. D	18. D	31. B	44. B
6. C	19. B	32. B	45. C
7. B	20. E	33. C	46. C
8. D	21. B	34. C	47. D
9. B	22. C	35. C	48. C
10. C	23. B	36. C	49. D
11. C	24. D	37. D	50. A
12. E	25. E	38. C	
13. D	26. E	39. C	

ANSWERS EXPLAINED

1. **(E)** Combine like terms: $4x = 10x - 24$
   Subtract $4x$ from and add 24 to each side: $24 = 6x$
   Divide both sides by 6: $x = 4$

   **You *could* use TACTIC 2 and backsolve. However, on a problem this easy, you should not.

2. **(C)** Do not FOIL. Just replace each w by 2:
   $$(w - 3)(w^2 - 3) = (2 - 3)(2^2 - 3) = (-1)(1) = -1.$$

3. **(A)** To divide by a fraction, multiply by its reciprocal:
   $$\frac{2}{\left(\frac{4}{3a^3}\right)} = 2\left(\frac{3a^3}{4}\right) = \frac{6a^3}{4} = \frac{3a^3}{2}.$$

   **Use TACTIC 3: Let $a = 2$. Then:
   $$\frac{2}{\left(\frac{4}{3a^3}\right)} = \frac{2}{\left(\frac{4}{3 \times 2^3}\right)} = \frac{2}{\left(\frac{4}{24}\right)} = \frac{2}{\left(\frac{1}{6}\right)} = 2 \times 6 = 12.$$
   Only Choice A is equal to 12 when $a = 2$.

4. **(C)** Since $m\angle AOB + m\angle AOC = 180°$, the measures of central angles AOB and COD are each 40°, as are the measures of their intercepted arcs. Therefore, arcs AB and CD are each $\frac{40}{360} = \frac{1}{9}$ of the circle. The sum of their lengths is $\frac{2}{9}$ of the circle.

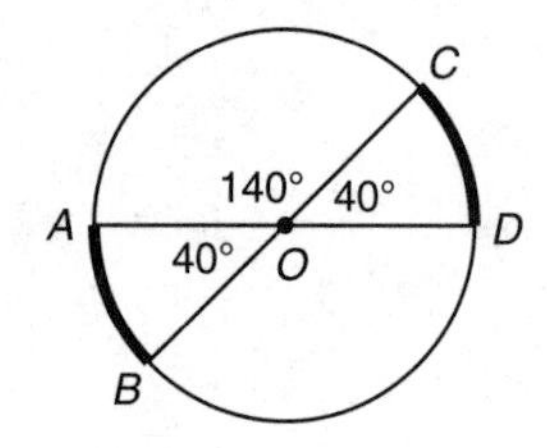

   **If you get stuck on a question like this, use TACTIC 5: since the diagram is drawn to scale, you can trust it. Clearly, $\frac{1}{9}$ and $\frac{1}{8}$ are too small, and $\frac{1}{3}$ is too big. Eliminate choices A, B, and E. Since $\frac{2}{9}$ and $\frac{1}{4}$ are very close ($\frac{2}{9} = 0.22$ and $\frac{1}{4} = 0.25$), you would have to guess between C and D.

5. **(D)** Subtract $2a$ and $2b$ from each side of the given equation:
   $$\begin{array}{rcl} 3a + 2b & = & 2a + 3b \\ \underline{-2a - 2b} & & \underline{-2a - 2b} \\ a & = & b \end{array}$$

   **Since this question has two variables but only one equation, there is one extra variable. By TACTIC 3, you can plug in any number for either variable. Say, $a = 2$. Then:
   $$3(2) + 2b = 2(2) + 3b \Rightarrow 6 + 2b = 4 + 3b \Rightarrow b = 2. \text{ So } b = a.$$

6. **(C)** Since $\angle QRS$ is an exterior angle of ΔPQR, its measure is the sum of the two opposite interior angles (see KEY FACT H2). So
$150 = a + 2a = 3a \Rightarrow a = 50$.
**In the diagram above, $150 + b = 180 \Rightarrow b = 30$. Then

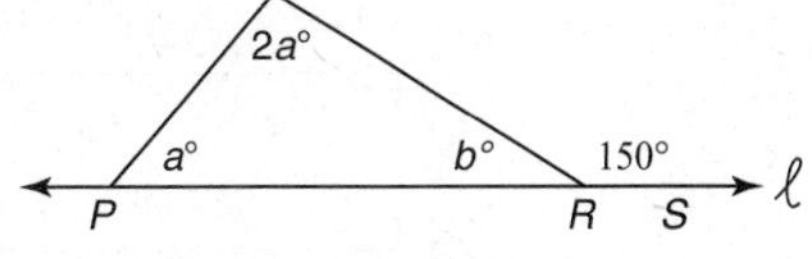

$$30 + a + 2a = 180 \Rightarrow 3a + 30 = 180 \Rightarrow 3a = 150 \Rightarrow a = 50.$$

7. **(B)** Rewrite the given equation in $y = mx + b$ form.

$$3x + 4y = 24 \Rightarrow 4y = -3x + 24 \Rightarrow y = -\frac{3}{4}x + 6.$$

So the slope, m, is $-\frac{3}{4}$ (see KEY FACT L8).

**Find two points on the given line and use the slope formula. For example, when $x = 0$, $y = 6$, and when $y = 0$, $x = 8$. Therefore, (0, 6) and (8, 0) are points on the line, and the slope of the line is $\frac{6-0}{0-8} = \frac{6}{-8} = -\frac{3}{4}$.

8. **(D)** Just replace x by -0.2:

$$g(-0.2) = \frac{100\,(-0.2)+12}{(-0.2)^3}.$$

Then enter this on your calculator, being sure to put parentheses around the entire numerator:

$$(100\,(-0.2) + 12) \div (-0.2)^3 = 1{,}000.$$

9. **(B)** Since $x^2 - y^2 = (x + y)(x - y)$, if $x^2 - y^2 = x - y$, then $x + y = 1$. So the average of x and y is $\frac{x+y}{2} = \frac{1}{2}$. Note that it is irrelevant that $x < y$.

**$x = 0$, $y = 1$ is a fairly obvious solution, in which case $\frac{x+y}{2} = \frac{0+1}{2} = \frac{1}{2}$.
This is not the only solution. In fact, there are infinitely many solutions. For example, if $x = -9$ and $y = 10$, $x^2 - y^2 = 81 - 100 = -19$ and $x - y = -9 - 10 = -19$. The average of x and y, however, is always the same.

10. **(C)** $f(x) = 4x^4 - 4 = 4 \Rightarrow 4x^4 = 8 \Rightarrow x^4 = 2 \Rightarrow x = \sqrt[4]{2} \approx 1.189$.
**Use TACTIC 2: backsolve. Try choice C. $f(1.19) = 4(1.19)^4 = 4.02$. That is so close to 4 that it must be the answer. The small difference is due to rounding in the answer choices. The real value of x is closer to 1.1892, and $4(1.1892)^4 - 4 = 3.9998$.

11. **(C)** In x years, John will have $2{,}000 + 150x$ cards and Bob will have $1{,}000 + 220x$ cards. Solve the inequality:

$$1{,}000 + 220x > 2{,}000 + 150x \Rightarrow 70x > 1{,}000 \Rightarrow x > 14.28.$$

So the smallest integer value for x is 15. Note that at the end of 2014, after 14 years, John will have $2{,}000 + 14(150) = 4{,}100$ cards and Bob will have $1{,}000 + 14(220) = 4{,}080$ cards. John still has more. In the following year, 2015, Bob will finally have more cards than John.

12. **(E)** Draw and label a diagram.

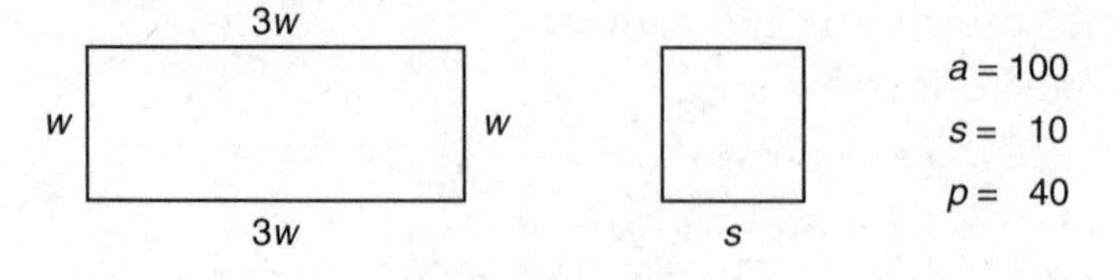

Since the area of the square is 100, each side is 10 and the perimeter is 40. Therefore, the perimeter of the rectangle, which equals $8w$, is $3 \times 40 = 120$. So $8w = 120 \Rightarrow w = 15$ and $3w = 45$. Finally, the area of the rectangle is $15 \times 45 = 675$.

13. **(D)** Only 12, choice D, is divisible by both 2 and 4 but not by 8. Choices A and C are not divisible by 4, and choices B and E are divisible by 8.

14. **(E)** Use the distance formula (KEY FACT L3):

$$d = \sqrt{(x_2 - x_1)^2 + (y_2 - y_1)^2}$$

$$d = \sqrt{(4-(-2))^2 + (-3-5)^2} = \sqrt{6^2 + (-8)^2} = \sqrt{36+64} = \sqrt{100} = 10$$

**Draw a diagram, and use the Pythagorean theorem.

$d^2 = 6^2 + 8^2 = 36 + 64 = 100.$

So $d = 10$.

15. **(A)** All scale drawings are direct proportions. Set up the ratio $\frac{\text{inches}}{\text{miles}}$ and cross multiply:

$$\frac{0.3}{75} = \frac{x}{420} \Rightarrow (0.3) \times (420) = 75x \Rightarrow 126 = 75x \Rightarrow x = 1.68.$$

16. **(E)** Since $\sin A = \frac{\text{opposite}}{\text{hypotenuse}} = \frac{BC}{AB}$ and since $\sin A = 0.6 = \frac{6}{10} = \frac{3}{5}$, let $BC = 3$ and $AB = 5$. Then $AC = 4$ ($\triangle ABC$ is a 3-4-5 right triangle), and $\sin B = \frac{4}{5} = 0.80$.

**By the Pythagorean identity (KEY FACT M2)

$$\sin^2 A + \cos^2 A = 1 \Rightarrow (0.6)^2 + \cos^2 A = 1 \Rightarrow 0.36 + \cos^2 A = 1 \Rightarrow$$

$$\cos^2 A = 0.64 \Rightarrow \cos A = 0.8$$

Since $\cos A = \sin B$, $\sin B = 0.8$.

17. **(A)** The 25 students earned a total of $25 \times 90 = 2{,}250$ points. Since the 15 girls earned a total of $15 \times 94 = 1{,}410$ points, the 10 boys earned $2{,}250 - 1{,}410 = 840$ points. So the average grade of the 10 boys was $\frac{840}{10} = 84$.

**Set this up as a weighted average (see KEY FACT O2). Let x be the average grade of the boys. Then:

$$\frac{(15)(94)+(10)(x)}{25} = 90 \Rightarrow 1{,}410 + 10x = 2{,}250 \Rightarrow 10x = 840 \Rightarrow x = 84.$$

18. **(D)** Since $AB = BC = AC$, $\triangle ACB$ is equilateral, and so the measure of each angle is 60°. Since $m\angle BCA + m\angle ACD = 180°$, $m\angle ACD = 120°$. Since $AC = CD$, $\triangle ACD$ is isosceles. Therefore, the congruent base angles each measure 30°. Finally, $m\angle ADC + m\angle CDE = 180°$, and so $m\angle CDE = 150°$.

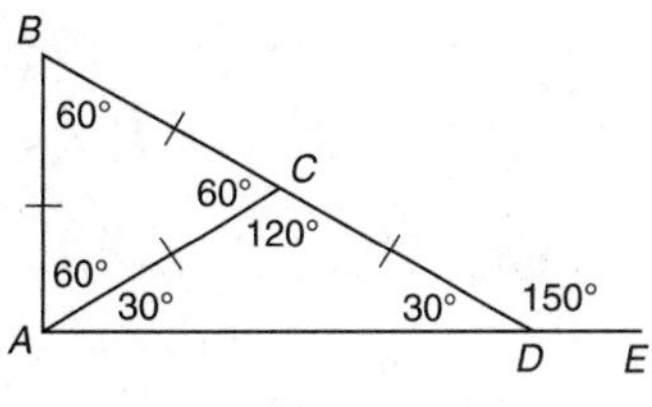

19. **(B)** In $\triangle ABC$, $\frac{b}{a}$ = tan 28° = 0.53.

**$\sin 28° = \frac{b}{11} \Rightarrow b = 11\sin 28° = 5.164.$

$\cos 28° = \frac{a}{11} \Rightarrow a = 11\cos 28° = 9.712.$

So $\frac{b}{a} = \frac{5.164}{9.712} = 0.53.$

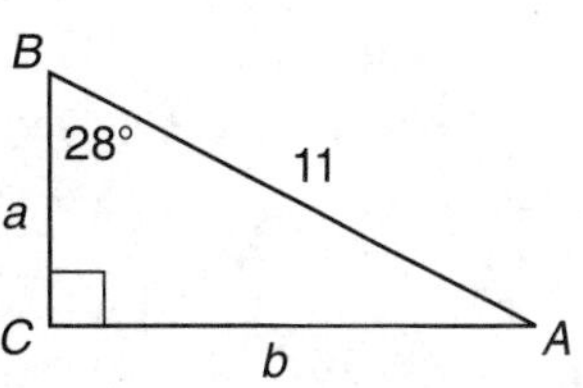

20. **(E)** The domain of a function that is an algebraic fraction consists of all real numbers except those for which the denominator equals 0.

$$3 - 2x = 0 \Rightarrow 3 = 2x \Rightarrow x = 1.5$$

**Use TACTIC 2: backsolve. Use your calculator to evaluate each choice until you get an error message. Be sure to enter the numbers correctly, putting the entire numerator and the entire denominator in parentheses.

When $x = 1.5$, $(1.5^2 - 1) \div (3 - 2(1.5))$ gives you an error message; none of the other choices do.

**On a graphing calculator, enter $y = \frac{(x^2 - 1)}{(3 - 2x)}$. Then go to TABLE and test the five choices.

When 1.5 is in the x-column, the y-column has ERR.

21. **(B)** In isosceles triangle ABC, height $\overline{BD}$ bisects base $\overline{AC}$, so $AD = 3$. You should immediately see that $\triangle ABD$ is a 3-4-5 right triangle, so $BD = 4$. Of course, if you do not notice that, you can use the Pythagorean theorem to find BD. Then area of $\triangle ABC = \frac{1}{2}(6)(4) = 12.$

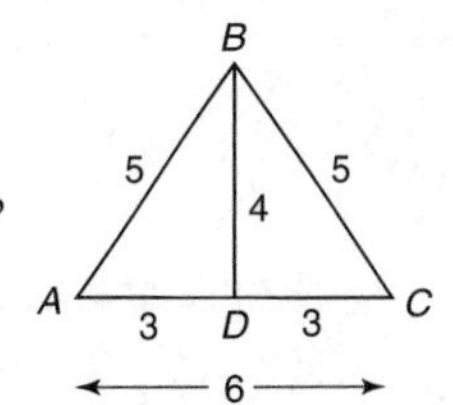

**If you happen to know Heron's formula for the area of a triangle, $A = \sqrt{s(s-a)(s-b)(s-c)}$, where a, b, and c are the lengths of the three sides and s is one-half the perimeter, you can use it here. Since

$$s = \frac{1}{2}(5 + 5 + 6) = 8;\ A = \sqrt{8(3)(3)(2)} = \sqrt{144} = 12.$$

22. **(C)** The coordinates are actually irrelevant. If *A* and *B* are any two points on a line, there are exactly 2 points *P*, *Q* that are twice as far from *A* as from *B*.

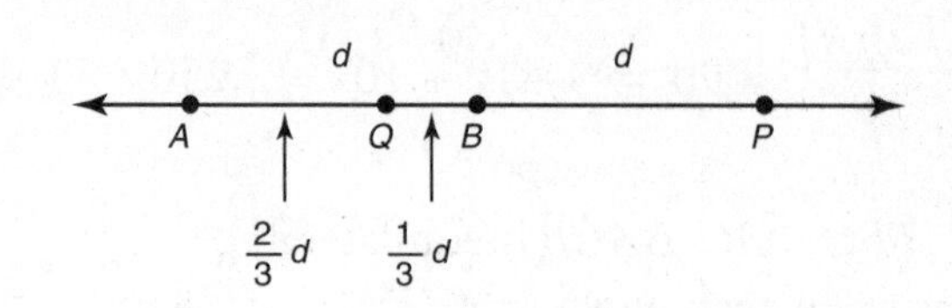

Assume $AB = d$. If *P* is *d* from *B* on the opposite side of *B* as *A*, then $BP = d$ and $AP = 2d$. So $AP = 2(BP)$. If *Q* is between *A* and *B* and if $AQ = \frac{2}{3}d$ and $QB = \frac{1}{3}d$, then $AQ = 2(BQ)$. In the actual example, *P* is (15, 25) and *Q* is (3, 9).

23. **(B)** If you draw segments $\overline{PW}$, $\overline{QX}$, $\overline{RY}$, and $\overline{SZ}$ perpendicular to the top edges of the cube, then it should be clear that quadrilaterals *PQRS* and *WXYZ* are congruent.

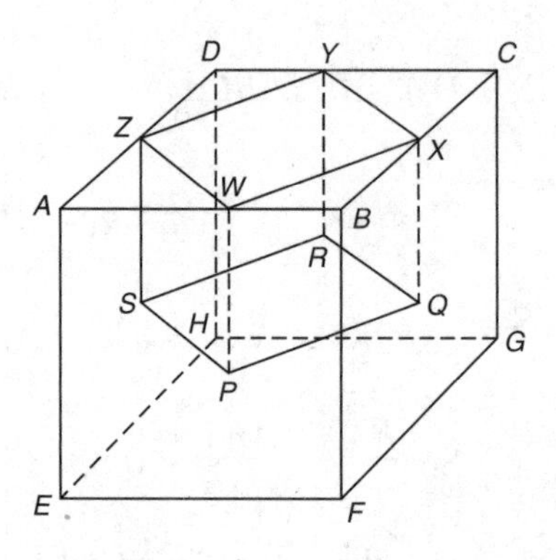

$\triangle XCY$ is an isosceles right triangle whose legs are 1 and whose hypotenuse, $\overline{XY}$, is $\sqrt{2}$. Similarly, each side of quadrilateral *XYZW* is $\sqrt{2}$, and so the perimeter is $4\sqrt{2}$.

24. **(D)** The easiest way to answer this question is to use the counting principle (see KEY FACT O2). The director can choose any of the 5 actors to play the first role, any of the 4 remaining actors for the second role, and any of the 3 remaining actors for the third role. Therefore, there are $5 \times 4 \times 3 = 60$ different ways to make the assignment.

**You can always systematically list the possibilities until you see a pattern. Assume the actors are *A*, *B*, *C*, *D*, and *E*, and list the possibilities in alphabetical order.

ABC	*ACB*	*ADB*	*AEB*
ABD	*ACD*	*ADC*	*AEC*
ABE	*ACE*	*ADE*	*AED*

These are the 12 possible ways to assign the actors if the first role is performed by actor *A*. Similarly, there are 12 ways to assign the actors if the first role is played by *B*, and so on. Each of the five actors could be assigned the first role, and so there are $5 \times 12 = 60$ possible assignments.

25. **(E)** Any positive integer, *n*, could be a factor of *z*. Simply let $x = y = n$. Then $z = 7n + 11n = 18n$, which is clearly divisible by *n*. Statements I, II, and III are all true.

Answers Explained–Model Test 2

**By trial and error or systematic reasoning, you can find values of x and y for each choice. For example,

x	y	$7x + 11y$	z	
2	1	7(2) + 11(1)	25	a multiple of 5
9	3	7(9) + 11(3)	96	a multiple of 48
11	7	7(11) + 11(7)	154	a multiple of 77

26. **(E)** By KEY FACT H15, since the ratio of similitude is $\frac{5}{7}$, the ratio of the areas of the two triangles is $\left(\frac{5}{7}\right)^2 = \frac{25}{49}$. So if x represents the area of triangle DEF, $\frac{25}{49} = \frac{10}{x} \Rightarrow 25x = 490 \Rightarrow x = 19.6$.

**Sketch and label the two triangles.

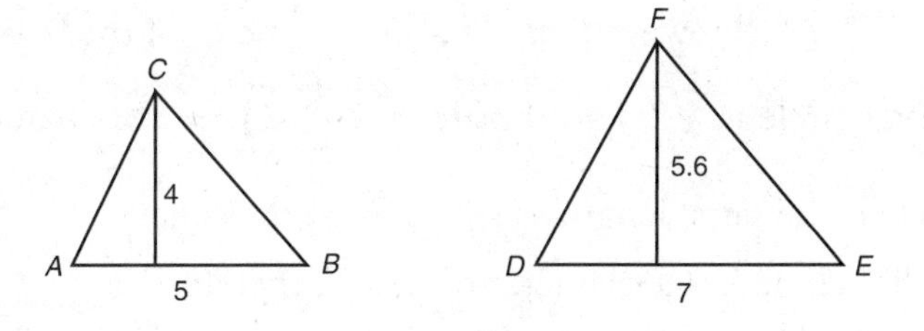

Since the area of ΔABC is 10, we have $10 = \frac{1}{2}(5)h \Rightarrow h = 4$. Since the heights of similar triangles are in the same ratio as their sides, the height of ΔDEF is $\left(\frac{7}{5}\right)(4) = 5.6$. So the area of ΔDEF is $\frac{1}{2}(7)(5.6) = 19.6$.

27. **(B)** By the laws of exponents (KEY FACT A11), $3^{100}4^{100}5^{100} = (3 \times 4 \times 5)^{100} = 60^{100} = 4^{100} \times 15^{100} = (2^2)^{100}15^{100} = 2^{200} \times 15^{100}$.

So $20a = 200 \Rightarrow a = 10$ and $50b = 100 \Rightarrow b = 2$. Therefore, $a + b = 12$.

28. **(D)** Use the counting principle (see KEY FACT O2). There are 8 choices for the thousands digit (any of the ten digits except 0 and 9). There are 9 choices each for the hundreds digit and for the ones digit (any digit except 9). Of course, there is only 1 choice for the tens digit (it must be 8). Therefore, there are $8 \times 9 \times 1 \times 9 = 648$ numbers that satisfy the given condition.

			8	
Number of choices	8	9	1	9

29. **(E)** Read carefully. None of these statements *must* be true, but the question asks which statement *could* be true. In any isosceles trapezoid, like $ABCD$, the diagonals are congruent (statement I is true).

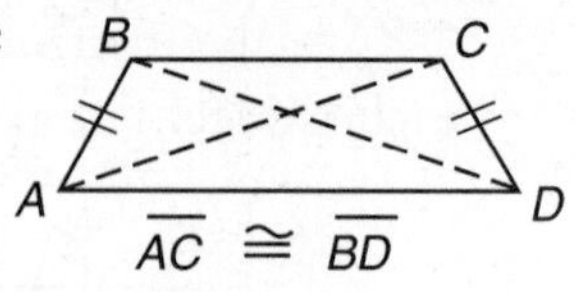

In trapezoid $RQST$, $\overline{RT} \cong \overline{RQ}$ and $\angle R \cong \angle Q$ (statements II and III are true). So statements I, II, and III are true.

30. **(E)** The x-intercepts of any graph are the points on the graph whose y-coordinates are 0 (see KEY FACT L1). To find where the line $y = 3x - 6$ crosses the x-axis, let $y = 0$:

$$0 = 3x - 6 \Rightarrow 3x = 6 \Rightarrow x = 2.$$

Of the five choices, only in E, $y = 2x - 4$, is $x = 2$ when $y = 0$.

**Graph $y = 3x - 6$ and each of the five choices on your calculator. Only $y = 2x - 4$ crosses the x-axis at the same point that $y = 3x - 6$ does.

31. **(B)** Pick a value for the side of the square, say 2. Then $\overline{BD}$, the diagonal of the square, is $2\sqrt{2}$. However, $\overline{BD}$ is also a diameter of the larger circle. So the radius of that circle is $\sqrt{2}$ and the area is $\pi\left(\sqrt{2}\right)^2 = 2\pi$. Diameter $\overline{EF}$ of the smaller circle is the same length as side $\overline{AB}$ of the square. So the smaller circle has a diameter of 2 and, therefore, a radius of 1. The area of the small circle is $\pi(1)^2 = \pi$. Finally, the ratio of the areas is $\frac{2\pi}{\pi} = 2$.

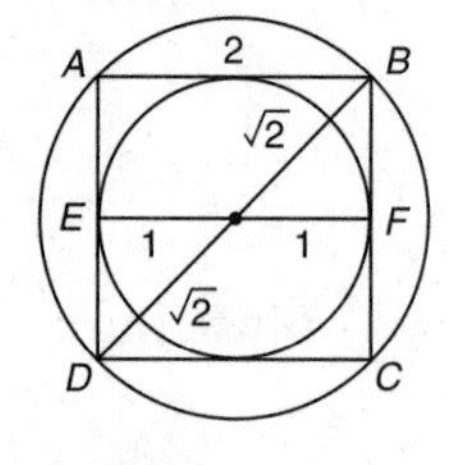

**If you get stuck on a question like this, do not leave it out. Trust the diagram, and make your best estimate. At the very least, eliminate absurd choices. For example, the ratio of the area of the larger circle to the area of the smaller circle must be a fraction greater than 1. However, choices D and E are less than 1.

32. **(B)** By KEY FACT L10, the equation of a circle whose center is at (h, k) and whose radius is r is $(x - h)^2 + (y - k)^2 = r^2$. So the equation is $(x - 2)^2 + (y - (-2))^2 = 2^2$ or $(x - 2)^2 + (y + 2)^2 = 4$.

**If you forgot the formula, you could have made a sketch and tested one or two points. Since (2, 0) is a point on the circle, $x = 2$, $y = 0$ must satisfy the equation. Only choice B is true when $x = 2$ and $y = 0$.

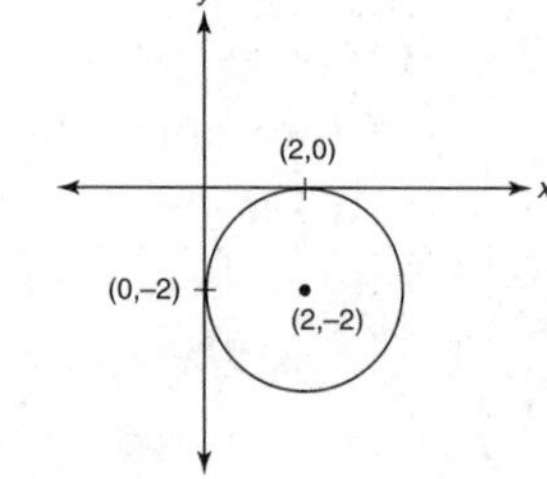

33. **(C)** Rewrite 4 as 2^2 and 8 as 2^3, and then use the laws of exponents (KEY FACT A11).

$$(2^{x^2-1})(4^{x+2}) = 8^{x+3} \Rightarrow (2^{x^2-1})((2^2)^{x+2}) = (2^3)^{x+3} \Rightarrow$$
$$(2^{x^2-1})(2^{2x+4}) = 2^{3x+9} \Rightarrow 2^{x^2-1+2x+4} = 2^{3x+9} \Rightarrow 2^{x^2+2x+3} = 2^{3x+9}.$$

So, $x^2 + 2x + 3 = 3x + 9 \Rightarrow x^2 - x - 6 = 0 \Rightarrow (x-3)(x+2) = 0$
Therefore, $x = 3$ or $x = -2$. Since it is given that $x > 0$, $x = 3$.

**Use TACTIC 2: backsolve starting with choice C. Replace x by 3 and use your calculator to verify that $(2^8)(4^5) = 8^6$.

34. **(C)** Simplifying the given expression requires the use of the only two identities you need to know (KEY FACT M1 and KEY FACT M2):

$$\tan^2\theta = \frac{\sin^2\theta}{\cos^2\theta} \text{ and } \sin^2\theta + \cos^2\theta = 1$$

$$3\cos^2\theta\tan^2\theta = 3\cancel{\cos^2\theta}\,\frac{\sin^2\theta}{\cancel{\cos^2\theta}} = 3\sin^2\theta$$

So $\sin^2\theta + 3\cos^2\theta\tan^2\theta + 4\cos^2\theta = \sin^2\theta + 3\sin^2\theta + 4\cos^2\theta = 4\sin^2\theta + 4\cos^2\theta.$
Now factor out a 4: $4(\sin^2\theta + \cos^2\theta) = 4(1) = 4$.

**Plug in any number for θ and carefully enter the entire expression in your calculator, making sure to use parentheses properly. For example, if $\theta = 30$:

$$(\sin(30))^2 + 3(\cos(30))^2(\tan(30))^2 + 4(\cos(30))^2 = 4.$$

35. **(C)** Remember that a logarithm is an exponent (see KEY FACT A15). $\log_8 16$ is the exponent, x, to which base 8 is raised to equal 16:

$$\log_8 16 = x \Rightarrow 8^x = 16$$

Since $8 = 2^3$ and $16 = 2^4$, we have $2^4 = 8^x = (2^3)^x = 2^{3x}$. So $4 = 3x \Rightarrow x = \frac{4}{3}$.

**If in the above solution you get that $8^x = 16$ but cannot solve the equation, you can estimate. Since $8^1 = 8$ and $8^2 = 64$, if $8^x = 16$, x must be between 1 and 2. Only choice C is between 1 and 2.

**If you know the change of base formula for logarithms (KEY FACT A16), you can evaluate this on your calculator:

$$\log_8 16 = \frac{\log 16}{\log 8} = 1.3333$$

You can even evaluate this without a calculator:

$$\frac{\log 16}{\log 8} = \frac{\log 2^4}{\log 2^3} = \frac{4\log 2}{3\log 2} = \frac{4}{3}$$

36. **(C)** Let the degree of arcs $\widehat{PQ}$, $\widehat{QR}$, and $\widehat{RP}$ be $3x$, $4x$, and $5x$, respectively. Since the sum of the measures of the three arcs is 360°:

$$3x + 4x + 5x = 360° \Rightarrow 12x = 360° \Rightarrow x = 30°.$$

So the measure of arc $\widehat{QR}$ is $4 \times 30 = 120°$. By KEY FACT J10, the measure of angle P is one-half the measure of arc $\widehat{QR}$: $\frac{1}{2}(120°) = 60°$.

37. **(D)** Remember that $g(x) = 2x + 3$ means that $g(\text{anything}) = 2(\text{that thing}) + 3$. Then $y = g(f(x)) = 2(f(x)) + 3 = 2(3x + 2) + 3 = (6x + 4) + 3 = 6x + 7$.

$y = 6x + 7$ is the equation of a line whose slope is 6 and whose y-intercept is 7.

**Let $x = 0$, then $f(0) = 3(0) + 2 = 2$, and $g(2) = 2(2) + 3 = 7$. So $(0, 7)$ is a point on the line whose equation is $y = g(f(x))$. In fact, it is the y-intercept.

38. **(C)** By KEY FACT E2, if the discriminant $b^2 - 4ac$ is positive, the quadratic equation $ax^2 + bx + c = 0$ has two different roots, and so the graph of $y = ax^2 + bx + c$ crosses the x-axis two times.

39. **(C)** The two given equations are each the equation of a line. If two lines do not intersect, they are parallel. So by KEY FACT L7, they have equal slopes. By solving for y to put $3x + 4y + 5 = 0$ into slope-intercept form, we get $y = -\frac{3}{4}x - \frac{5}{4}$. So the slope is $-\frac{3}{4}$. Similarly, by rewriting $kx + 6y + 7 = 0$ as $y = -\frac{k}{6}x - \frac{7}{6}$, we see that its slope is $-\frac{k}{6}$. So

$$-\frac{3}{4} = -\frac{k}{6} \Rightarrow -4k = -18 \Rightarrow k = 4.5.$$

40. **(E)** By KEY FACT E3, if $ax^2 + bx + c = 0$, the sum of the roots is $\frac{-b}{a}$ and the product of the roots is $\frac{c}{a}$. In choice E, the product of the roots is $\frac{c}{a} = \frac{5}{1} = 5$ and the sum of the roots is $\frac{-b}{a} = \frac{-(-10)}{1} = \frac{10}{1} = 10$. In none of the other choices is the sum of the roots twice the product of the roots.

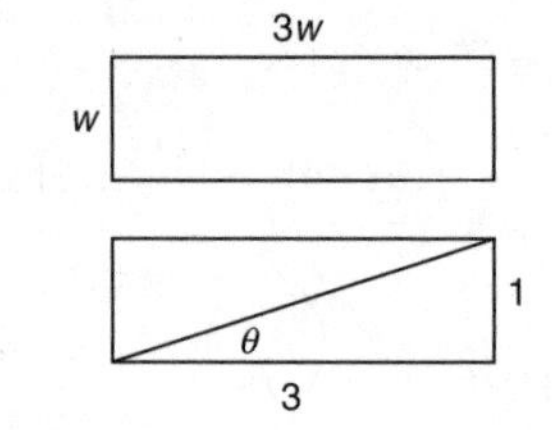

41. **(A)** Of course, you should draw a diagram and label it. You could label the sides w and $3w$, but it is even easier to label them 1 and 3. Now you have two choices:

(1) Since $\tan\theta = \frac{1}{3}$, $\theta = \tan^{-1}\left(\frac{1}{3}\right) = 18.430$. Then $\sin\theta = \sin(18.430) = 0.316$.

(2) Use the Pythagorean theorem to find AC.

$(AC)^2 = 3^2 + 1^2 = 10 \Rightarrow AC = \sqrt{10}$. Then $\sin\theta = \frac{1}{\sqrt{10}} = 0.316$.

42. **(A)** $g(4)$ is the y-component of the point on the graph of $y = g(x)$ where $x = 4$. Since the point (4, 3) is on the graph of $y = g(x)$, $g(4) = 3$. So $h(g(4)) = h(3) = -2$. Similarly, since (4, −3) is a point on the graph of $y = h(x)$, $h(4) = -3$. Then $g(h(4)) = g(-3) = 1$. Finally, $h(g(4)) - g(h(4)) = -2 - 1 = -3$.

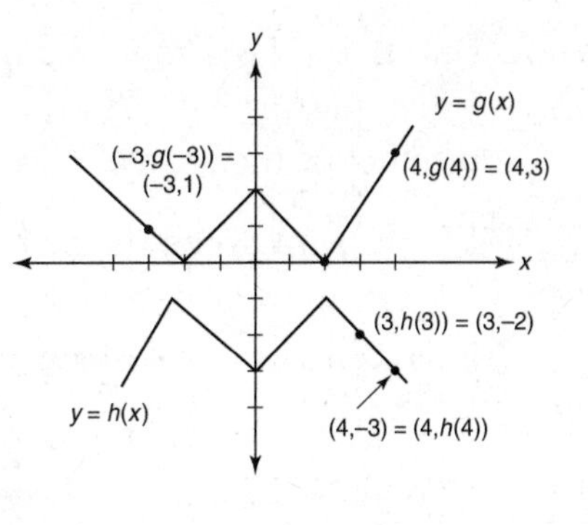

43. **(D)** The area of circle O is $\pi(10^2) = 100\pi$. Since sector OPR is $\frac{50}{360}$ of the circle, its area is $\frac{50}{360}(100\pi) = 43.63$.

44. **(B)** By using the laws of logarithms (KEY FACT A18), we get $\log_b 72 = \log_b(8)(9) = \log_b 8 + \log_b 9 = \log_b 2^3 + \log_b 3^2 = 3\log_b 2 + 2\log_b 3 = 3x + 2y$.

**Let $b = 10$ and use your calculator. $\log 2 = 0.301$, $\log 3 = 0.477$, and $\log 72 = 1.857$. Which of the answer choices is 1.857 when $x = 0.301$ and $y = 0.477$? Only choice B.

45. **(C)** If after adding the red marbles, the probability of drawing a red one is $\frac{3}{4}$, the probability of drawing a blue one will be $\frac{1}{4}$. This means that $\frac{1}{4}$ of the marbles are blue. Since there are 75 blue marbles in the jar, the jar must contain $4 \times 75 = 300$ marbles. So $300 - 75 = 225$ of the marbles are red. Since there were originally 40 red marbles, $225 - 40 = 185$ red marbles had to be added to the jar.

46. **(C)** By KEY FACT K8, the formula for the surface area of a sphere is $S = 4\pi r^2$. (Remember that this is one of the five formulas given to you on the test.)

$$S_1 = 4\pi(3)^2 = 4\pi(9) = 36\pi.$$

$$S_2 = 4\pi(9)^2 = 4\pi(81) = 324\pi.$$

So the average of the surface areas of spheres 1 and 2 is $\frac{36\pi + 324\pi}{2} = \frac{360\pi}{2} = 180\pi$. Then the surface area of sphere 3 is 180π, and so $4\pi r^2 = 180\pi \Rightarrow 4r^2 = 180 \Rightarrow r^2 = 45$. Finally, $r = \sqrt{45} = 6.71$.

47. **(D)** Statement I contradicts the given true statement that if Adam wins the lottery, he will buy Jane a car. So I is false. Neither statement II nor statement III *must* be true, but each of them *could* be true. The original statement did not say what would happen if Adam did not win the lottery, only what would happen if he did win it. Even without winning the lottery, Adam might choose to buy Jane a car. Only statements II and III are true.

48. **(C)** If $1 + i$ is a root of a quadratic equation with real coefficients, then so is its conjugate, $1 - i$. So the sum of the roots is $(1 + i) + (1 - i) = 2$ and the product of the roots is $(1 + i)(1 - i) = 1 - i^2 = 1 - (-1) = 2$. By KEY FACT E3, the sum of the roots of $ax^2 + bx + c = 0$ is $\frac{-b}{a}$ and the product of the roots is $\frac{c}{a}$. Since $a = 1$, we have $\frac{-b}{a} = 2 \Rightarrow b = -2$ and $\frac{c}{a} = 2 \Rightarrow c = 2$. So $b + c = 0$.

**As above, if $1 + i$ is a root of $x^2 + bx + c = 0$, then so is $1 - i$. Therefore, $(x - (1 + i))$ and $(x - (1 - i))$ are the factors of $x^2 + bx + c$.

$$(x - (1 + i))\,(x - (1 - i)) = ((x - 1) - i)\,((x - 1) + i) =$$
$$(x - 1)^2 - i^2 = (x^2 - 2x + 1) - (-1) = x^2 - 2x + 2$$

So $b = -2$ and $c = 2$, and therefore $b + c = 0$.

49. **(D)** The formula for the volume of a sphere is $V = \frac{4}{3}\pi r^3$ (KEY FACT K8). (Remember that this formula is given to you on the first page of the test.) The formula for the volume of a cube is $V = e^3$ (KEY FACT K1). So:

$$e^3 = \frac{4}{3}\pi r^3 \Rightarrow \frac{e^3}{r^3} = \frac{4}{3}\pi \Rightarrow \frac{e}{r} = \sqrt[3]{\frac{4}{3}\pi} = 1.61$$

50. **(A)** If the graph of $y = g(x)$ is the result of shifting the graph of $y = f(x)$ 3 units right and 2 units up, shifting the graph of $y = g(x)$ 2 units down and 3 units left will bring us back to the graph of $y = f(x)$. So by KEY FACT N4:

$$f(x) = g(x + 3) - 2 =$$
$$((x + 3)^2 - 2(x + 3) + 2) - 2 = ((x^2 + 6x + 9) - 2x - 6 + 2 - 2) = x^2 + 4x + 3.$$

Model Test 3

50 Questions / 60 MINUTES

Directions: For each question, determine which of the answer choices is correct and fill in the oval on the answer sheet that corresponds to your choice.

Notes:

1. You will need to use a scientific or graphing calculator to answer some of the questions.
2. Be sure your calculator is in degree mode.
3. Each figure on this test is drawn as accurately as possible unless it is specifically indicated that the figure has not been drawn to scale.
4. The domain of any function f is the set of all real numbers x for which $f(x)$ is also a real number, unless the question indicates that the domain has been restricted in some way.
5. The box below contains five formulas that you may need to answer one or more of the questions.

Reference Information. This box contains formulas for the volumes of three solids and the areas of two of them.

For a sphere with radius r:

- $V = \frac{4}{3}\pi r^3$
- $A = 4\pi r^2$

For a right circular cone with radius r, circumference c, height h, and slant height l:

- $V = \frac{1}{3}\pi r^2 h$
- $A = \frac{1}{2}cl$

For a pyramid with base area B and height h:

- $V = \frac{1}{3}Bh$

Answer Sheet

MODEL TEST 3

1 Ⓐ Ⓑ Ⓒ Ⓓ Ⓔ	14 Ⓐ Ⓑ Ⓒ Ⓓ Ⓔ	27 Ⓐ Ⓑ Ⓒ Ⓓ Ⓔ	40 Ⓐ Ⓑ Ⓒ Ⓓ Ⓔ
2 Ⓐ Ⓑ Ⓒ Ⓓ Ⓔ	15 Ⓐ Ⓑ Ⓒ Ⓓ Ⓔ	28 Ⓐ Ⓑ Ⓒ Ⓓ Ⓔ	41 Ⓐ Ⓑ Ⓒ Ⓓ Ⓔ
3 Ⓐ Ⓑ Ⓒ Ⓓ Ⓔ	16 Ⓐ Ⓑ Ⓒ Ⓓ Ⓔ	29 Ⓐ Ⓑ Ⓒ Ⓓ Ⓔ	42 Ⓐ Ⓑ Ⓒ Ⓓ Ⓔ
4 Ⓐ Ⓑ Ⓒ Ⓓ Ⓔ	17 Ⓐ Ⓑ Ⓒ Ⓓ Ⓔ	30 Ⓐ Ⓑ Ⓒ Ⓓ Ⓔ	43 Ⓐ Ⓑ Ⓒ Ⓓ Ⓔ
5 Ⓐ Ⓑ Ⓒ Ⓓ Ⓔ	18 Ⓐ Ⓑ Ⓒ Ⓓ Ⓔ	31 Ⓐ Ⓑ Ⓒ Ⓓ Ⓔ	44 Ⓐ Ⓑ Ⓒ Ⓓ Ⓔ
6 Ⓐ Ⓑ Ⓒ Ⓓ Ⓔ	19 Ⓐ Ⓑ Ⓒ Ⓓ Ⓔ	32 Ⓐ Ⓑ Ⓒ Ⓓ Ⓔ	45 Ⓐ Ⓑ Ⓒ Ⓓ Ⓔ
7 Ⓐ Ⓑ Ⓒ Ⓓ Ⓔ	20 Ⓐ Ⓑ Ⓒ Ⓓ Ⓔ	33 Ⓐ Ⓑ Ⓒ Ⓓ Ⓔ	46 Ⓐ Ⓑ Ⓒ Ⓓ Ⓔ
8 Ⓐ Ⓑ Ⓒ Ⓓ Ⓔ	21 Ⓐ Ⓑ Ⓒ Ⓓ Ⓔ	34 Ⓐ Ⓑ Ⓒ Ⓓ Ⓔ	47 Ⓐ Ⓑ Ⓒ Ⓓ Ⓔ
9 Ⓐ Ⓑ Ⓒ Ⓓ Ⓔ	22 Ⓐ Ⓑ Ⓒ Ⓓ Ⓔ	35 Ⓐ Ⓑ Ⓒ Ⓓ Ⓔ	48 Ⓐ Ⓑ Ⓒ Ⓓ Ⓔ
10 Ⓐ Ⓑ Ⓒ Ⓓ Ⓔ	23 Ⓐ Ⓑ Ⓒ Ⓓ Ⓔ	36 Ⓐ Ⓑ Ⓒ Ⓓ Ⓔ	49 Ⓐ Ⓑ Ⓒ Ⓓ Ⓔ
11 Ⓐ Ⓑ Ⓒ Ⓓ Ⓔ	24 Ⓐ Ⓑ Ⓒ Ⓓ Ⓔ	37 Ⓐ Ⓑ Ⓒ Ⓓ Ⓔ	50 Ⓐ Ⓑ Ⓒ Ⓓ Ⓔ
12 Ⓐ Ⓑ Ⓒ Ⓓ Ⓔ	25 Ⓐ Ⓑ Ⓒ Ⓓ Ⓔ	38 Ⓐ Ⓑ Ⓒ Ⓓ Ⓔ	
13 Ⓐ Ⓑ Ⓒ Ⓓ Ⓔ	26 Ⓐ Ⓑ Ⓒ Ⓓ Ⓔ	39 Ⓐ Ⓑ Ⓒ Ⓓ Ⓔ	

USE THIS SPACE FOR SCRATCH WORK

1. If $\frac{3x}{5} = \frac{4x+2}{7}$, then $x =$

(A) -10
(B) -2
(C) 0
(D) 2
(E) 10

2. If $x = -3$ and $y = 2$, what is the value of $2x^2 - 3y^2$?

(A) -30
(B) -6
(C) 0
(D) 6
(E) 18

3. What is the value of $3|-5| + 5|-3|$?

(A) -30
(B) -15
(C) 0
(D) 15
(E) 30

4. If $f(x) = x^4 - x^2$ and $g(x) = x^3 - 3x^2 + 2x$, then for all values of x for which $g(x) \neq 0$, $\frac{f(x)}{g(x)} =$

(A) $\frac{2(x+1)}{x-2}$
(B) $\frac{x(x+1)}{x-2}$
(C) $\frac{x(x-1)}{x-2}$
(D) $\frac{x(x-1)}{x+2}$
(E) $\frac{x^2+1}{x-2}$

USE THIS SPACE FOR SCRATCH WORK

5. If $x \neq 0$, then $\frac{x}{\left(\frac{4}{x}\right)} =$

(A) $\frac{1}{4}$
(B) 4
(C) $\frac{x}{4}$
(D) $\frac{x}{2}$
(E) $\frac{x^2}{4}$

6. If Lior flips five fair coins and if, for each of them, Ezra guesses whether it landed heads or tails, what is the probability that Ezra makes at least one correct guess?

(A) .16
(B) .50
(C) .80
(D) .97
(E) 2.5

7. If $y = 3x$ and $2x + 3y = 22$, then $y =$

(A) 2
(B) 3
(C) 6
(D) $\frac{22}{9}$
(E) $\frac{66}{9}$

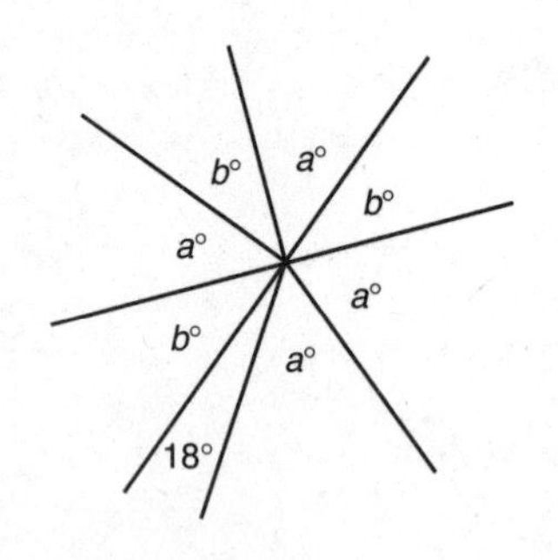

Note: Figure not drawn to scale.

8. If in the figure above, the ratio of a to b is 3 to 2, then what is the value of $a + b$?

(A) 38
(B) 57
(C) 76
(D) 95
(E) 113

9. Which of the following statements about the solution set of $\frac{w+3}{3}=\frac{w+12}{w}$ is true?

(A) The solution set is the empty set.
(B) The solution set has exactly one member.
(C) The solution set has exactly two positive members.
(D) The solution set has exactly two negative members.
(E) The solution set has one positive member and one negative member.

10. What is the reciprocal of the sum of the reciprocals of 2 and 3?

(A) $\frac{1}{6}$
(B) $\frac{1}{5}$
(C) $\frac{5}{11}$
(D) $\frac{5}{6}$
(E) $\frac{6}{5}$

11. If $\sqrt[4]{20^2-12^2}=\sqrt[n]{64}$, then $n=$

(A) $\frac{3}{4}$
(B) $\frac{3}{2}$
(C) 2
(D) 3
(E) 4

12. If in quadrilateral $ABCD$, $m\angle A : m\angle B : m\angle C : m\angle D = 1:1:2:2$, which of the following statements could be true?

I. $ABCD$ is a parallelogram.
II. $ABCD$ is a rhombus.
III. $ABCD$ is a trapezoid.

(A) I only
(B) III only
(C) I and II only
(D) I and III only
(E) I, II, and III

USE THIS SPACE FOR SCRATCH WORK

USE THIS SPACE FOR SCRATCH WORK

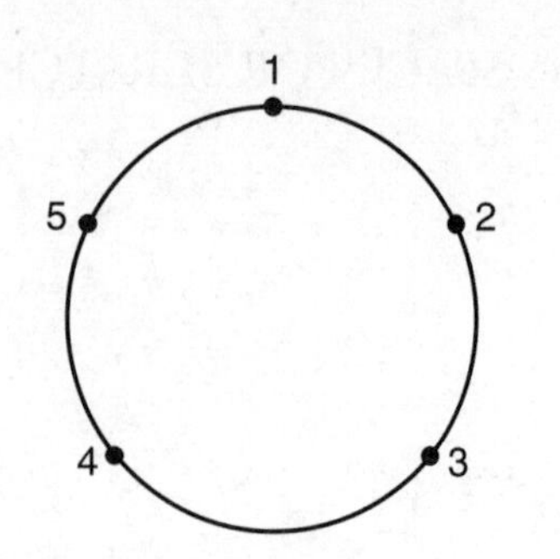

13. In how many ways can Hillary, Ira, Jack, Ken, and Lou sit around the table above if either Ira or Jack must sit in seat 1 and neither Ken nor Lou can sit in seat 5?

(A) 6
(B) 12
(C) 24
(D) 48
(E) 120

14. Which of the following is the equation of a line that is perpendicular to the line whose equation is $y = 2x$ and that has a negative x-intercept?

(A) $y = 2x - 1$
(B) $y = -2x + 1$
(C) $y = \frac{1}{2}x - 1$
(D) $y = -\frac{1}{2}x + 1$
(E) $y = -\frac{1}{2}x - 1$

15. For what value(s) of n is $\sqrt{n+1} = n - 1$?

(A) 0
(B) 3
(C) 0 and 3
(D) 1 and −1
(E) No values

16. If $f(x) = x^2 - 5$ and $g(x) = x^2 + 5$, what is the value of $f(g(5))$?

(A) 0
(B) 50
(C) 405
(D) 600
(E) 895

17. If the average of 5 positive integers is 70, what is the largest possible value of their median?

(A) 70
(B) 114
(C) 116
(D) 201
(E) 346

18. If ℓ is parallel to the line whose equation is $x + y = 11$, then ℓ must pass through which quadrants?

(A) I and II
(B) I and IV
(C) II and IV
(D) I, II, and IV
(E) II, III and IV

19. If angles P and Q are supplementary, angles Q and R are complementary, and x is the sum of the measure of the three angles, then which of the following must be true?

(A) $x = 90°$
(B) $90° < x < 180°$
(C) $x = 180°$
(D) $180° < x < 270°$
(E) $x = 270°$

20. If $|x + 3| < 2$, which of the following must be true?

I. x is negative
II. $x + 3 < 2$
III. $-x - 3 < -2$

(A) I only
(B) II only
(C) I and II only
(D) II and III only
(E) I, II, and III

USE THIS SPACE FOR SCRATCH WORK

USE THIS SPACE FOR SCRATCH WORK

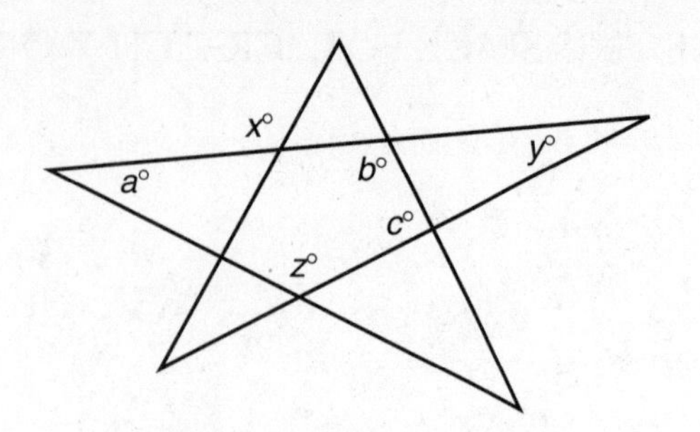

Note: Figure not drawn to scale.

21. In the figure above, if $x = 110$, $y = 40$, and $z = 130$, what is the value of $a + b + c$?

    (A) 170
    (B) 230
    (C) 290
    (D) 320
    (E) 410

22. If A is the point (5, 4) and B is the point (−3, −2), what is the area of the circle that has $\overline{AB}$ as a diameter?

    (A) 8π
    (B) 10π
    (C) 20π
    (D) 25π
    (E) 100π

23. What is the value of $\log_9 27$?

    (A) $\frac{2}{3}$
    (B) 1
    (C) $\frac{3}{2}$
    (D) 2
    (E) 3

24. The perimeter of a square is a inches, and the area of the square is b square inches. If $a + b = 45$, what is the length in inches of one of the sides?

    (A) 4.5
    (B) 5
    (C) 9
    (D) $3\sqrt{5}$
    (E) $5\sqrt{3}$

25. In ΔABC, C is a right angle and $\tan A = 1$. What is the value of $\sin A + \cos A$?

(A) $\frac{\sqrt{2}}{2}$
(B) 1
(C) $\sqrt{2}$
(D) 2
(E) $2\sqrt{2}$

26. If $f(x) = \sqrt[3]{x+5} - 3$, then $f^{-1}(x) =$

(A) $(x+3)^3 - 5$
(B) $(x+5)^3 + 3$
(C) $(x-3)^3 + 5$
(D) $(x+5)^3 - 3$
(E) $f^{-1}(x)$ does not exist

27. There are 25 students in Mrs. Wang's first-period algebra class. On Monday, five students were absent and the other 20 students took a test. The average grade for those students was 86. The next day after the five absent students took the test, the class average was 88. What was the average of those five students' grades?

(A) 90
(B) 92
(C) 94
(D) 96
(E) 98

28. What is the sum of the x- and y-coordinates of the point where the lines $y = 3x + 2$ and $y = 2x + 3$ intersect?

(A) 1
(B) 5
(C) 6
(D) 11
(E) 12

USE THIS SPACE FOR SCRATCH WORK

29. For what value of k will the equation $5x^2 + 6x + 3k = 0$ have exactly one solution?

(A) 0
(B) $\frac{3}{5}$
(C) $\frac{5}{6}$
(D) 1
(E) $\frac{6}{5}$

30. On January 1, 1990, the value of a certificate of deposit (CD) was $765. If the value of the CD increased by 6% each year, what was its value, in dollars, on January 1, 2002?

(A) 1,316
(B) 1,370
(C) 1,498
(D) 1,539
(E) 1,611

31. Which of the following lines has the greatest y-intercept?

(A) $y = 3x + 10$
(B) $2y = x + 15$
(C) $x + y = 12$
(D) $\frac{1}{2}y - \frac{1}{2}x = 10$
(E) $x - y - 25 = 0$

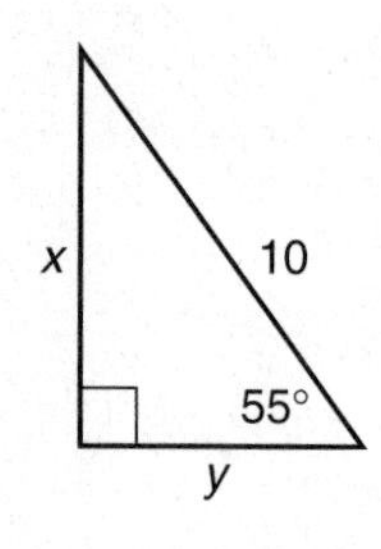

32. In the figure above, what is the value of $x - y$?

(A) 0.14
(B) 1.43
(C) 2.46
(D) 6.09
(E) 8.55

USE THIS SPACE FOR SCRATCH WORK

USE THIS SPACE FOR SCRATCH WORK

33. What is the area of an equilateral triangle whose perimeter is 10?

(A) 4.81
(B) 5.56
(C) 7.86
(D) 43.30
(E) 50

34. If a cube whose edges are x has the same volume as a rectangular solid whose length, width, and height are x, $x-3$, and $x+4$, respectively, then $x=$

(A) 9
(B) 12
(C) 16
(D) 144
(E) 1,728

35. What is the greater of the two y-intercepts of the circle whose center is at (3, −2) and whose radius is 4?

(A) −4.65
(B) 0.65
(C) 1.65
(D) 2.65
(E) 4.65

36. If $f(x)=\frac{x^2-16}{x^2+9}$, which of the following is a number that is NOT in the domain of f?

(A) −4
(B) −3
(C) 3
(D) 4
(E) All real numbers are in the domain.

37. If $0° < x < 90°$, then $\frac{\sin^4 x - \cos^4 x}{\sin x - \cos x} =$

(A) $\sin x + \cos x$
(B) $\sin x - \cos x$
(C) $\sin^2 x - \cos^2 x$
(D) $\sin^3 x - \cos^3 x$
(E) $\sin^3 x + \cos^3 x$

38. If $i^2 = -1$, which of the following are real numbers?

I. i^{21}
II. $i^{21} + i^{22}$
III. $i^{21} + i^{22} + i^{23}$

(A) None
(B) I only
(C) II only
(D) III only
(E) I, II, and III

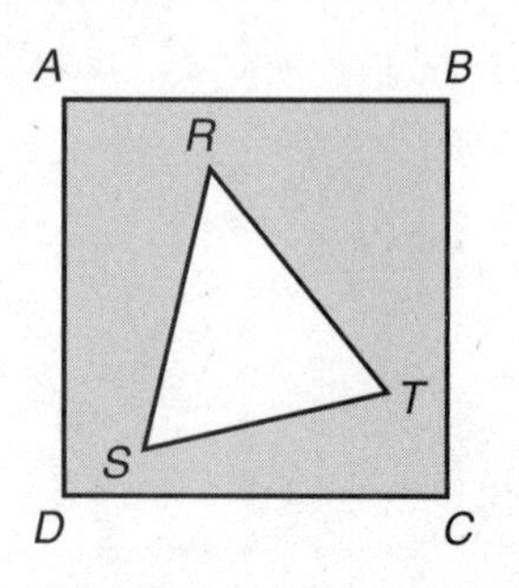

Note: Figure not drawn to scale.

39. In the figure above, *ABCD* is a square and *RST* is an equilateral triangle. If the area of the shaded region is 3 times the area of the white triangle and if $RT = 2$, then what is *AB*?

(A) 2
(B) $\sqrt{3}$
(C) $4\sqrt{3}$
(D) $2\sqrt[4]{3}$
(E) $4\sqrt[4]{3}$

USE THIS SPACE FOR SCRATCH WORK

USE THIS SPACE FOR SCRATCH WORK

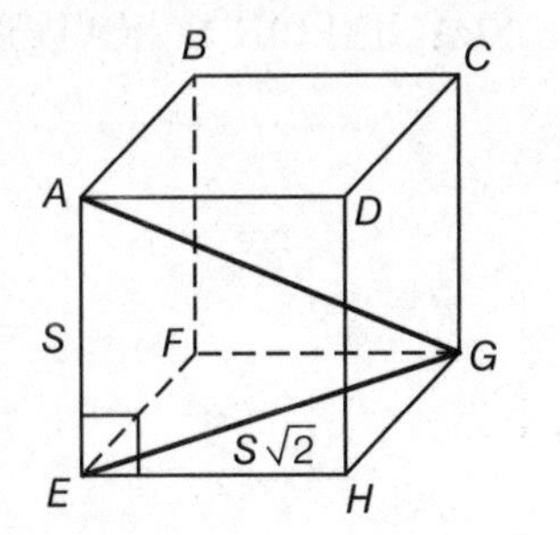

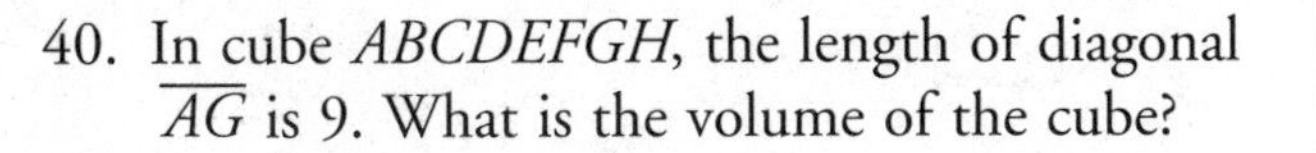

40. In cube *ABCDEFGH*, the length of diagonal $\overline{AG}$ is 9. What is the volume of the cube?

(A) $3\sqrt{3}$
(B) $9\sqrt{3}$
(C) $27\sqrt{3}$
(D) $81\sqrt{3}$
(E) 729

41. If b and c are real numbers, $i^2 = -1$, and $r + si$ is a root of the equation $2x^2 + bx + c = 0$, which of the following must be true?

I. $r - si$ is a root of the equation
II. $b = -4r$
III. $c = 2r^2 + 2s^2$

(A) None
(B) I only
(C) I and II only
(D) I and III only
(E) I, II, and III

42. What is the area of a circle that is inscribed in an equilateral triangle whose sides are 2?

(A) $\frac{\pi}{3}$
(B) $\frac{\pi}{2}$
(C) π
(D) $\frac{4}{3}\pi$
(E) 3π

Model Test 3

USE THIS SPACE FOR SCRATCH WORK

43. If $f(x) = 7x + 3$ and $g(x) = 2x - 5$, which of the following statements must be true?

   I. The lines $y = f(g(x))$ and $y = g(f(x))$ are parallel.
   II. The lines $y = f(g(x))$ and $y = g(f(x))$ are perpendicular.
   III. The line $y = f(g(x)) - g(f(x))$ is horizontal.

   (A) None
   (B) I only
   (C) II only
   (D) III only
   (E) I and III only

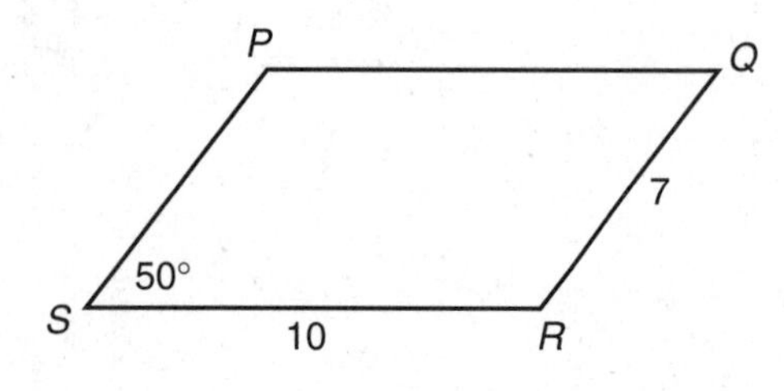

44. What is the area of parallelogram *PQRS* above?

   (A) 45.0
   (B) 53.6
   (C) 61.8
   (D) 70.0
   (E) 83.4

45. "If London is the capital of England, then Moscow is the capital of France" is logically equivalent to which of the following statements:

   (A) London is the capital of England and Moscow is not the capital of France.
   (B) If London is the capital of England, then Moscow is not the capital of France.
   (C) If Moscow is the capital of France, then London is the capital of England.
   (D) If Moscow is not the capital of France, then London is not the capital of England.
   (E) Moscow is not the capital of France, and London is the capital of England.

46. If $f(x) = 4x + 7$, $g(x) = 7x + 4$, and $h(x) = (f \circ g)(x)$, then which of the following is $h^{-1}(x)$?

(A) $11x + 11$
(B) $28x^2 + 65x + 28$
(C) $4x - 7$
(D) $7x - 4$
(E) $\frac{x - 23}{28}$

USE THIS SPACE FOR SCRATCH WORK

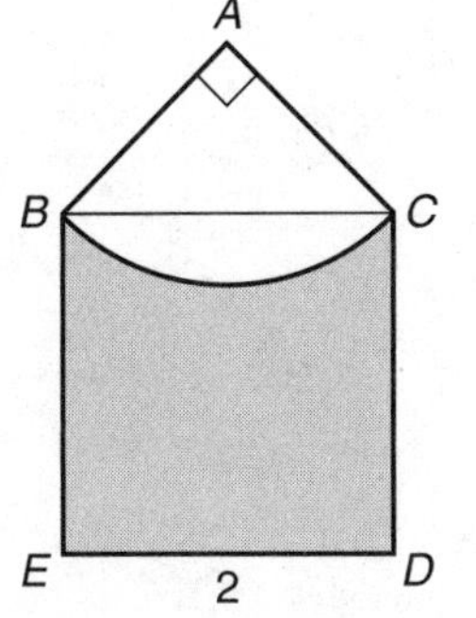

47. In the diagram above, $BCDE$ is a square and ΔABC is a right isosceles triangle. Arc $\widehat{BC}$ is the arc of a circle whose center is at A. If $ED = 2$, what is the area of the shaded region?

(A) $6 - \frac{1}{2}\pi$
(B) $5 - \frac{1}{2}\pi$
(C) $6 - 2\pi$
(D) $2\pi - 6$
(E) $5 - \pi$

48. How many pounds of peanuts must be added to a mixture of 20 pounds of peanuts and 50 pounds of cashews if the resulting mixture is to be 60% peanuts by weight?

(A) 12
(B) 30
(C) 42
(D) 52
(E) 55

USE THIS SPACE FOR SCRATCH WORK

49. If a sphere and a right circular cone have the same radius and equal volumes, what is the ratio of the height of the cylinder to its radius?

(A) $\frac{1}{4}$
(B) 4
(C) $\frac{1}{3}\pi$
(D) $\frac{3}{4}\pi$
(E) $\frac{4}{3}\pi$

50. The first two terms of a sequence are 4 and 5, and each subsequent term is the sum of the two preceding terms. (For example, the third term is $4 + 5 = 9$, and the fourth term is $5 + 9 = 14$.) How many of the first 1,000 terms are odd?

(A) 333
(B) 334
(C) 500
(D) 666
(E) 667

Answer Key

MODEL TEST 3

1. E
2. D
3. E
4. B
5. E
6. D
7. C
8. D
9. E
10. E
11. D
12. B
13. C
14. E
15. B
16. E
17. C
18. C
19. D
20. C
21. B
22. D
23. C
24. B
25. C
26. A
27. D
28. C
29. B
30. D
31. D
32. C
33. A
34. B
35. B
36. E
37. A
38. D
39. D
40. D
41. E
42. A
43. E
44. B
45. D
46. E
47. B
48. E
49. B
50. D

ANSWERS EXPLAINED

For many of the questions in this model test, an alternative solution, indicated by two asterisks (**), follows the first solution. Almost always the first solution is the direct mathematical one and the other is based on one of the tactics discussed in the "Tactics" chapter.

1. **(E)** Use TACTIC C2: solve proportions by cross multiplying:

$$\frac{3x}{5}=\frac{4x+2}{7}\Rightarrow 7(3x)=5(4x+2)\Rightarrow$$
$$21x=20x+10\Rightarrow x=10$$

**Of course, anytime you need to solve for a single variable, you can backsolve (TACTIC 2).

2. **(D)** $2x^2-3y^2=2(-3)^2-3(2^2)=2(9)-3(4)=18-12=6$

3. **(E)** $3|-5|+5|-3|=3(5)+5(3)=15+15=30$

4. **(B)** Factor the numerator and the denominator and simplify:

$$\frac{f(x)}{g(x)}=\frac{x^4-x^2}{x^3-3x^2+2x}=\frac{x^2(x^2-1)}{x(x^2-3x+2)}$$
$$=\frac{x^2\cancel{(x-1)}(x+1)}{x\cancel{(x-1)}(x-2)}=\frac{x(x+1)}{x-2}$$

**You can avoid the algebra by using TACTIC 3: plug in a number for x. For example, if $x=3$:

$$\frac{x^4-x^2}{x^3-3x^2+2x}=\frac{3^4-3^2}{3^3-3(3)^2+2(3)}=\frac{72}{6}=12$$

Only choice B equals 12 when $x=3$.

5. **(E)** $x\div\frac{4}{x}=x\left(\frac{x}{4}\right)=\frac{x^2}{4}$

**Use TACTIC 3: replace x with a number, say $x=2$. Then $\frac{2}{\left(\frac{4}{2}\right)}=\frac{2}{2}=1$. Only choices D and E are equal to 1 when $x=2$. So eliminate A, B, and C and pick another number for x, say 4. $\frac{4}{\left(\frac{4}{4}\right)}=\frac{4}{1}=4$. Of choices D and E, only choice E equals 4 when $x=4$.

6. **(D)** For each coin, the probability that Ezra guesses wrong is $\frac{1}{2}$. So the probability that he guesses wrong 5 times in a row is $\frac{1}{2} \times \frac{1}{2} \times \frac{1}{2} \times \frac{1}{2} \times \frac{1}{2} = \frac{1}{32}$. The probability that he does not guess incorrectly each time (and hence the probability that he is correct at least once) is $1 - \frac{1}{32} = \frac{31}{32} \approx 0.97$.

7. **(C)** $y = 3x \Rightarrow 3y = 9x \Rightarrow 2x + 3y = 2x + 9x = 11x$. So $11x = 22$. Therefore, $x = 2$ and $y = 3(2) = 6$.

   **Use TACTIC 2: backsolve. Start with 6, choice C, which happens to work.

8. **(D)** Use KEY FACT C1 and let $a = 3x$ and $b = 2x$. By KEY FACT G3, the sum of all the angles in the given figure is 360°, so:

$$4a + 3b + 18 = 360 \Rightarrow 4(3x) + 3(2x) + 18 = 360$$
$$\text{So, } 12x + 6x = 342 \Rightarrow 18x = 342 \Rightarrow x = 19$$
$$\text{Therefore, } a = 57,\ b = 38, \text{ and } a + b = 57 + 38 = 95$$

9. **(E)** Use TACTIC C2: cross multiply and solve:

$$w(w + 3) = 3(w + 12) \Rightarrow w^2 + 3w = 3w + 36$$
$$\text{So, } w^2 = 36 \Rightarrow w = 6 \text{ or } w = -6$$

   The solution set has one positive member and one negative member.

10. **(E)** The reciprocals of 2 and 3 are $\frac{1}{2}$ and $\frac{1}{3}$. Their sum is $\frac{1}{2} + \frac{1}{3} = \frac{5}{6}$. The reciprocal of $\frac{5}{6}$ is $\frac{6}{5}$.

11. **(D)** $\sqrt[4]{20^2 - 12^2} = \sqrt[4]{400 - 144} = \sqrt[4]{256} = 4$. Since $4 = \sqrt[3]{64}$, $n = 3$.

12. **(B)** Since the sum of the measures of the four angles in any quadrilateral is 360° (KEY FACT I1) and since the measures of the angles are in the ratio of 1 : 1 : 2 : 2, by KEY FACT C1, for some number *x*, the measures of the angles are *x*, *x*, 2*x*, and 2*x*. Therefore:

$$x + x + 2x + 2x = 360 \Rightarrow 6x = 360 \Rightarrow x = 60.$$

   So $m\angle A = 60°$, $m\angle B = 60°$, $m\angle C = 120°$, and $m\angle D = 120°$. In a parallelogram (and hence a rhombus), by KEY FACT I4 the sum of the measures of any two adjacent angles (for example, angles *A* and *B*) is 180°, so I and II are false and you can eliminate choices A, C, D, and E. The answer must be B. In fact, *ABCD* has to be an isosceles trapezoid.

13. **(C)** Use KEY FACT O2 (the counting principle). There are 2 choices for seat 1: Ira or Jack. After one of them is assigned to seat 1, there are 2 choices for seat 5: Hillary and whichever of Ira or Jack has not been assigned to seat 1. Then, since there are no more restrictions, there are 3 ways to assign seat 2, 2 ways to assign seat 3, and only 1 way to assign seat 4. There are a total of $2 \times 2 \times 3 \times 2 \times 1 = 24$ ways for the five people to be seated.

14. **(E)** By KEY FACT L8, the slope of $y = 2x$ is 2, and by KEY FACT L7, the slope of any line perpendicular to it is $-\frac{1}{2}$. Eliminate choices A, B, and C. Now check choices D and E to see which has a negative x-intercept. To find the x-intercept, let $y = 0$.

Choice D: $0 = -\frac{1}{2}x + 1 \Rightarrow \frac{1}{2}x = 1 \Rightarrow x = 2$

Choice E: $0 = -\frac{1}{2}x - 1 \Rightarrow \frac{1}{2}x = -1 \Rightarrow x = -2$

The answer is E.

**On a graphing calculator, graph $y = 2x$ and each choice.

15. **(B)** Square both sides of the equation $\sqrt{n+1} = n - 1$:

$n + 1 = (n - 1)^2 = n^2 - 2n + 1 \Rightarrow n^2 - 3n = 0 \Rightarrow n(n - 3) = 0 \Rightarrow n = 0$ or $n = 3$

Check both values:

- $\sqrt{3+1} = 2 = 3 - 1$, so 3 works
- $\sqrt{0+1} = 1 \neq 0 - 1$, so 0 does not work

The only solution is 3.

**Use TACTIC 2: backsolve. Test 0. Since it doesn't work, eliminate choices A and C. Test 3. Since it works, eliminate D and E.

16. **(E)** $g(5) = 5^2 + 5 = 30$. So, $f(g(5)) = f(30) = 30^2 - 5 = 900 - 5 = 895$

17. **(C)** By TACTIC O1, if the average of 5 numbers is 70, their sum is $5 \times 70 = 350$. The median of the 5 numbers is the middle one when they are listed in increasing order. To make the third one as large as possible, make the others as small as possible. Since the numbers are not necessarily different positive integers, the two smallest could both be 1: 1, 1, ___, ___, ___. That leaves 348 for the remaining 3 integers. Since the fourth and fifth numbers must be at least as large as the third, the most the median could be is $348 \div 3 = 116$. The five numbers could be 1, 1, 116, 116, 116.

18. **(C)** By rewriting $x + y = 11$ as $y = -x + 11$, you see that by KEY FACT L8, the slope of the given line is -1. Therefore, by KEY FACT L7, the slope of line ℓ is also -1. Any line whose slope is negative must pass through quadrants II and IV. The line $y = -x + 2$ passes through quadrant I, but not quadrant III; the line $y = -x - 2$ passes through quadrant III, but not quadrant I; and the line $y = -x$ passes through neither quadrant I nor quadrant III.

19. **(D)** Since angles P and Q are supplementary, $m\angle P + m\angle Q = 180°$. Since angles Q and R are complementary, $m\angle Q + m\angle R = 90° \Rightarrow m\angle R < 90°$. Therefore, $180° < m\angle P + m\angle Q + m\angle R < 180° + 90° = 270°$.

20. **(C)** By KEY FACT A2, $|x + 3| < 2 \Rightarrow -2 < x + 3 < 2$ (II is true). Also, $x + 3 < 2 \Rightarrow x < -1$, so x must be negative (I is true). Statement III is equivalent to the inequality $x + 3 > 2$, which is false. Only I and II are true.

21. **(B)** In problems such as this, just fill in every angle you can until you have all the angles you need. For easy reference, we have labeled several of the angles.

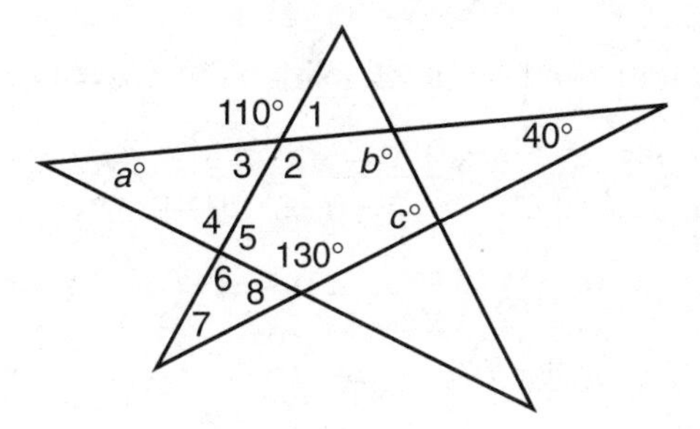

Since the sum of the measures of the two angles in a linear pair is 180° (KEY FACT G2), $m\angle 1 = 70°$ and $m\angle 8 = 50°$.

Since vertical angles are congruent (KEY FACT G4), $m\angle 2 = 110°$ and $m\angle 3 = 70°$.

Since the sum of the measures of the three angles in a triangle is 180°,

$$40° + m\angle 2 + m\angle 7 = 180° \Rightarrow 40° + 110° + m\angle 7 = 180° \Rightarrow m\angle 7 = 30°.$$

Similarly, $m\angle 6 + m\angle 7 + m\angle 8 = 180° \Rightarrow m\angle 6 = 100°$.

Then $m\angle 4 = 100° \Rightarrow a = 10°$. Also $m\angle 5 = 80°$. Finally, since by KEY FACT I2 the sum of the measures of the 5 angles in a pentagon is $(5 - 2)(180°) = 540°$:

$$\begin{aligned} 110 + 80 + 130 + b + c &= 540 \Rightarrow \\ 320 + b + c &= 540 \Rightarrow \\ b + c &= 220 \Rightarrow a + b + c = 230 \end{aligned}$$

22. **(D)** By KEY FACT L3, AB, the distance from A to B, is:

$$\sqrt{(-3-5)^2 + (-2-4)^2} = \sqrt{(-8)^2 + (-6)^2} = \sqrt{64+36} = \sqrt{100} = 10$$

So the diameter of the circle is 10, the radius is 5, and by KEY FACT J8, the area is $\pi(5)^2 = 25\pi$.

23. **(C)** By KEY FACT A15:

$$x = \log_9 27 \Rightarrow 9^x = 27 \Rightarrow \left(3^2\right)^x = 3^3$$

$$\text{So, } 3^{2x} = 3^3 \Rightarrow 2x = 3 \Rightarrow x = \frac{3}{2}$$

**By KEY FACT A16:

$$\log_9 27 = \frac{\log 27}{\log 9} = 1.5 = \frac{3}{2}$$

24. **(B)** Let s represent the length of a side. Then $a = 4s$ and $b = s^2$. So:

$$a + b = 45 \Rightarrow 4s + s^2 = 45 \Rightarrow s^2 + 4s - 45 = 0 \Rightarrow$$
$$(s + 9)(s - 5) = 0 \Rightarrow s = -9 \text{ or } s = 5$$

Since s must be positive, $s = 5$.

**Use TACTIC 2: backsolve.

25. **(C)** Draw right triangle ABC.

Since $\tan A = 1$, and by KEY FACT M1,

$$\tan A = \frac{\text{opposite}}{\text{adjacent}} = \frac{BC}{AC}, \text{ then } AC = BC.$$

A
C
B

Assume AC and BC are each 1; then by KEY FACT H7, $AB = \sqrt{2}$. So, again by KEY FACT M1:

$$\sin A + \cos A = \frac{1}{\sqrt{2}} + \frac{1}{\sqrt{2}} = \frac{2}{\sqrt{2}}$$

A
1
$\sqrt{2}$
C
1
B

Since $\frac{2}{\sqrt{2}}$ is not an answer choice, you can either:

- Rationalize the denominator:

$$\frac{2}{\sqrt{2}} \times \frac{\sqrt{2}}{\sqrt{2}} = \frac{2\sqrt{2}}{2} = \sqrt{2}, \text{ or}$$

- Use your calculator: $\frac{2}{\sqrt{2}} = 1.414$.

Only choice C, $\sqrt{2}$, equals 1.414.

**Use your calculator to find A: $A = \tan^{-1}(1) = 45°$. Then:

$$\sin 45° + \cos 45° = 0.7071 + 0.7071 = 1.414 \approx \sqrt{2}$$

26. **(A)** Use KEY FACT N8. In the equation $y = \sqrt[3]{x+5} - 3$, switch x and y and solve for y:

$$x = \sqrt[3]{y+5} - 3 \Rightarrow x + 3 = \sqrt[3]{y+5} \Rightarrow (x+3)^3 = y + 5 \Rightarrow y = (x+3)^3 - 5$$

27. **(D)** The original 20 students earned a total of $20 \times 86 = 1{,}720$ points. The total number of points earned by all 25 students was $25 \times 88 = 2{,}200$. Therefore, the five students who took the test late earned a total of $2{,}200 - 1{,}720 = 480$ points. So their average was $480 \div 5 = 96$.

**Let x be the average of the five students, and treat this as a weighted average problem:

$$\frac{20(86) + 5x}{25} = 88 \Rightarrow 1720 + 5x = 2200 \Rightarrow 5x = 480 \Rightarrow x = 96$$

28. **(C)** Since $y = 3x + 2$ and $y = 2x + 3$, then $3x + 2 = 2x + 3 \Rightarrow x = 1$. So the x-coordinate of the point of intersection is 1. To find the y-coordinate, plug $x = 1$ into either equation, say $y = 3x + 2$: $y = 3(1) + 2 = 5$. So the point of intersection of the two lines is (1, 5), and the sum of the x- and y-coordinates is $1 + 5 = 6$.

**Graphing the two lines on a graphing calculator would easily get you the answer but probably no faster than would the algebraic solution.

**If you don't have a graphing calculator and want to avoid the algebra, you could make a quick sketch, which would surely eliminate choices A, D, and E.

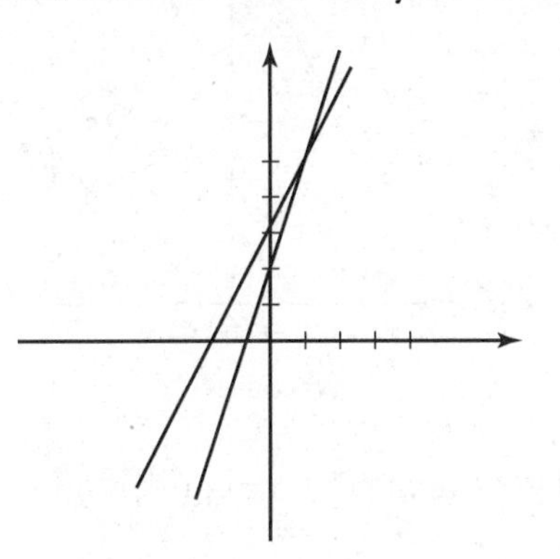

29. **(B)** By KEY FACT E2, a quadratic equation has exactly one solution if and only if its discriminant equals 0. In the equation $5x^2 + 6x + 3k = 0$, the discriminant, D, is:

$$b^2 - 4ac, \text{ where } a = 5, b = 6, \text{ and } c = 3k$$

So, $D = 6^2 - 4(5)(3k) = 36 - 60k$, and

$$36 - 60k = 0 \Rightarrow 36 = 60k \Rightarrow k = \frac{36}{60} = \frac{6}{10} = \frac{3}{5}$$

30. **(D)** On January 1, 1991, the value of the CD was $765 + 0.06(765) = (1.06)(765)$. On January 1 of each year, the value of the CD was (1.06) times its value on January 1 of the preceding year. After 12 years, the value of the CD was $(1.06)^{12}(765) = 1{,}539$.

31. **(D)** Since the *x*-coordinate of any point on the *y*-axis is 0, replace *x* by 0 in each of the choices and determine which has the greatest *y*-value. In choice D, when $x = 0$, $y = 20$. In each of the other choices, the value of *y* is less than 20.

    **Rewrite each equation in the form $y = mx + b$. Choice D becomes $y = x + 20$. So by KEY FACT L8, its *y*-intercept is 20. The other equations all have values of *b* less than 20.

32. **(C)** By KEY FACT M1

    $\sin 55° = \frac{x}{10} \Rightarrow x = 10 \sin 55°$ and

    $\cos 55° = \frac{y}{10} \Rightarrow y = 10 \cos 55°$.

    So $x - y = 10 \sin 55° - 10 \cos 55° = 2.456$.

    You do not have to evaluate *x* and *y* separately and then subtract. You can do it all on one line on your calculator.

33. **(A)** Use TACTIC 4: draw and label a diagram. Since the perimeter of the triangle is 10, each side is $\frac{10}{3}$. Since ΔPSQ in a 30-60-90 right triangle, by KEY FACT H8, $QS = \frac{5}{3}$ and $PS = \frac{5}{3}\sqrt{3}$.

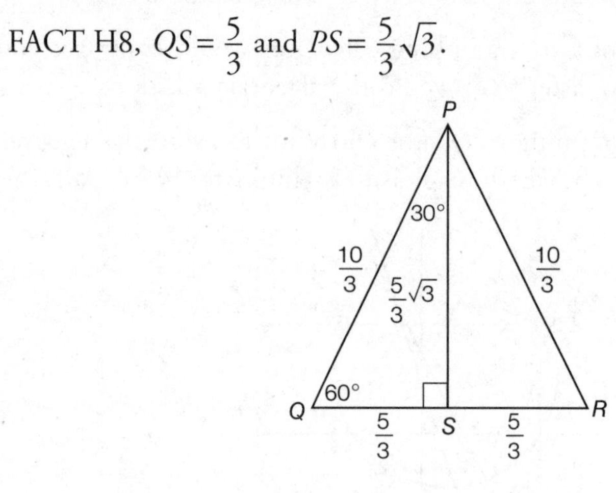

By KEY FACT H11, the area of an equilateral triangle of side *s* is $\frac{s^2\sqrt{3}}{4}$, so

$$A = \frac{\left(\frac{10}{3}\right)^2 \sqrt{3}}{4} = 4.81.$$

**If you don't remember the formula for the area of an equilateral triangle, use the formula $A=\frac{1}{2}bh$: $b=\frac{10}{3}$ and by KEY FACT H8, $h=\frac{5}{3}\sqrt{3}$. So

$A=\frac{1}{2}\left(\frac{10}{3}\right)\left(\frac{5}{3}\right)\sqrt{3}=4.81$.

34. **(B)** By KEY FACT K1, the volume of the cube is x^3, and the volume of the rectangular solid is:

$$x(x-3)(x+4)=x(x^2+x-12)=x^3+x^2-12x.$$

$$\text{So } x^3=x^3+x^2-12x \Rightarrow 0=x^2-12x \Rightarrow$$

$$x(x-12)=0 \Rightarrow x=12$$

**Use TACTIC 2: backsolve. If $x=16$, the volume of the cube is $16^3=4{,}096$ and the volume of the rectangular solid is $(16)(13)(20)=4{,}160$. Since 4096 and 4160 are close, try a number close to 16, say 12. It works.

35. **(B)** By KEY FACT L10, the equation of the circle is $(x-3)^2+(y+2)^2=16$. Since any graph crosses the y-axis when $x=0$, we get:

$$(0-3)^2+(y+2)^2=16 \Rightarrow 9+(y+2)^2=16$$

$$\text{So, } (y+2)^2=7 \Rightarrow y+2=\pm\sqrt{7} \Rightarrow y=-2\pm\sqrt{7}$$

The two y-intercepts are $-2+\sqrt{7}$ and $-2-\sqrt{7}$. The greater of the two is $-2+\sqrt{7} \approx -2+2.65=0.65$.

**Use TACTIC 4: draw a diagram. A quick sketch of the circle is sufficient to eliminate choices A, D, and E and maybe even C.

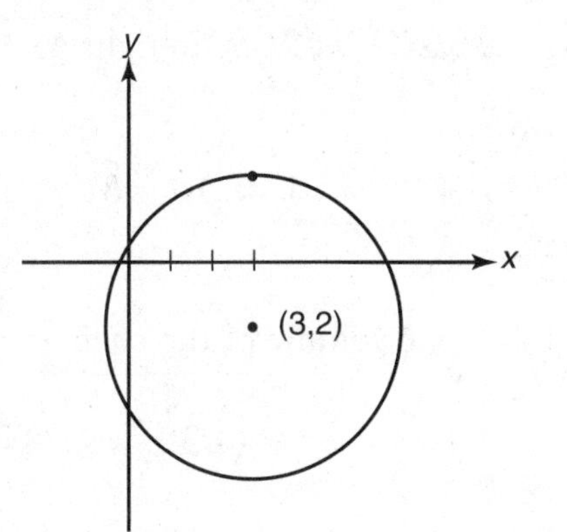

36. **(E)** The domain of f is the set of all real numbers for which $f(x)$ is defined. The only potential problem is that the denominator could be 0. However, for any real number x, x^2+9 is positive. The denominator of $f(x)$ is never equal to 0, and the domain of f is the set of all real numbers.

**If you graph $y=\frac{x^2-16}{x^2+9}$ on a graphing calculator, you can see that all four of the answer choices (–4, –3, 3, and 4) are in the domain.

37. **(A)**

$$\frac{\sin^4 x - \cos^4 x}{\sin x - \cos x} = \frac{(\sin^2 x + \cos^2 x)(\sin^2 x - \cos^2 x)}{\sin x - \cos x} =$$

$$\frac{(1)(\sin x - \cos x)(\sin x + \cos x)}{\sin x - \cos x} = \sin x + \cos x$$

**Use TACTIC 3. Pick any number for x and use your calculator to evaluate the given expression and each of the answer choices.

38. **(D)** When 21, 22, and 23 are divided by 4, the remainders are 1, 2, and 3, respectively.

Therefore, by KEY FACT P2, $i^{21} = i$, $i^{22} = i^2 = -1$, and $i^{23} = i^3 = -i$.

So i^{21} is not a real number (I is false);

$i^{21} + i^{22} = i + (-1)$, which is also not a real number (II is false);

$i^{21} + i^{22} + i^{23} = i + -1 + (-i) = -1$, which is a real number (III is true).

Only III is true.

39. **(D)** By KEY FACT H11, the area of ΔRST is $\frac{(2)^2\sqrt{3}}{4} = \sqrt{3}$. Since the area of the shaded region is 3 times as great, its area is $3\sqrt{3}$ and the total area of the square is $\sqrt{3} + 3\sqrt{3} = 4\sqrt{3}$. Finally, the length of side AB is $\sqrt{4\sqrt{3}} = \sqrt{4} \times \sqrt{\sqrt{3}} = 2\sqrt[4]{3}$.

40. **(D)** If s is the side of the cube, then by KEY FACT H7, the length of $\overline{EG}$, a diagonal of the square base, is $s\sqrt{2}$. So in right ΔAEG, by the Pythagorean theorem, we have:

$s^2 + (s\sqrt{2})^2 = 9^2 \Rightarrow s^2 + 2s^2 = 81 \Rightarrow 3s^2 = 81 \Rightarrow$

$s^2 = 27 \Rightarrow s = \sqrt{27} = 3\sqrt{3}$

So by KEY FACT K1, the volume of the cube is:

$$V = s^3 = (3\sqrt{3})^3 = (3^3)(\sqrt{3})^3 = 27(3\sqrt{3}) = 81\sqrt{3}$$

41. **(E)** If $r + si$ is a root of any quadratic equation with real coefficients, then $r - si$ is also a root (I is true). Since $(r + si) + (r - si) = 2r$, the sum of the roots is $2r$. However, by KEY FACT E3, the sum of the roots of a quadratic equation is $\frac{-b}{a} = \frac{-b}{2}$. So $2r = \frac{-b}{2} \Rightarrow -b = 4r \Rightarrow b = -4r$ (II is true). Since $(r + si)(r - si) = r^2 - s^2i^2 = r^2 - (s^2)(-1) = r^2 + s^2$, the product of the roots is $r^2 + s^2$. However, by KEY FACT E3, the product of the roots of a quadratic equation is $\frac{c}{a} = \frac{c}{2}$. So $r^2 + s^2 = \frac{c}{2} \Rightarrow c = 2r^2 + 2s^2$ (III is true). I, II, and III are true.

42. **(A)** Use TACTIC 4: draw and label a diagram.

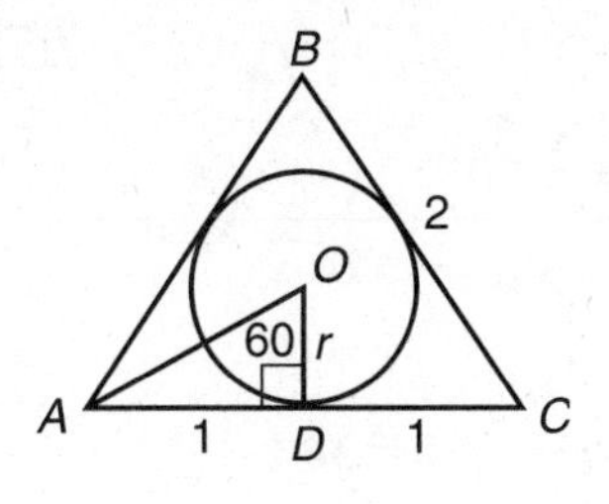

Then ΔAOD is a 30-60-90 right triangle whose shorter leg is r, the radius of the circle, and whose longer leg, AD, is 1. But by KEY FACT H8, $AD = r\sqrt{3}$

$$1 = r\sqrt{3} \Rightarrow r = \frac{1}{\sqrt{3}} \Rightarrow \pi r^2 = \pi\left(\frac{1}{\sqrt{3}}\right)^2 = \pi\left(\frac{1}{3}\right) = \frac{\pi}{3}$$

43. **(E)** $y = f(g(x)) = f(2x - 5) = 7(2x - 5) + 3 = 14x - 35 + 3 = 14x - 32$

$y = g(f(x)) = g(7x + 3) = 2(7x + 3) - 5 = 14x + 6 - 5 = 14x + 1$

By KEY FACT L8, the slope of each line is equal to 14. So by KEY FACT L7, the lines are parallel and, therefore, not perpendicular. (I is true and II is false).

$y = f(g(x)) - g(f(x)) = (14x - 32) - (14x + 1) = -33$

By KEY FACT L8, $y = -33$ is the equation of a horizontal line (III is true). Only I and III are true.

44. **(B)** By KEY FACT I9, the formula for the area of a parallelogram is $A = bh$. Use side $\overline{SR}$ as the base and draw in height $\overline{PT}$.

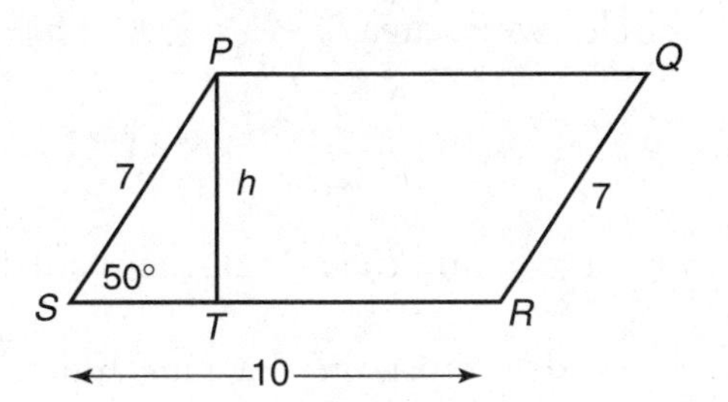

Then by KEY FACT M1, $\sin 50° = \frac{h}{7} \Rightarrow h = 7\sin 50° = 5.36$. Finally:

$$A = bh = 10(5.36) = 53.6$$

45. **(D)** By KEY FACT R4, a conditional statement $(p \rightarrow q)$ is equivalent to its contrapositive $(\sim p \rightarrow \sim q)$. Choice D is the contrapositive of the given statement. Note that whether or not the original statement is true is irrelevant. In this case it isn't and, therefore, neither is its contrapositive.

46. **(E)** By KEY FACT N6:

$$h(x) = (f \circ g)(x) = f(g(x)) = f(7x + 4) =$$
$$4(7x + 4) + 7 = 28x + 16 + 7 = 28x + 23$$

To find $h^{-1}(x)$, use KEY FACT N8. Write $y = 28x + 23$, switch x and y, and solve for y:

$$x = 28y + 23 \Rightarrow 28y = x - 23 \Rightarrow y = h^{-1}(x) = \frac{x-23}{28}$$

47. **(B)** Since each side of square $BCDE$ is 2, hypotenuse $\overline{BC}$ of isosceles right ΔABC is 2, and by KEY FACT H7, $AB = AC = \frac{2}{\sqrt{2}} = \sqrt{2}$.

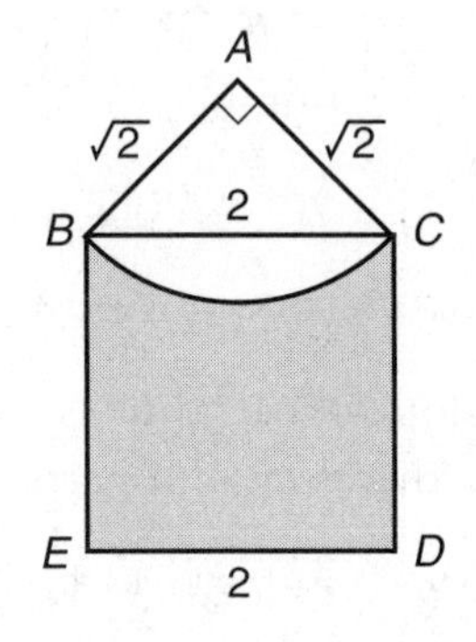

Since $\overline{AB}$ is a radius of the circle whose center is A and that passes through B and C, the area of the circle is $\pi(\sqrt{2})^2 = 2\pi$. Since $m\angle A = 90°$, sector ABC is $\frac{90}{360} = \frac{1}{4}$ of the circle. So its area is $\frac{2\pi}{4} = \frac{1}{2}\pi$. The area of square $ABCD$ is $2^2 = 4$ and the area of $\Delta ABC = \frac{1}{2}(\sqrt{2})(\sqrt{2}) = 1$. So the total area of the original figure is 4 + 1 = 5, and the area of the shaded region is $5 - \frac{1}{2}\pi$.

**Use TACTICS 5 and 8: trust the diagram and eliminate absurd choices. Since the area of square $BCDE$ is 4, the area of the shaded region must be slightly less than 4. Immediately eliminate choice A, which is greater than 4, and choice C, which is negative. Choices B, D, and E are all positive numbers less than 4, but choices D and E are much too small. The answer must be B.

48. **(E)** If x pounds of peanuts are added to the existing mixture, the result will be a mixture whose total weight will be $(70 + x)$ pounds, of which $(20 + x)$ pounds will be peanuts. Then, by expressing 60% as $\frac{6}{10}$, we have

$$\frac{20+x}{70+x} = \frac{6}{10} \Rightarrow 200 + 10x = 420 + 6x \Rightarrow 4x = 220 \Rightarrow x = 55$$

**Use TACTIC 2: backsolve. If 42 pounds of peanuts are added, peanuts will be $\frac{42+20}{42+70} = \frac{62}{112}$ = 55% of the mixture. Since that is not enough, eliminate choice C and choices A and B, which are even smaller. Test D or E.

49. **(B)** By KEY FACTS K8 and K6, the volume of a sphere is $\frac{4}{3}\pi r^3$ and the volume of a right circular cone is $\frac{1}{3}\pi r^2 h$. (Remember that both of these formulas are given to you on the first page of the test.) Then:

$$\frac{4}{3}\pi r^3 = \frac{1}{3}\pi r^2 h \Rightarrow 4r^3 = r^2 h \Rightarrow 4r = h \Rightarrow \frac{h}{r} = 4$$

**Use TACTIC 3. Plug in a number for the radius, say $r = 1$. Then the volume of the sphere is $\frac{4}{3}\pi(1)^3 = \frac{4}{3}\pi$, and the volume of the cone is $\frac{1}{3}\pi(1)^2 h = \frac{1}{3}\pi h$. So $\frac{4}{3}\pi = \frac{1}{3}\pi h \Rightarrow h = 4$.

50. **(D)** Write out enough terms of the sequence until you see a pattern: 4, 5, 9, 14, 23, 37, 60, 97, 157, . . . The sequence of odds and evens repeats indefinitely in groups of three: $\boxed{\text{even, odd, odd}}$, $\boxed{\text{even, odd, odd}}$, . . . Since each group contains one even and two odds, the first 333 groups of 3 contain 999 terms—333 of which are even and 666 of which are odd. The 1,000th term is the first term of the next group and is even. So in all, there are 334 even terms and 666 odd terms.

Index

THE FOLLOWING DOCUMENTATION APPLIES IF YOU PURCHASED
***SAT Subject Test Math Level 1, 3rd Edition* with CD-ROM.**
Please disregard this information if your edition does not contain the CD-ROM.

How to Use the CD-ROM

The software is not installed on your computer; it runs directly from the CD-ROM. Barron's CD-ROM includes an "autorun" feature that automatically launches the application when the CD is inserted into the CD-ROM drive. In the unlikely event that the autorun feature is disabled, follow the manual launching instructions below.

Windows®

Insert the CD-ROM and the program should launch automatically. If the software does not launch automatically, follow the steps below.

1. Click on the Start button and choose "My Computer."
2. Double-click on the CD-ROM drive, which will be named **SATmath1**.
3. Double-click **SATmath1.exe** application to launch the program.

Macintosh®

1. Insert the CD-ROM.
2. Double-click the CD-ROM icon.
3. Double-click the **SATmath1** icon to start the program.

SYSTEM REQUIREMENTS

The program will run on a PC with:
Windows® Intel® Pentium II 450 MHz
or faster, 128MB of RAM
1024 X 768 display resolution
Windows 2000, XP, Vista
CD-ROM Player

The program will run on a Macintosh® with:
PowerPC® G3 500 MHz
or faster, 128MB of RAM
1024 X 768 display resolution
Mac OS X v.10.1 through 10.4
CD-ROM Player